LAPLACE CIRCUIT ANALYSIS AND ACTIVE FILTERS

LAPLACE CIRCUIT ANALYSIS AND ACTIVE FILTERS

Don A. Meador
DeVry Institute of Technology

Prentice Hall, Englewood Cliffs, New Jersey 07632

Library of Congress Cataloging-in-Publication Data

Meador, Don A. (*date*)
Laplace circuit analysis and active filters / Don A. Meador.
p. cm.
Includes bibliographical references and index.
ISBN 0-13-523481-6
1. Electric circuit analysis. 2. Laplace transformation.
3. Electric filters, Active. I. Title.
TK454.M44 1991
621.319'2—dc20 90-48503
CIP

Editorial/production supervision and
interior design: **Kathryn Pavelec**
Cover design: **Wanda Lubelska Design**
Manufacturing buyer: **Mary McCartney and Ed O'Dougherty**

Printed in the United States of America

10 9 8 7 6 5 4 3 2 1

ISBN 0-13-523481-6

Prentice-Hall International (UK) Limited, *London*
Prentice-Hall of Australia Pty. Limited, *Sydney*
Prentice-Hall Canada Inc., *Toronto*
Prentice-Hall Hispanoamericana, S.A., *Mexico*
Prentice-Hall of India Private Limited, *New Delhi*
Prentice-Hall of Japan, Inc., *Tokyo*
Simon & Schuster Asia Pte. Ltd., *Singapore*
Editora Prentice-Hall do Brasil, Ltda., *Rio de Janeiro*

to Joan, Angie, and Abbie

Contents

Preface

The student is assumed to have a working knowledge of dc and ac steady-state time-domain circuit analysis, and to have had a beginning course in calculus. (Since calculus is not crucial to understanding the book, it is possible to skip around those parts without adverse problems.) The student should also have a working knowledge of op-amps for the filter section.

The book is intended for either engineers or technologists. The material is appropriate for a one- or two-semester course in a mid-level engineering or an upper-level technology program.

Laplace transforms are presented as a circuit analysis tool rather than as a math course. Although the mathematical equations are included, the proofs and theorems are minimized, and their use and applications are emphasized.

The book is self-contained, so other sources are unnecessary when going into greater depth than the text presentation—advanced equation and derivations are given in this book. This makes it possible to teach at many levels. The chapters do not go through long dissertation on the equations, but are thorough in examples.

Special attention to the details of how to work a problem are used to eliminate a weakness found in other books. Most texts give you an idea of how to work a problem, but avoid showing the details of more complex problems.

Some textbooks show several ways to work special case problems. This book uses techniques that apply equally well to typical and special case problems. For example, the derivative of $tu(t)$ and $u(t)$ in most texts are handled as uniquely different problems. Here the derivatives are handled in the same way for both cases.

A major goal of this book is to use a minimum of formulas that will apply to a maximum of cases. The student should not be barraged with special cases when first confronted with Laplace transforms. The student should have a holistic understanding of all cases.

In this text a new teaching/learning technique is used. The different topics are approached in a similar or parallel manner. This allows the student to apply techniques learned on one topic to another possibly unrelated topic. This is most clearly seen in the approach of complex waveforms in Chapter 2 and Bode plots in Chapter 5.

We may call this the "parallel learning curve" method. This method significantly reduces the learning curve. Instead of starting a new learning curve with each new topic, we use part of the previous learning curve on the next topic. With the explosion of information we must somehow become capable of learning more information faster. This method allows us to do that.

The inverse Laplace transform in this book has a new formula that eliminates the need for extensive inverse tables. This formula finds the inverse of multiple-order complex-conjugate roots. This finally makes it possible to write a computer program to find the inverse Laplace of any practical electronic circuit.

Following is a brief summary of the book.

Chapter 1 is a brief overview of the book and some study hints.

Chapter 2 defines complex waveforms. It goes step by step from the basic parts of waveforms with discontinuities to complex waveforms. This chapter also prepares the student for Bode plots in Chapter 5 since both chapters use the same technique.

Chapter 3 covers the Laplace and the inverse Laplace transform. The Laplace transform technique in this book makes extensive use of Laplace operations, reducing the amount of material to memorize. Only equations that work on the general and special cases are used. This chapter is unique due to the development of a new inverse transform formula. This formula makes it possible to find the inverse Laplace of any practical electronic problem with only two techniques.

Chapter 4 is written to show the student how to use Laplace transforms to solve electronic circuits. It is organized so that the student will be given a review of circuit analysis at the same time. Most texts hop around to different topics of circuit analysis as if trying to humiliate the student into reading other texts to brush up on the topics. Here the circuit analysis rules are clearly stated and used in the examples. Even the very basics of circuit analysis (what is series and parallel, Ohm's law, etc.) are included in an appendix.

Chapter 5 is written so that the drawing of Bode plots are similar to the way complex waveforms are drawn in Chapter 2. This reduces the difficulty of teaching this topic since the student will have an insight to the drawing technique. Most other texts use the gosh-and-by-golly technique to find *x*-axis crossing of these graphs. In this book everything has a formula to calculate any point on the Bode plot. This allows the student to start seeing the construction of Bode plots as a whole instead of as a collection of special cases. The construction of the Bode plot for breaks occurring at the same frequency has been simplified and explained so that the student will not be in the dark on this subject any more.

Chapter 6 is an introduction to filters which takes the student from the general use of Laplace transforms to a specific use. This chapter builds on Chapter 5 to show how frequency plots can be used to define filter specifications. The different types of filters (LP, HP, BP, and notch) are defined. Also, in this chapter the approximate op-amp is defined to refresh the student's memory before the op-amp is used as an active filter.

Chapter 7 show the student how to build normalized low-pass filters. First the admittance Laplace transfer function is found from basic topologies. The admittance transfer functions are then transformed to circuit values using coefficient matching. At this point the Butterworth, Chebyshev, and elliptic filters are defined. Here, unlike other textbooks, the Butterworth filter is normalized in the same way as the Chebyshev and elliptic. This makes frequency shifting of all three filters the same process.

Chapter 8 shows the student how to shift the normalized filter's frequency, gain, and impedance to make a practical filter. This chapter also shows how to design high-pass, band-pass, and notch filters by transforming the equations in Chapter 7.

An IBM compatible disk that is described in Appendix G can be obtained through:

Millennial Marketing
Educational Software
P.O. Box 2123
Lee's Summit, Mo. 64063

I would like to thank my colleagues and friends for their encouragement and help, my students for what they taught me about learning, but most of all my thanks to my wife, daughters, and family for their unquestioning understanding and support in the long hours required to create this child.

Don A. Meador

LAPLACE CIRCUIT ANALYSIS AND ACTIVE FILTERS

CHAPTER 1

Introduction

1-1 WHAT'S IN THIS BOOK

This book covers the mathematical tools required to work electronic circuits having complex waveforms and reactive components. At the center of these mathematical tools are Laplace transforms. Using Laplace transforms, we are able to analyze a circuit having complex waveforms at multiple frequencies for the complete response (from $t = 0$ to $t = \infty$). In addition, these tools apply to steady-state responses of single-frequency sinusoidal circuits.

The first half of the book is devoted to learning these mathematical tools, the second half to applying them to active filters. In Chapter 2 we will learn how to describe complex waveforms. These complex waveforms will not be emphasized in later chapters, so the mathematical tools will be clearly seen, but the complex waveforms will apply equally. Although we are learning Laplace transforms, we will be using impractical circuits simply to learn the techniques—playing games with electronic circuits. In the second half we apply these tools to active filter design.

1-2 LAPLACE TRANSFORMS

The Laplace transform is a technique that transforms a differential equation from the time domain (equations expressed as a function of time) to the s-domain (equations expressed as a function of a complex variable s). This makes the process of solving the differential equation an algebraic process. Solving a differential equation is somewhat difficult, and in electronics, simultaneous differential equations are common. When the process is algebraic, simultaneous equations become much easier to solve.

In the beginning, for electronics, it is not important to understand exactly what Laplace transforms are. We are more interested in how to use the tool than how the tool was "developed and manufactured." However, when Laplace transforms are introduced in the following chapters, we begin with how the

Laplace transforms are derived but we should not be overly concerned with this. Understanding Laplace transforms is much easier when we know how they are used.

1-3 NOTATIONS

In this book we must be very careful to recognize function notation. We will be concerned primarily with two types of functions, $f(t)$ and $F(s)$. These are read "f of t" and "F of s." They are not "f times t" or "f times s," but represent a function of time and a function of s. These two functions are the same except that one is in the time domain and the other is in the s-domain (or Laplace domain). The functions "$f(t)$ and $F(s)$" and "$h(t)$ and $H(s)$" are generic names and do not refer to a specific function. The function "$e(t)$ and $E(s)$" usually refers to a source voltage, and the function "$v(t)$ and V(s)" usually refers to a voltage drop across a component.

The typical electronic notations are used in this book. These notations are typical of most textbooks, but some may seem unusual, depending on the texts used for basic circuit analysis. We should use the notations presented in this book while learning the technique to avoid confusion and the use of "translation sheets."

1-4 CALCULATIONS

An important aspect of this book is the way the examples are calculated. Full computer/calculator accuracy is used to calculate the answers, rounded to five significant digits. If an intermediate result is shown, subsequent calculation will be based on the full computer/calculator number rather than the five significant digits shown. For calculations based on tabulated values, only the values in the tables will be used, but after that point full computer/calculator accuracy will be maintained.

We should develop the habit of carrying the full-digit accuracy of our computer/calculator to prevent becoming obsolete in our accuracy. As time progresses, the accuracy of manufactured components increases and the need for carrying more significant digits in our calculations increases. If we learn to carry the full accuracy of our computer/calculator, our calculations will never become obsolete.

1-5 HOW TO STUDY

The purpose of this book is learning. Therefore, this is the most important section in the book. If we fail here, we will learn nothing in the following chapters. The rules are simple to follow, but must be followed consistently. Learning Laplace transforms enough to work the problems in this book means very little. The main idea is to know Laplace transforms as well as we know Ohm's law. Then a door to many new worlds will be opened.

RULE 1: Study in short, frequent intervals of time.

Laplace transforms are easy to watch and understand when someone else is working the problems, but difficult when we wait too long to start working problems. We must work problems as soon as we are exposed to any new idea, no matter how small. More can be learned in one hour spread over several days than in two hours of unbroken time. The longer hours at a stretch will only make us feel noble—we won't retain much.

RULE 2: Always write the equation being used in its original form with variables instead of numbers.

There are always equations that must be memorized. The easiest way is to write the equation in its original form first (without copying from the text or notes). Second, use algebra to solve for the variable required. Last, substitute in the numbers. The second and last step may be reversed, depending on personal preference. Many of the equations will be similar, and this will help to keep us from combining different equations.

RULE 3: Work problems without the text or notes.

When we have to use references to work a problem, the only thing we learn is that we do not know how to work the problem. We must work the problems without the text. When we are not sure what to do, try anything, but try something. (A good thing to try is to make a list of known and unknown variables and values.) When we have an answer, then and only then should we use our notes and the text to determine if we did something wrong.

CHAPTER 2

Analysis of Waveforms

2-0 INTRODUCTION

Not only does the Laplace transform provide a method to analyze complex circuits, but it also provides a method to analyze circuits excited by complex waveforms. A method is therefore required to describe complex waveforms. In this chapter we define basic functions used to describe complex waveforms mathematically.

2-1 STEP FUNCTION

Ideally, a switch will supply a voltage or current to a circuit instantaneously. This instantaneousness transition is shown in Fig. 2-1. The mathematical function we will use is called the unit step function and is denoted by

$$u(t) \tag{2-1}$$

Multiplying the unit step by a constant is used to simulate a dc source of any value being switched on. The general form of a function with magnitude K being switched on at $t = 0$ is

$$f(t) = Ku(t) \tag{2-2}$$

An examination of a simple function and how it contrasts with the $u(t)$ function will be helpful for a full understanding of this function. We will use the function

$$g(t) = 3t + 4 \tag{2-3}$$

for our discussion. The part of $g(t)$ enclosed in parentheses is called the argument. For each value of t (the argument) there exists a unique solution of $g(t)$. For example, if the argument is 2, then $g(2) = 3(2) + 4 = 10$, and with an argument of -3, $g(-3) = -5$. In contrast, $u(t)$ does not have a unique value for each t.

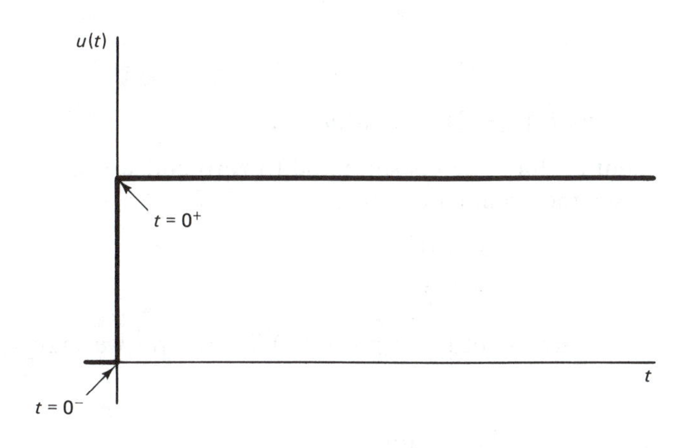

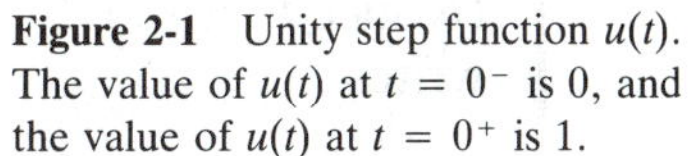

Figure 2-1 Unity step function $u(t)$. The value of $u(t)$ at $t = 0^-$ is 0, and the value of $u(t)$ at $t = 0^+$ is 1.

There are only two possible values for $u(t)$, 0 or 1. Whenever the $u(t)$ argument is greater than zero, the value of $u(t)$ is 1. When the $u(t)$ argument is less than zero, the value of $u(t)$ is 0. Therefore, $u(t)$ cannot be defined by a polynomial but must be defined piecewise as follows:

$$u(t) = 1 \qquad \text{for } t > 0 \tag{2-4a}$$

$$= 0 \qquad \text{for } t < 0 \tag{2-4b}$$

We also need to know how to handle $u(t)$ at $t = 0$ since it is undefined at this time. Figure 2-1 shows that the function $u(t)$ has two values at $t = 0$. We can split $t = 0$ into two parts. The time at an infinitely small distance to the right of $t = 0$ is written as $t = 0^+$, and the time at an infinitely small distance to the left of $t = 0$ is written as $t = 0^-$. Therefore,

$$u(0^+) = 1$$

$$u(0^-) = 0$$

We can now define two rules for $u(t)$ at $t = 0$.

> **RULE 1:** When we are working with the function to the right of $t = 0$, we use $u(0^+)$ for $u(t)$ at $t = 0$.
> **RULE 2:** When we are working with the function to the left of $t = 0$, we use $u(0^-)$ for $u(t)$ at $t = 0$.

There are two important items to keep in mind about the argument of $u(t)$. First, the argument may be a function of t rather than just t. In this case the algebraic distributive law cannot be applied since this is a function notation rather than an algebraic expression. Second, the step function's value goes through the zero-to-one transition when the value of the argument is zero, not just when t is zero. Therefore, the value of this function is 1 when the argument is positive, and the value of this function is 0 when the argument is negative.

EXAMPLE 2-1

Draw a graph of the functions $4u(t - 3)$ and $2u(5t - 25)$.

SOLUTION The argument of $4u(t - 3)$ is set equal to zero and we solve for the value of t that causes the argument to be zero.

$$t - 3 = 0$$

$$t = 3$$

(Notice that $t > 3$ causes the argument to be positive.) Therefore, we may write

$$4u(t - 3) = \begin{cases} 4 & \text{for } t > 3 \\ 0 & \text{for } t < 3 \end{cases}$$

The graph is shown in Fig. 2-2a.

The argument of $2u(5t - 25)$ is set equal to zero and t is solved.

$$5t - 25 = 0$$

$$t = 5$$

In this case, when t is greater than 5, the argument of $2u(5t - 25)$ is greater than zero. The result is shown in Fig. 2-2b. Note that the magnitude of the function is 2 since we cannot use the distributive law on a function notation.

The step function is sometimes referred to as the switch function. When we multiply the step function by another function, the product of the two functions will be zero until the argument of the step function is greater than zero. This gives us a way to represent mathematically, for instance, a sinusoidal wave "switched on" at any particular time.

Figure 2-3 shows an example of how $u(t)$ is used to switch on a sinusoidal function at $t = 0$. Figure 2-3c shows what we want to represent mathematically. We start with the sinusoidal function shown in Fig. 2-3a. Since we want the sinusoidal function switched on at $t = 0$, we use $u(t)$ as shown in Fig. 2-3b. When we multiply these two functions together, the values to the left of $t = 0$ will be zero since $u(t) = 0$ for $t < 0$. To the right of $t = 0$ the sinusoidal function is multiplied by 1 and therefore the product is $\sin(\omega t)$. Figure 2-3c shows the graph of $g(t) = \sin(\omega t)\, u(t)$.

2-2 RAMP FUNCTION

The ramp function is a straight line with a value of zero at and before the origin increasing or decreasing with a slope of K. Figure 2-4 shows the graph of a straight line. The slope is found by

$$K = \frac{\Delta f(t)}{\Delta t} = \frac{f_2(t) - f_1(t)}{t_2 - t_1} \tag{2-5}$$

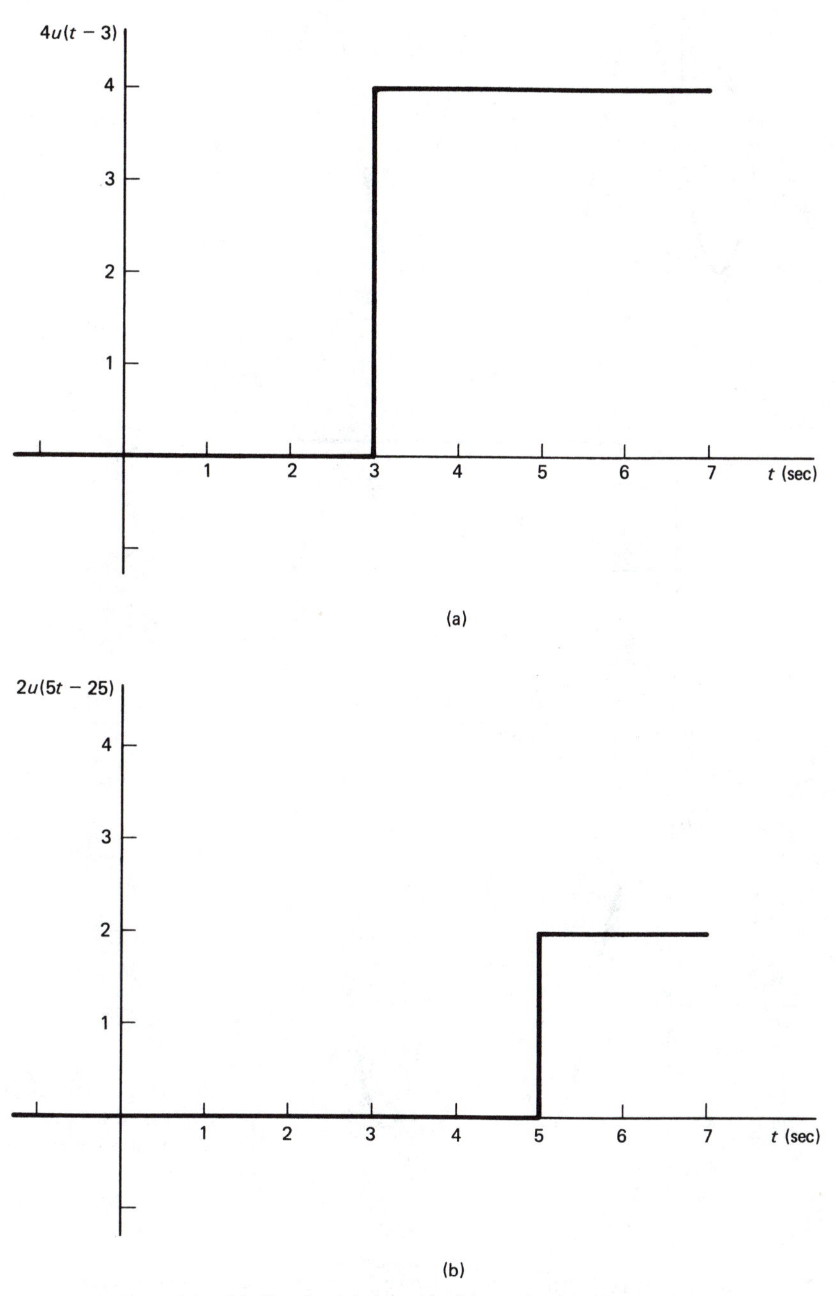

Figure 2-2 (a) Graph of $4u(t - 3)$; (b) graph of $2u(5t - 25)$.

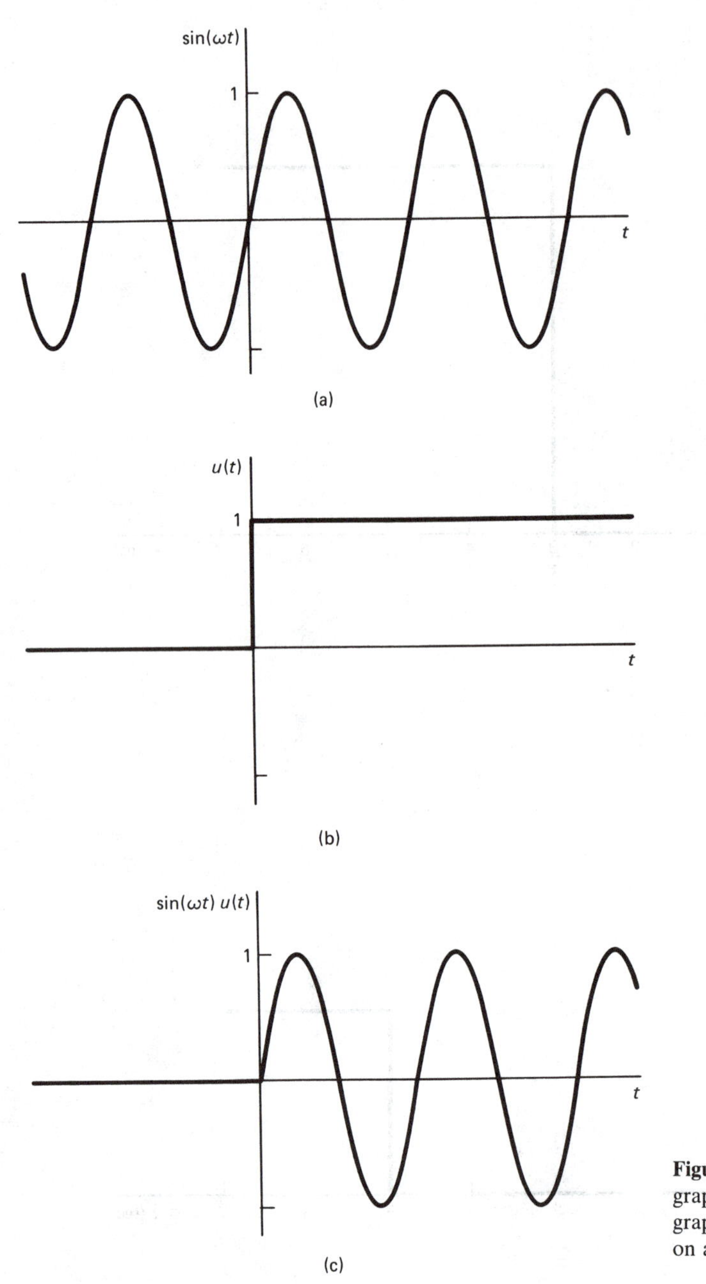

Figure 2-3 (a) Graph of sin (ωt); (b) graph of the switch function $u(t)$; (c) graph of a sinusoidal function switched on at $t = 0$.

Since the slope is K and the intercept is zero, the function is written $f(t) = Kt$. If this function is voltage, the units are volts/second, and if the function is a current, the units are amperes/second.

Since the ramp function is zero before the origin, the straight-line equation is multiplied by $u(t)$ in order to form the ramp function. Therefore, the general

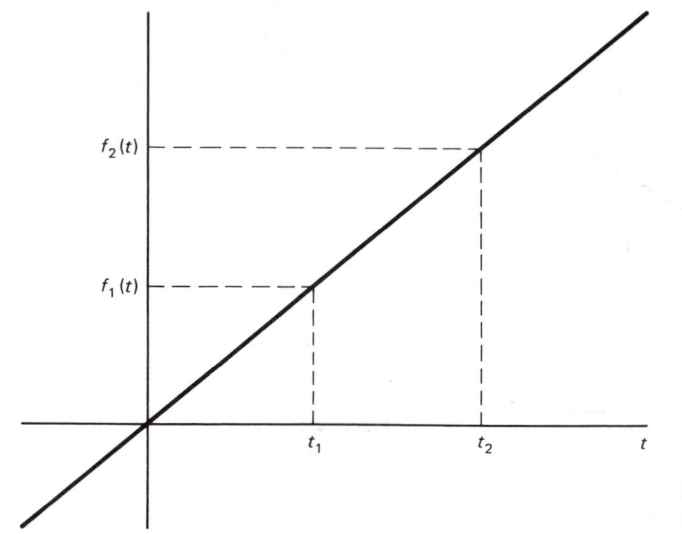

Figure 2-4 Graph of a straight line used to make a ramp function.

equation for the ramp function is

$$f(t) = Ktu(t) \tag{2-6}$$

EXAMPLE 2-2

Draw the graph for $g(t) = 2tu(t)$ for $-1 < t < 3$ sec.

SOLUTION The result is shown in Fig. 2-5.

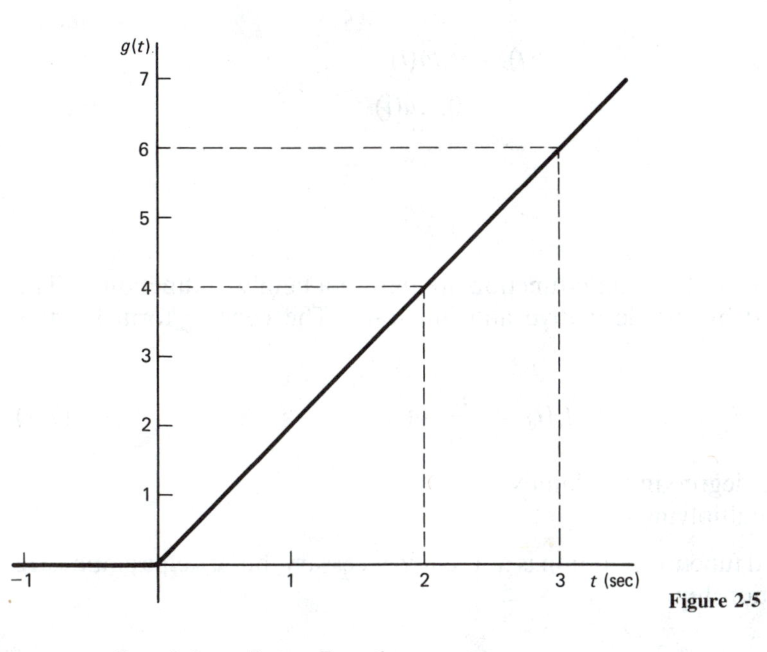

Figure 2-5

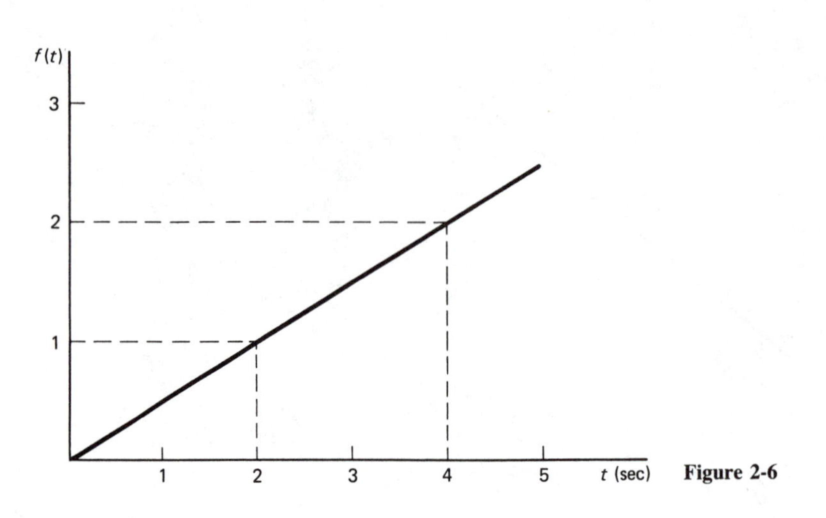

Figure 2-6

EXAMPLE 2-3

Write the equation for Fig. 2-6.

SOLUTION Since this is a straight line, the slope can be determined at any two points using Eq. (2-5).

$$K = \frac{f_2(t) - f_1(t)}{t_2 - t_1}$$

$$= \frac{2 - 1}{4 - 2} = \frac{1}{2}$$

From Eq. (2-6),

$$f(t) = Ktu(t)$$

$$= 0.5tu(t)$$

2-3 t^n FUNCTION

The step function and the ramp function are part of a family of functions. The family is related by the derivative and integral. The general form for this family is

$$f_n(t) = \frac{At^n}{n!} u(t) \tag{2-7}$$

where n = the degree in the family
A = a multiplying constant

(The subscripted function notation is not really necessary, but it helps emphasize the family relationship.)

For the step function with a height of 3, n is equal to 0 and A is equal to 3.

$$f_n(t) = \frac{At^n}{n!} u(t)$$

$$f_0(t) = \frac{3t^0}{0!} u(t)$$

Note that zero factorial is by definition equal to 1.

$$= 3u(t)$$

For a ramp with a slope of 3, n is equal to 1 and A is equal to 3.

$$f_n(t) = \frac{At^n}{n!} u(t)$$

$$f_1(t) = \frac{3t^1}{1!} u(t)$$

$$= 3tu(t)$$

The two functions $f_0(t)$ and $f_1(t)$ are related to each other through the derivative and the integral. The derivative of the ramp is the slope of the ramp, which is a constant value starting at $t = 0$.

$$f_0(t) = \frac{d}{dt} f_1(t)$$

$$= \frac{d}{dt} Atu(t)$$

$$= Au(t)$$

The integral of the step must therefore be the ramp function. This is an important point that we will use later.

2-4 IMPULSE FUNCTION

In Section 2-3 we did not take the derivative of the step function. This was because $u(t)$ has a discontinuity at $t = 0$. The derivative of $u(t)$ is the impulse function and is denoted by

$$\delta(\mathrm{t}) \tag{2-8}$$

where δ is the Greek lowercase letter delta. The general form with a multiplying constant is

$$K\delta(t) \tag{2-9}$$

This impulse function has zero width, infinite height, and a finite area. In order to understand this function we will start with an imperfect step function and find the derivative.

In the next few paragraphs, we will work with the derivative in an electronic context. An electronic differentiator's output is not really the derivative of the input signal, but usually has an output voltage proportional to the derivative of the input. Since this is the case, the units of the differentiator's output will not be the same units as the mathematical derivative. We read the output voltage of the differentiator and supply the correct units if the mathematical units are required.

Figure 2-7 shows an imperfect step function and its derivative. The derivative has a value only when the original function has a nonzero slope. In this case the only time the derivative has a value is between $t = 0$ and $t = t_1$ seconds. The derivative, shown in Fig. 2-7b, has a value in this time interval equal to the slope of the ramp and is therefore K/t_1. The area under the derivative curve in Fig. 2-7b is

$$\text{area} = t_1 \left(\frac{K}{t_1}\right)$$

$$= \text{K}$$

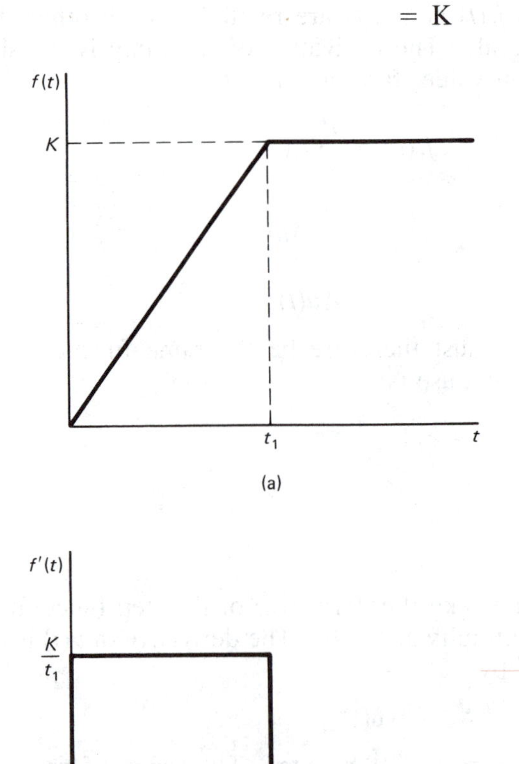

Figure 2-7 (a) Imperfect step function; (b) derivative of the imperfect step function.

Figure 2-8a is closer to being a perfect step function. The derivative in Fig. 2-8b has a larger amplitude, but the area under Fig. 2-8b is the same area that we found in Fig. 2-7b.

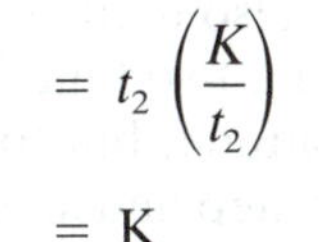

$$= t_2 \left(\frac{K}{t_2} \right)$$

$$= \text{K}$$

As we keep decreasing the amount of time allowed for the ramp function, the derivative will increase in height, but the area will remain constant.

$$\text{area} = \lim_{t \to 0} \frac{K}{t} t = \lim_{t \to 0} K = K$$

$$\text{height} = \lim_{t \to 0} \frac{K}{t} = \infty$$

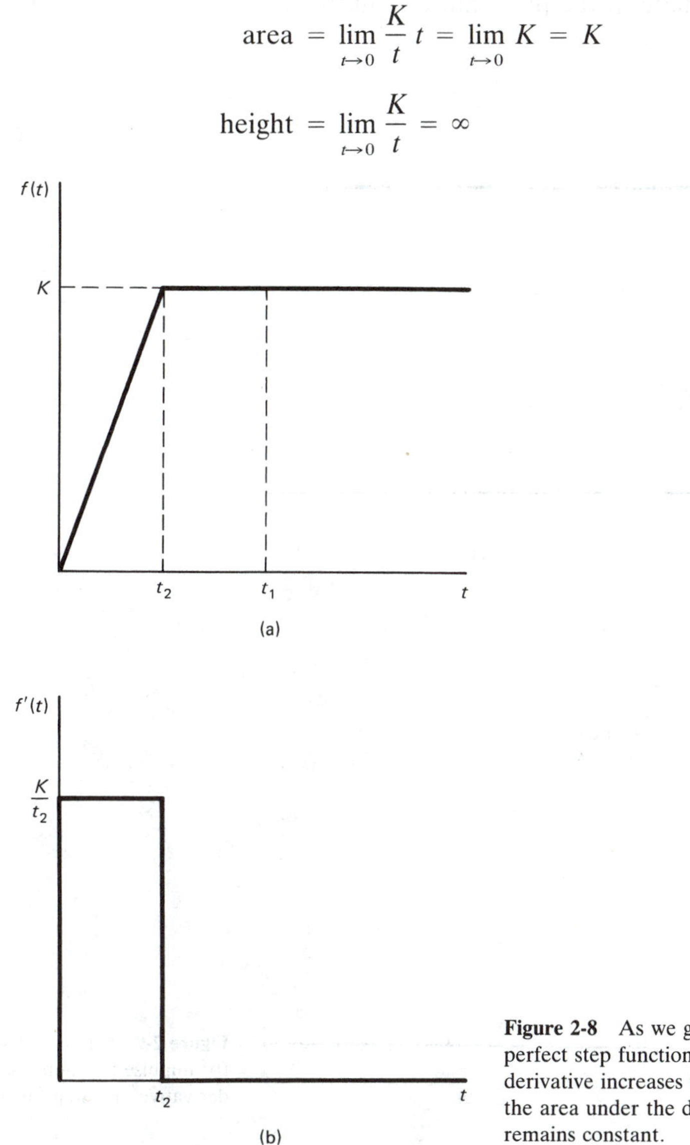

Figure 2-8 As we get closer to a perfect step function the height of the derivative increases toward infinity, but the area under the derivative curve remains constant.

Thus we have defined the impulse function of area K. The impulse function has infinite height, zero width, and finite area.

In electronic circuits it is unlikely that we will have a voltage or current of infinite height. It is also unlikely that we will be able to have a perfect step function. In reality these two functions will be an imperfect step and a narrow pulse with a large amplitude, but to simplify the mathematics we will approximate these as a perfect step and an impulse. ("Narrow" in this context means narrow when compared to the shortest time constant of the circuit.) The area of the impulse will equal the area of the narrow pulse. The impulse may be assigned any time within the limits of the original pulse since to make this approximation, the pulse must be narrow.

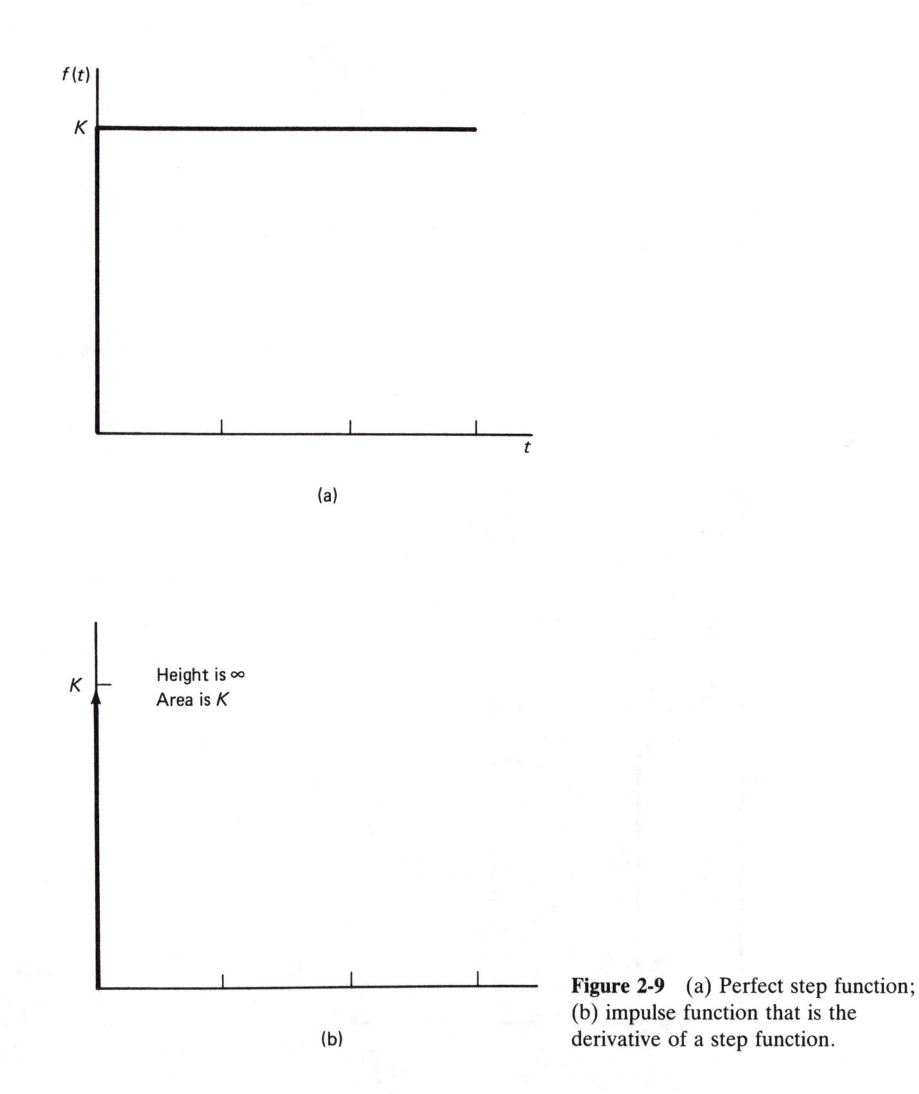

Figure 2-9 (a) Perfect step function; (b) impulse function that is the derivative of a step function.

Figure 2-9 shows the step function, $Ku(t)$, and its derivative, the impulse function. In equation form,

$$f(t) = Ku(t) \tag{2-10a}$$

$$f'(t) = \frac{d}{dt} f(t) = K\delta(t) \tag{2-10b}$$

Since the height of the impulse is infinity, an arrow is used. We may draw the height of the arrow to correspond to the area, but we must never forget that the height of the impulse is infinity. Another way to show the impulse function is to draw all impulses the same height and write the value of the area next to the impulse symbol.

EXAMPLE 2-4

Graph the derivative of the function shown in Fig. 2-10.

SOLUTION From the graph

$$g(t) = -2u(t)$$

Therefore,

$$g'(t) = -2\delta(t)$$

The graph of the derivative is shown in Fig. 2-11.

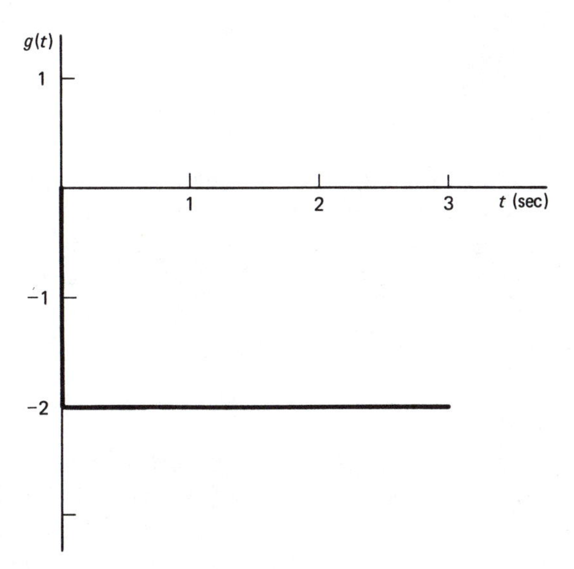

Figure 2-10

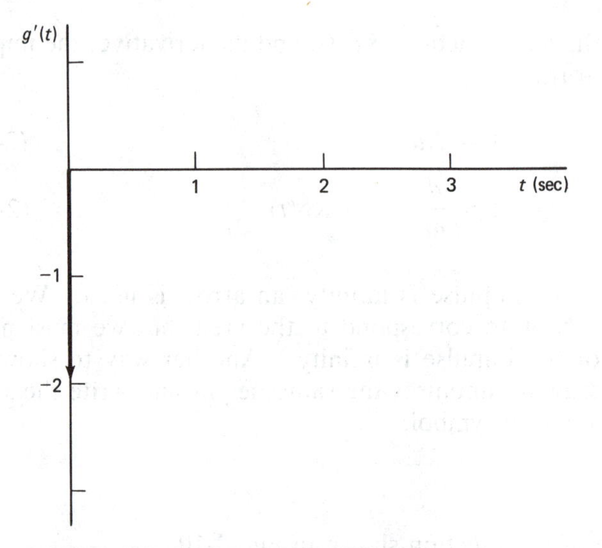

Figure 2-11

2-5 EXPONENTIAL FUNCTION

There are two forms of the exponential:

$$e^{-\alpha t} \tag{2-11a}$$

$$e^{-t/\tau} \tag{2-11b}$$

where $e = 2.7183...$
α = the damping constant
τ = the time constant

The relationship between α and τ is found by setting the two exponentials equal to each other.

$$e^{-\alpha t} = e^{-t/\tau} \tag{2-12a}$$

$$\ln(e^{-\alpha t}) = \ln(e^{-t/\tau})$$

$$\alpha = \frac{1}{\tau} \tag{2-12b}$$

Figure 2-12 shows a graph of the exponential function. We will always assume that at five time constants the value of the function is 0. This is an assumption universally used. Five time constants is the value of time that causes the exponential function to equal e^{-5} (the exact value is $e^{-5} = 0.0067379...$), and one time constant is the value of time that causes the exponential function to equal e^{-1} ($= 0.36788...$).

EXAMPLE 2-5

For the function shown, find (a) the amount of time for two time constants, (b) the time when the function goes to zero, and (c) α and τ. (d) Graph the function.

$$g(t) = 14e^{-20t}u(t)$$

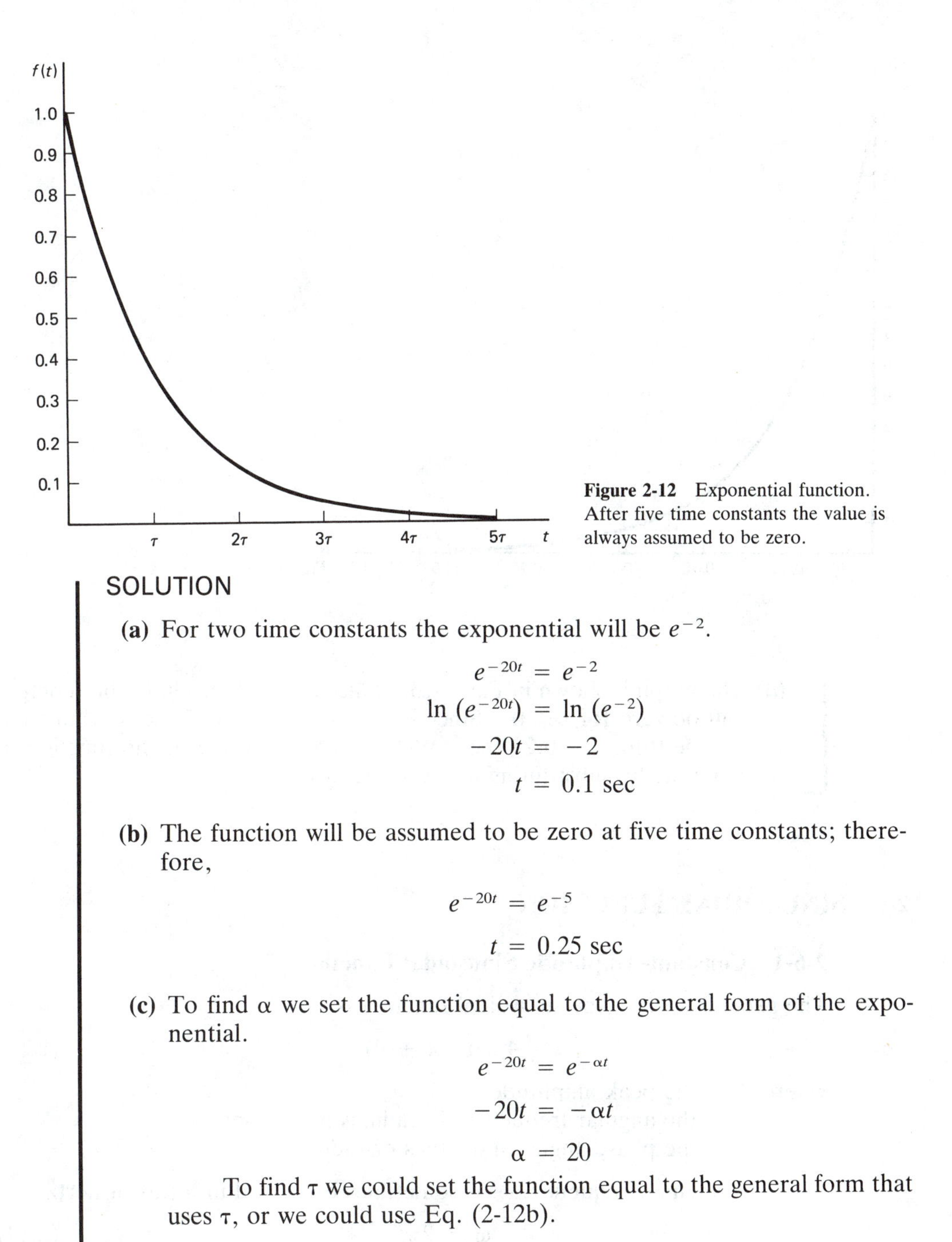

Figure 2-12 Exponential function. After five time constants the value is always assumed to be zero.

SOLUTION

(a) For two time constants the exponential will be e^{-2}.

$$e^{-20t} = e^{-2}$$
$$\ln (e^{-20t}) = \ln (e^{-2})$$
$$-20t = -2$$
$$t = 0.1 \text{ sec}$$

(b) The function will be assumed to be zero at five time constants; therefore,

$$e^{-20t} = e^{-5}$$
$$t = 0.25 \text{ sec}$$

(c) To find α we set the function equal to the general form of the exponential.

$$e^{-20t} = e^{-\alpha t}$$
$$-20t = -\alpha t$$
$$\alpha = 20$$

To find τ we could set the function equal to the general form that uses τ, or we could use Eq. (2-12b).

$$\alpha = \frac{1}{\tau}$$
$$\tau = \frac{1}{\alpha} = \frac{1}{20} = 0.05$$

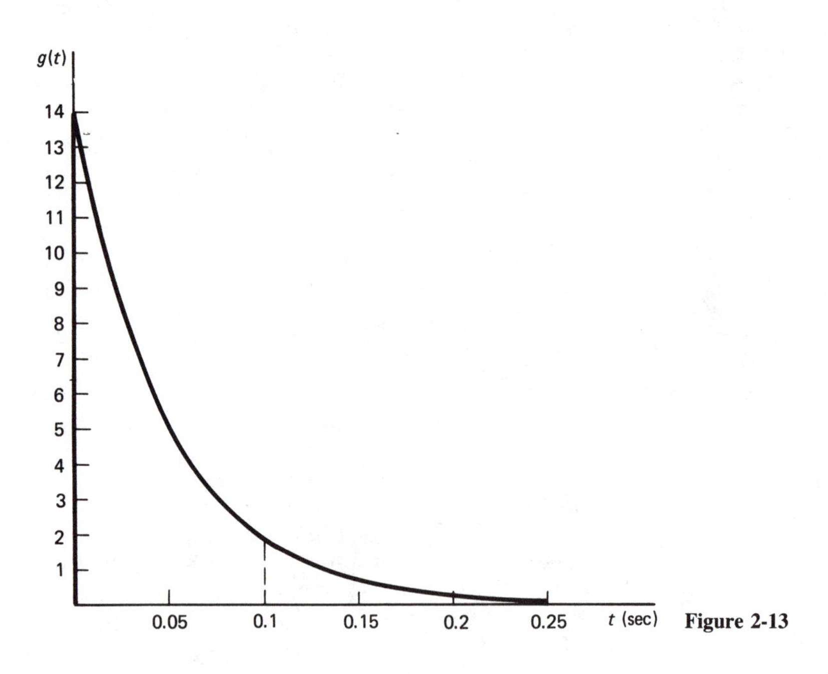

Figure 2-13

(d) The graph is shown in Fig. 2-13. Since $u(t)$ is 0 for $t < 0$, the function will be zero for $t < 0$. Since e^{-20t} is 0 for $t > 0.25$ sec, the function will be 0 for $t > 0.25$ sec. We can find the value of the function at any time by substituting in the value of t desired.

2-6 SINUSOIDAL FUNCTION

2-6-1 Constant-Amplitude Sinusoidal Function

The general form of the sinusoidal function is

$$A \sin(\omega t + \theta) \tag{2-13}$$

where A = the peak amplitude
ω = the angular frequency, in radians per second
θ = the phase angle, in degrees or radians

The angular frequency, ω, is sometimes expressed in terms of hertz.

$$\omega = 2\pi f \tag{2-14}$$

where f is the frequency in hertz (Hz).

The phase angle, θ, is usually expressed in degrees when phasors are being used, but ωt is always expressed in radians. When evaluating the sine function at a specific time with θ in degrees, we must first convert θ to radians before we can add it to ωt. We may therefore find it more convenient to express θ in radians. The relation between radians and degrees is

$$\pi \text{ radians} = 180° \tag{2-15}$$

EXAMPLE 2-6

Put the following function (a) in a form suitable for evaluation on a calculator using radians and (b) in a form using degrees. (c) Evaluate the function at $t = 0.5$ sec.

$$g(t) = 10 \sin (5t + 35°)$$

SOLUTION

(a) The $5t$ is in radians; therefore, convert the 35°:

$$35°\left(\frac{\pi \text{ rad}}{180°}\right) = \frac{7\pi}{36} \text{ rad}$$

The function can then be expressed as

$$g(t) = 10 \sin\left(5t + \frac{7\pi}{36}\right)$$

(b) The 35° part is in degrees; therefore, convert $5t$.

$$(5t \text{ rad})\left(\frac{180°}{\pi \text{ rad}}\right) = \frac{900}{\pi} t$$

$$g(t) = 10 \sin\left(\frac{900}{\pi} t° + 35°\right)$$

(c) In both cases the value of $g(t)$ at $t = 0.5$ sec will be the same. Being sure that the calculator is in the correct mode (radians or degrees), we should get

$$g(0.5) = 10 \sin\left[5(0.5) + \frac{7\pi}{36}\right]$$

or

$$g(0.5) = 10 \sin\left[\frac{900}{\pi}(0.5)° + 35°\right]$$

In both cases

$$g(0.5) = 0.30723$$

There are two methods of evaluating a sinusoidal function. The first way is to choose a value of time and solve for the value of the function at that time.

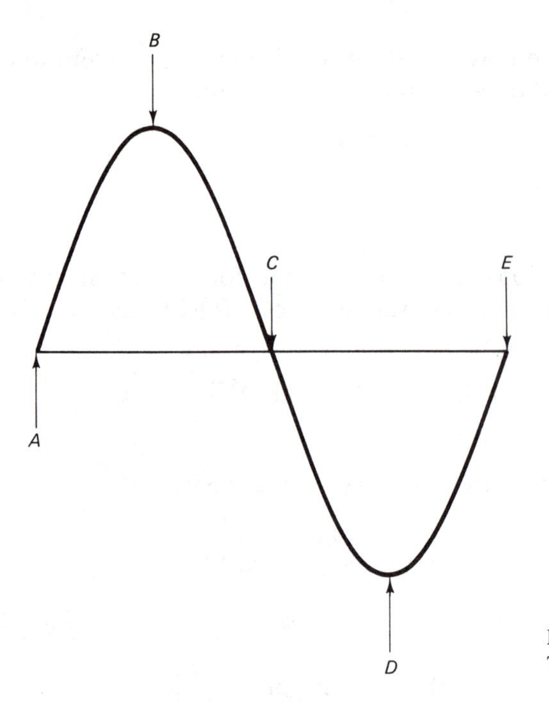

Figure 2-14 Points used to construct Table 2-1.

TABLE 2-1

Location	Equation (rad)
A	$\omega t + \theta = 0$
B	$\omega t + \theta = \pi/2$
C	$\omega t + \theta = \pi$
D	$\omega t + \theta = 3\pi/2$
E	$\omega t + \theta = 2\pi$

This method was used in Example 2-6. The second method is to choose a known value of the sinusoidal function and solve for the time when this value will occur. In the second method we choose a location on the sinusoidal function where we know the function's value. Figure 2-14 shows the most frequently used locations of sin $(\omega t + \theta)$. At each location we know what $\omega t + \theta$ is equal to, as shown in Table 2-1. We can now combine these two methods to draw the sinusoidal function. This is best shown by example.

EXAMPLE 2-7

Find (a) the angular frequency, (b) the frequency in hertz, and (c) the period of the waveform. (d) Sketch the graph.

$$v(t) = 5 \sin\left(4t - \frac{\pi}{4}\right)$$

SOLUTION

(a) The angular frequency is

$$\omega = 4 \text{ rad/sec}$$

(b) The frequency is found from the relationship

$$\omega = 2\pi f$$

Solving for f yields

$$f = \frac{\omega}{2\pi} = \frac{4}{2\pi}$$

$$= 0.63662 \text{ Hz}$$

(c) The period is the reciprocal of the frequency and therefore

$$T = \frac{1}{f}$$

$$= \frac{1}{0.63662}$$

$$= 1.5708 \text{ sec}$$

(d) To sketch the graph, first convert the 45° to radians.

$$(45°)\left(\frac{\pi}{180°}\right) = \frac{\pi}{4} \text{ rad}$$

Therefore,

$$v(t) = 5 \sin\left(4t - \frac{\pi}{4}\right)$$

Next we find the time of the peaks and x-axis crossing. Using the second method and Table 2-1 gives

$$4t - \frac{\pi}{4} = 0 \qquad \Rightarrow t = 0.19635 \text{ sec}$$

$$4t - \frac{\pi}{4} = \frac{\pi}{2} \qquad \Rightarrow t = 0.58905 \text{ sec}$$

$$4t - \frac{\pi}{4} = \pi \qquad \Rightarrow t = 0.98175 \text{ sec}$$

$$4t - \frac{\pi}{4} = \frac{3\pi}{2} \qquad \Rightarrow t = 1.3744 \text{ sec}$$

$$4t - \frac{\pi}{4} = 2\pi \qquad \Rightarrow t = 1.7671 \text{ sec}$$

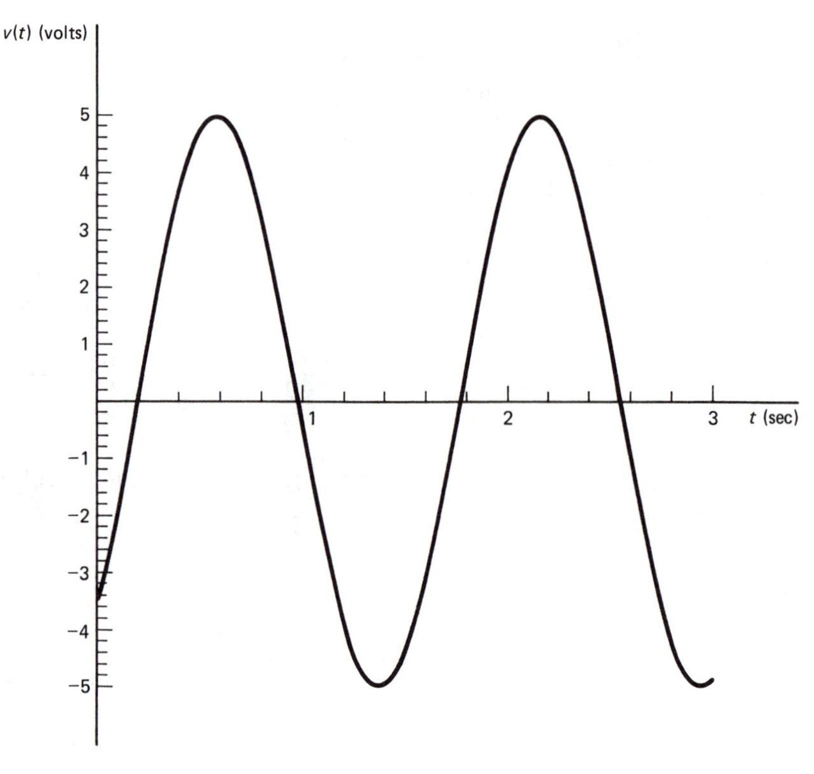

Figure 2-15

Using the first method, we can find the value of the sinusoidal function at $t = 0$.

$$v(0) = 5 \sin\left[4(0) - \frac{\pi}{4}\right] = -3.5355$$

The function is graphed in Fig. 2-15.

2-6-2 Exponential Amplitude Sinusoidal Function

The amplitude of the sinusoidal function does not have to be a constant. The amplitude may be a time-dependent function. The most important time-dependent function used for the sinusoidal amplitude is the exponential function. The general form is

$$f(t) = Ae^{-\alpha t} \sin(\omega t + \theta) \tag{2-16}$$

Figure 2-16 shows the general shape of the function. The time-dependent amplitude is $Ae^{-\alpha t}$ and is shown by the dashed line. The sine function is therefore confined within the limits of $+Ae^{-\alpha t}$ and $-Ae^{\alpha t}$. The amplitude and consequently the function will be zero when the exponential function reaches five time constants ($e^{-\alpha t} = e^{-5}$).

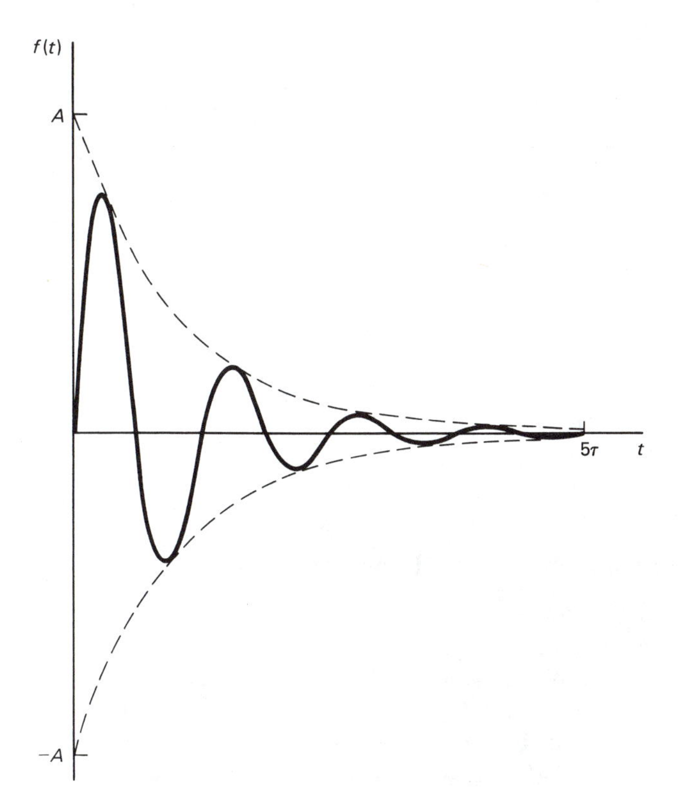

Figure 2-16 Graph of the general function $Ae^{-\alpha t}\sin(\omega t + \theta)$.

EXAMPLE 2-8

Find the sine's peaks and zero crossings for the function from $t = 0$ to when the function goes to zero.

$$f(t) = 4e^{-0.5t}\sin\left(0.4\pi t + \frac{\pi}{3}\right)$$

SOLUTION First find when the function goes to zero. This is controlled by the exponential function.

$$e^{-0.5t} = e^{-5}$$

Therefore,

$$t = 10 \text{ sec}$$

Next we set the argument of the sinusoidal function equal to the angles where we get peaks and zero crossings and we solve for t. Since the amplitude of the sinusoidal function is time dependent, we must substitute back into the whole function to find the value of the entire function. This is summarized in Table 2-2.

TABLE 2-2

$0.4\pi t + (\pi/3) =$	t	$4e^{-0.5t} \sin [0.4\pi t + (\pi/3)]$
0	< 0	—
$\pi/2$	0.41667	3.2477
π	1.6667	0
$3\pi/2$	2.9167	−0.93049
2π	4.1667	0
$2\pi + (\pi/2)$	5.4167	0.26659
$2\pi + \pi$	6.6667	0
$2\pi + (3\pi/2)$	7.9167	−0.076380
4π	9.1667	0
$4\pi + (\pi/2)$	10.417	0

2-6-3 Splitting and Combining Sinusoidal Functions

It is often more convenient to express sinusoidal functions containing a phase angle as the sum of a sine and cosine function without a phase angle. A method similar to polar-to-rectangular and rectangular-to-polar conversions may be used to split and combine sine and cosine functions. When converting a sine function to a phasor, the root-mean-square (rms) value is typically used. However, to make the math more direct, we will use the peak value. In any case the frequencies of sinusoidal functions must be the same.

Figure 2-17a shows $6 \sin (\omega t)$ drawn as a phasor (using the peak value). Figure 2-17b shows $6 \sin (\omega t + 90°)$, which may also be expressed as $6 \cos (\omega t)$. With this idea we may think of the x-axis as the sine axis and measure positive angles counterclockwise from the x-axis. We may think of the y-axis as the cosine axis measuring positive angles counterclockwise from the y-axis.

To split $A \sin (\omega t + \theta)$, we project the vector lengths onto the x-axis and the y-axis as shown in Fig. 2-18. The vector length projected onto the x-axis is the magnitude of the sine part, and the vector length projected onto the y-axis is the magnitude of the cosine part. Therefore,

$$A \sin (\omega t + \theta) = [A \cos (\theta)] \sin (\omega t) + [A \sin (\theta)] \cos (\omega t) \tag{2-17}$$

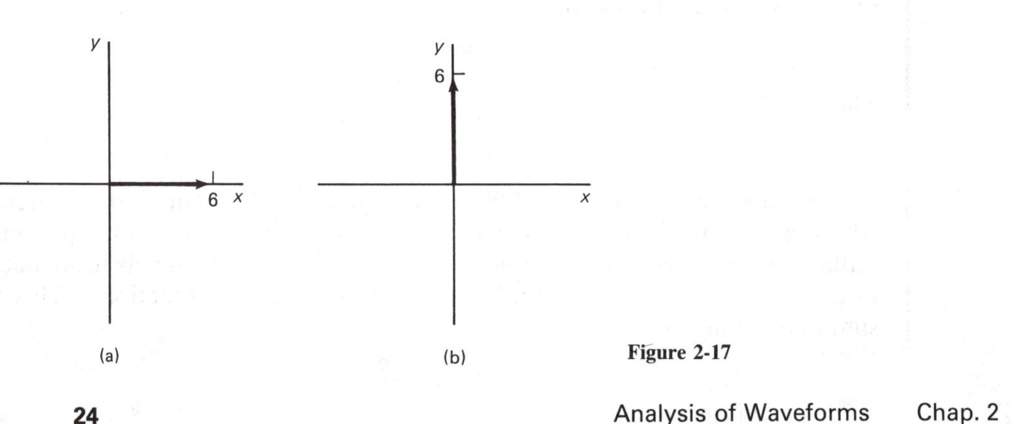

Figure 2-17

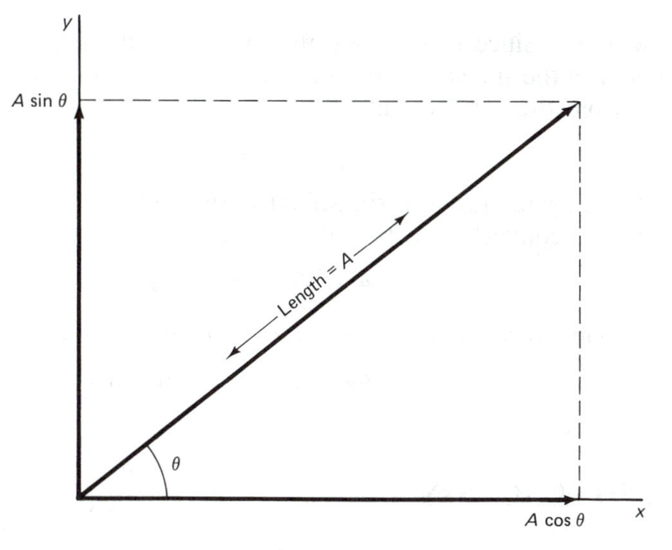

Figure 2-18 $A \sin(\omega t + \theta)$ being split into a sine and a cosine component.

Since this is exactly what is done in polar-to-rectangular conversions, a polar-to-rectangular operation on a calculator may be used.

$$A \angle\theta = A\cos(\theta) + jA\sin(\theta) \tag{2-18}$$

We multiply the first term or the real part by $\sin(\omega t)$. We then remove the j and multiply the second term by $\cos(\omega t)$.

EXAMPLE 2-9

Split into sine and cosine parts.

$$5\sin(3t + 70°)$$

SOLUTION

$$5 \angle 70° = 1.7101 + j4.6985$$

Therefore,

$$5\sin(3t + 70°) = 1.7101\sin(3t) + 4.6985\cos(3t)$$

EXAMPLE 2-10

Combine into a single sine function.

$$4e^{-3t}\sin(2t) - 3e^{-3t}\cos(2t)$$

SOLUTION If the exponentials or the frequency of both sinusoidal functions had been different, we would be unable to combine the sine and cosine functions. Since in this case the powers of the exponentials and the frequency of the sinusoidal are the same, we may combine the functions. First factor out the exponential

$$e^{-3t}\,[4 \sin (2t) - 3 \cos (2t)]$$

and work with just the sinusoidal parts. We put the function in a polar form and convert to rectangular form.

$$4 - j3 = 5 \angle -36.870^\circ$$

This leads us to the single sine function with a phase angle.

$$5e^{-3t} \sin (2t - 36.870^\circ)$$

2-7 SHIFTED FUNCTION

All functions do not start at $t = 0$. We need a way to shift or delay a function. In this section we develop the shifting theorem. Figure 2-19 shows the function $f_1(t) = 2tu(t)$ shifted by 2 seconds.

Table 2-3 shows the values of $f_1(t)$ unshifted and $f_1(t)$ shifted, which we will call $f_2(t)$. What we notice is that the values are exactly the same except that they occur at times that differ by exactly 2 seconds. To shift this function to $t = 2$ sec, it means the function starting at $t = 2$ will have the same values as it did starting at $t = 0$. This can be done by changing all of the t's to $(t - 2)$.

$$f_2(t) = f_1(t - 2) = 2(t - 2)u(t - 2)$$

Notice that $u(t - 2)$ will be zero until we reach $t = 2$ sec. Substituting into $f_2(t)$, we will get the values shown in Table 2-3.

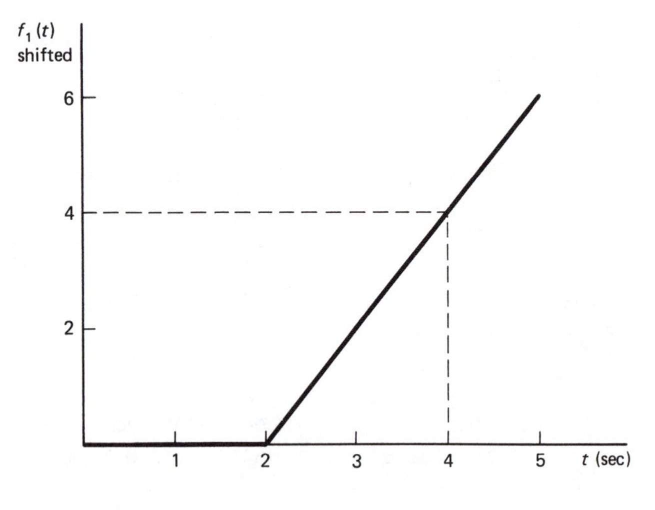

Figure 2-19 Function $2tu(t)$ shifted to the right by 2 seconds.

TABLE 2-3

t	$f_1(t)$	$f_2(t)$
0	0	0
1	2	0
2	4	0
3	6	2
4	8	4
5	10	6

The shifting theorem can be summarized as follows: To shift a function to start at $t = a$ seconds, we replace each t with $t - a$. Therefore, the function $f_1(t)$

$$f_1(t) = f(t)u(t) \tag{2-19a}$$

becomes a new function, $f_2(t)$,

$$f_2(t) = f_1(t - a) = f(t - a)u(t - a) \tag{2-19b}$$

If a is positive, the function will be shifted to the right of $t = 0$ and delay the function by a seconds. If a is negative, the function will be shifted to the left.

EXAMPLE 2-11

For the following functions, shift them so that they start at 4 seconds.

(a) $tu(t) + 4u(t)$
(b) $4e^{-2t} \sin (7t + \pi/2)\, u(t)$

SOLUTION

(a) $(t - 4)u(t - 4) + 4u(t - 4)$
(b) $4e^{-2(t-4)} \sin [7(t - 4) + \pi/2]\, u(t - 4)$

EXAMPLE 2-12

Shift each function back to the origin.

(a) $7(t - 3)^2u(t - 3)$
(b) $5 \sin (2t - 10)\, u(t - 4)$

SOLUTION

(a) Looking at the step function we see that the function was delayed by 3 seconds. We want to shift the function the opposite direction; therefore, let $t = t + 3$.

$$7[(t + 3) - 3]^2u[(t + 3) - 3] = 7t^2u(t)$$

(b) Again we begin by looking at the step function, and we see that this function has been shifted by 4 seconds. The sine function must first be put in the proper form of a shifted function.

$$\sin(2t - 10)$$

$$\sin(2t - 8 - 2)$$

$$\sin[2(t - 4) - 2]$$

Therefore, the function in a proper form is

$$5\sin[2(t - 4) - 2]\,u(t - 4)$$

We now substitute $t = t + 4$.

$$5\sin\{2[(t + 4) - 4] - 2\}\,u((t + 4) - 4) = 5\sin(2t - 2)\,u(t)$$

2-8 PUTTING IT ALL TOGETHER

Most complex waveforms can be expressed using the impulse function, step function, ramp function, sinusoidal function, and the shifting theorem. In this section we use all of these concepts together to describe complex waveforms.

When dealing with complex waveforms, we either need to graph a function from a mathematical expression, or we need to write a mathematical expression from a given graph. In each case there are some general guidelines to follow. We will look first at graphing a function from a mathematical expression.

There are two methods used when you are given a mathematical expression to graph. The first is to graph each term separately and then add the functions point by point until you know what the total function looks like. The other method is to evaluate the function at points of time where a new term in the mathematical expression becomes active. The first method is used when the function is complex or we are unsure of how to graph it. The second method is used more frequently since it is faster to use.

EXAMPLE 2-13

Graph the function using the two methods just discussed.

$$f(t) = 5u(t) - 7u(t - 2) + 3u(t - 3)$$

SOLUTION Using the first method, we graph each term as shown in Fig. 2-20a–c. We then add these at each location in time and plot the sum as shown in Fig. 2-20d. The second method is basically the same as the first method except that we are seeing things change mathematically instead of graphically. The significant times and values are shown in Table 2-4. As you become more experienced, this will become a mental process not requiring a written table.

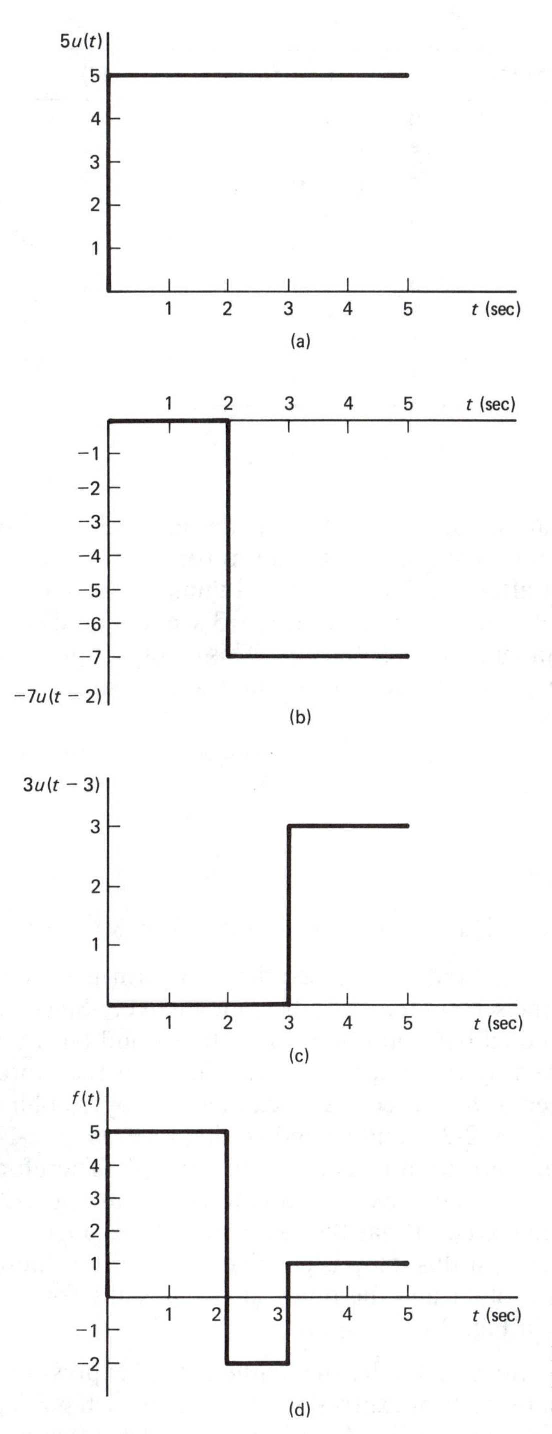

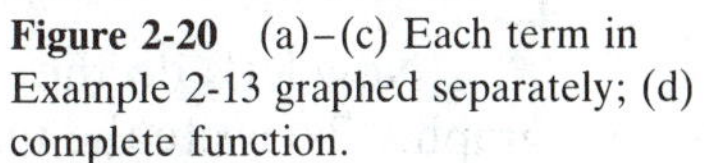

Figure 2-20 (a)–(c) Each term in Example 2-13 graphed separately; (d) complete function.

TABLE 2-4

t	$5u(t) - 7u(t-2) + 3u(t-3)$	$f(t)$
$t = 0^-$	$0 - 0 + 0$	0
$t = 0^+$	$5 - 0 + 0$	5
$0 < t < 2$	$5 - 0 + 0$	5
$t = 2^-$	$5 - 0 + 0$	5
$t = 2^+$	$5 - 7 + 0$	-2
$2 < t < 3$	$5 - 7 + 0$	-2
$t = 3^-$	$5 - 7 + 0$	-2
$t = 3^+$	$5 - 7 + 3$	1
$t > 3$	$5 - 7 + 3$	1

To simplify calculation when several terms are active at the same time, we may combine them into a single effective term (or a few effective terms). Notice in Table 2-4 that after $t = 2$ sec, we could think of the $5u(t) - 7u(t - 2)$ as a single function of $-2u(t - 2)$. After $t = 3$ sec we could combine the first three terms and think of them as $1u(t - 3)$. This is very helpful in functions of several terms since we will in effect be working with a smaller number of terms at any given time.

EXAMPLE 2-14

Graph the function.

$$g(t) = 4tu(t) - 4(t - 2)u(t - 2) - 8(t - 5)u(t - 5) + 8(t - 6)u(t - 6)$$

SOLUTION From $t = 0$ until $t = 2$ sec the first term is the only active term. At $t = 2$ sec the second term also becomes active. Since both terms are ramps and have equal but opposite slopes, the second term counteracts further changes caused by the first term. The function therefore remains at the level it obtained at $t = 2$ sec. This can be seen by graphing the two functions separately (Fig. 2-21a and b) and adding them (Fig. 2-21c).

At $t = 5$ sec the third term adds a negative ramp. Therefore, at $t = 5$ sec we get a ramp in the negative direction as shown in Fig. 2-22a. At $t = 6$ sec another ramp starts. It has the same magnitude slope as the ramp that starts at $t = 5$ sec, but this ramp is positive. Therefore, these last two ramps counteract each other and the function remains at 0 for $t > 6$. The final result is shown in Fig. 2-22b.

Now let's do the reverse and write the mathematical expression from the graph. To write the mathematical expression, we look at each point the graph changes its current direction or level. At each point we try different functions

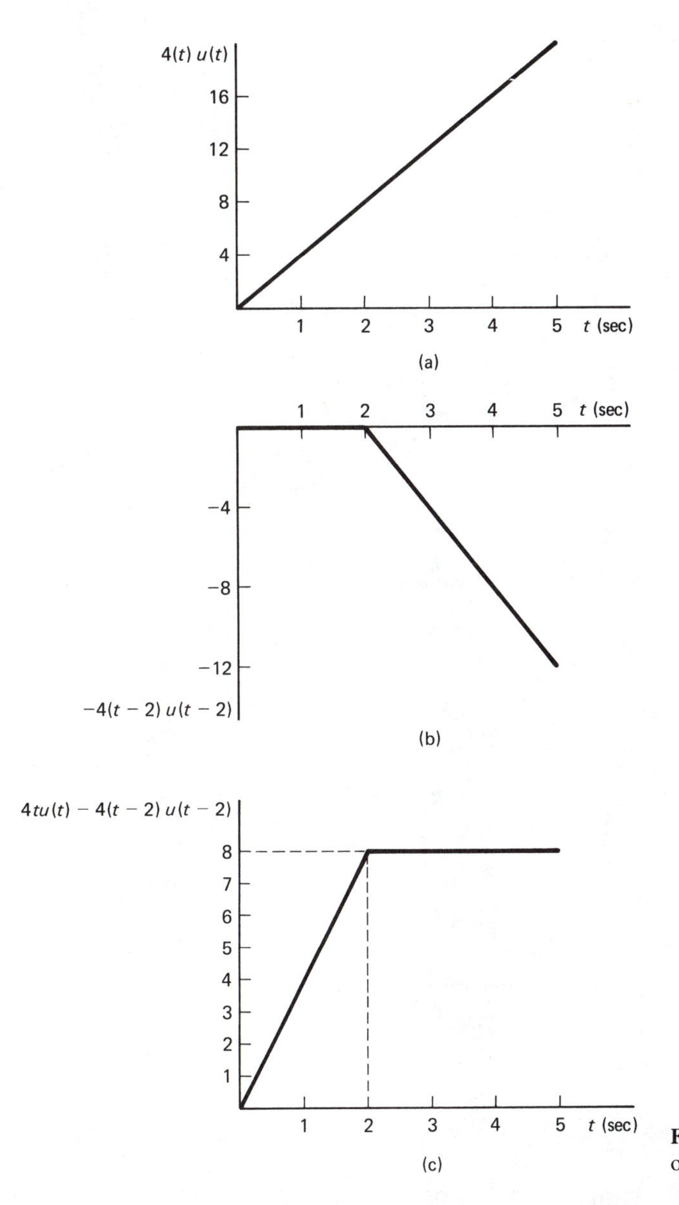

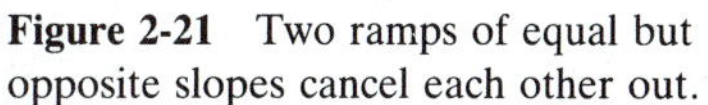

Figure 2-21 Two ramps of equal but opposite slopes cancel each other out.

until we find the one that gives us the desired result, but it is not all guesswork. We can make some general rules on common function changes that will occur.

The first common change is a step change. Figure 2-23 shows a step change occurring at $t = a$. When there are two values at one time, there will always be a step function involved. The change in the height is $f(a^+) - f(a^-)$ whether it is a positive or negative change. Therefore, the term to add to the function is

$$[f(a^+) - f(a^-)]u(t - a) \tag{2-20}$$

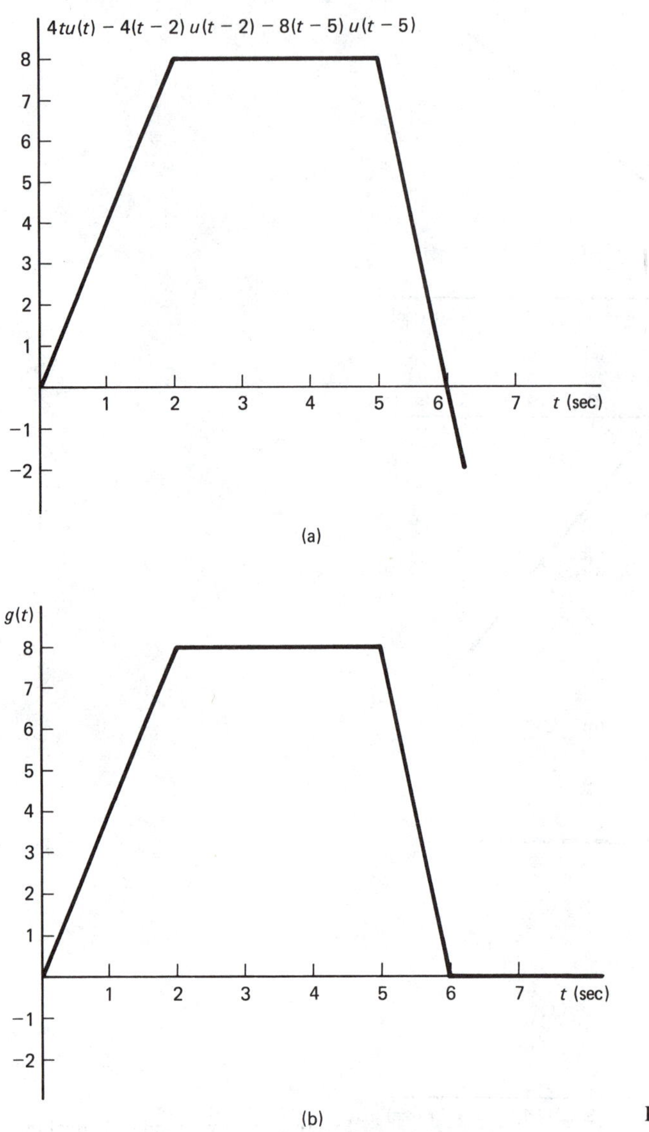

Figure 2-22

Another common change is the slope of a ramp. Figure 2-24 shows a function with a positive change in the slope at $t = a$. The slope of A becomes a slope of B when a term with a slope of C becomes active in the function.

$$B = A + C$$

Therefore,

$$C = B - A \tag{2-21a}$$

The term to add to the function is then

$$C(t - a)u(t - a) \tag{2-21b}$$

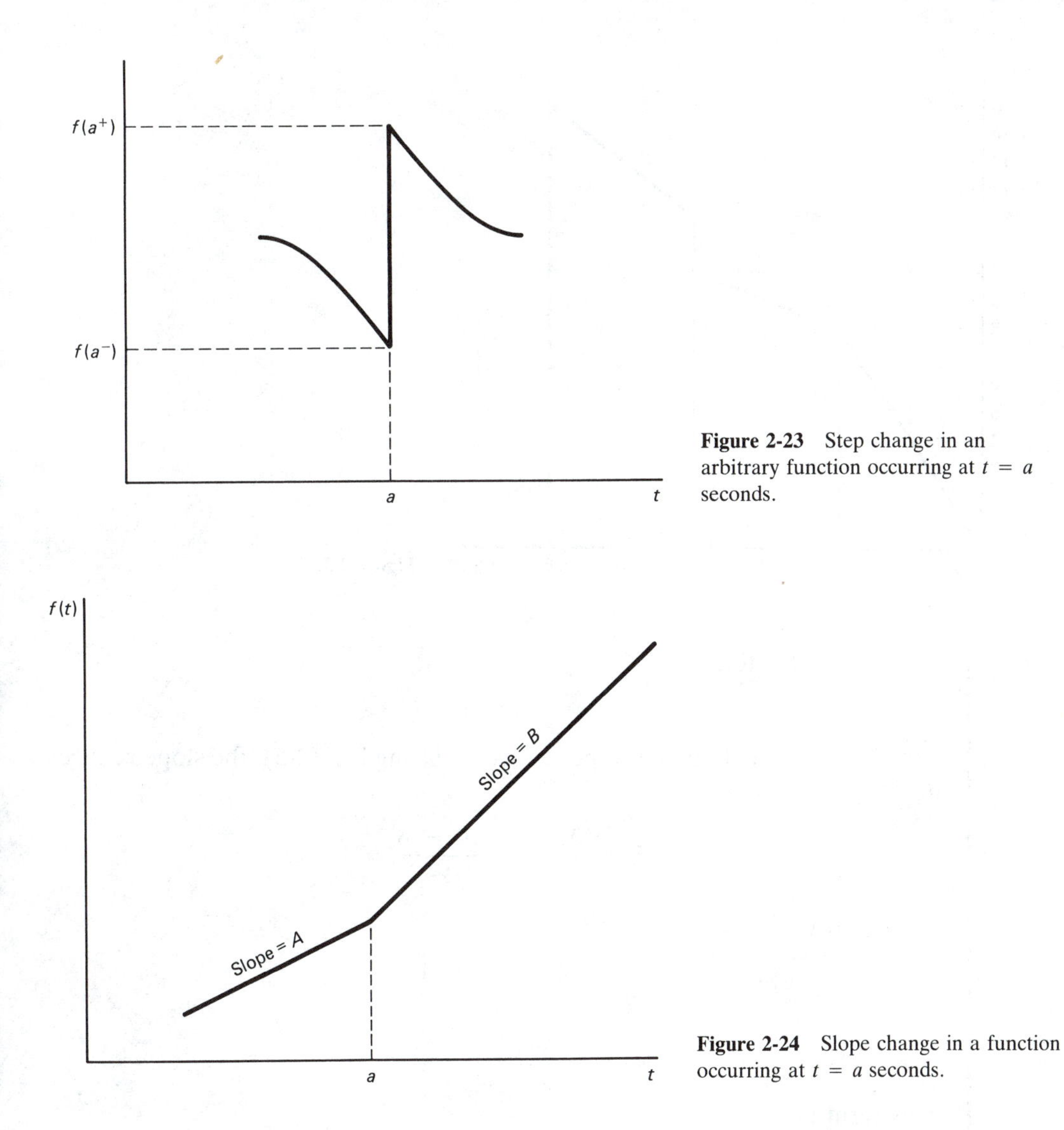

Figure 2-23 Step change in an arbitrary function occurring at $t = a$ seconds.

Figure 2-24 Slope change in a function occurring at $t = a$ seconds.

EXAMPLE 2-15

Write the function for Fig. 2-25.

SOLUTION Starting at $t = 0$ there is an impulse with an area of 2; therefore, the first term is

$$2\delta(t)$$

For $t = 0$ to 2 sec there is a ramp. Using Eq. (2-5), we can find the slope.

$$K = \frac{\Delta f(t)}{\Delta t} = \frac{8 - 0}{2 - 0} = 4$$

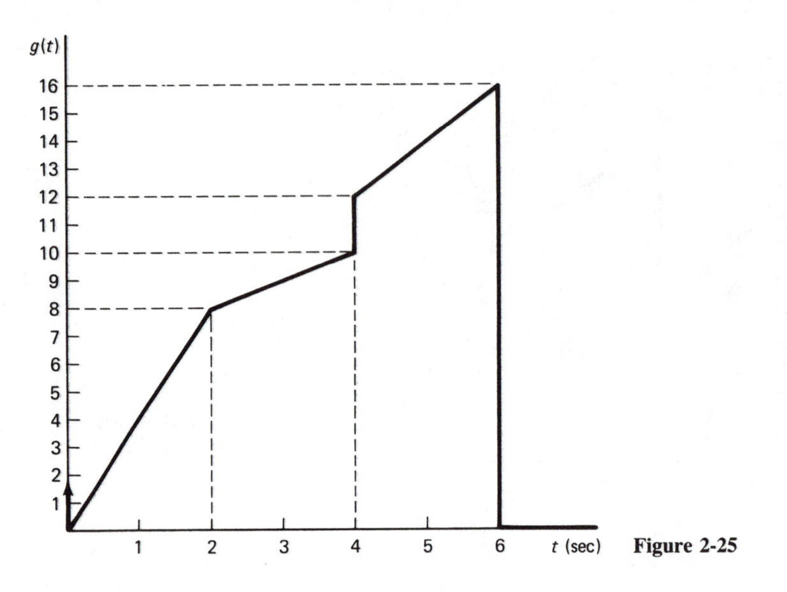

Figure 2-25

Therefore, this term is

$$4tu(t)$$

For $t = 2$ to 4 sec the slope reduces. Using Eq. (2-5), the slope reduces to

$$K = \frac{\Delta f(t)}{\Delta t} = \frac{10 - 8}{4 - 2} = 1$$

From Eq. (2-21a),

$$\begin{aligned} C &= B - A \\ &= 1 - 4 \\ &= -3 \end{aligned}$$

This term becomes

$$-3(t - 2)u(t - 2)$$

At $t = 4$ sec we have two terms to add to our function: a step change and a ramp change. First we will find the step change using Eq. (2-20).

$$[f(a^+) - f(a^-)]u(t - a)$$
$$(12 - 10)u(t - 4)$$
$$2u(t - 4)$$

Also at $t = 4$ the slope changes to

$$K = \frac{\Delta f(t)}{\Delta t} = \frac{16 - 12}{6 - 4} = 2$$

and the previous slope was 1. Using Eq. (2-21a) gives us

$$C = B - A$$
$$= 2 - 1$$
$$= 1$$
$$C(t - a)u(t - a)$$
$$(t - 4)u(t - 4)$$

At $t = 6$ sec the function has a step and a change in the ramp slope. For the step

$$[f(a^+) - f(a^-)]u(t - a)$$
$$(0 - 16)u(t - 6)$$
$$-16u(t - 6)$$

The new ramp has a slope of 0 and the old slope was 2; therefore,

$$C = B - A$$
$$= 0 - 2$$
$$= -2$$
$$C(t - a)u(t - a)$$
$$-2(t - 6)u(t - 6)$$

The complete equation is the sum of the terms.

$$g(t) = 2\delta(t) + 4tu(t) - 3(t - 2)u(t - 2) + 2u(t - 4) + (t - 4)u(t - 4) - 16u(t - 6) - 2(t - 6)u(t - 6)$$

EXAMPLE 2-16

Write the equation for Fig. 2-26. The function goes to 5 V at $t = 10$ sec.

SOLUTION All of the terms of the function become active at $t = 0$. The function will be in the form of

$$v(t) = K_1u(t) + K_2e^{-\alpha t}\sin(\omega t + \theta)$$

The exponential sine function is shifted up by a dc voltage of 5 V; therefore, $K_1 = 5$.

Before we find K_2 we must find α. We know that five time constants are over at $t = 10$ sec.

$$e^{-\alpha(10)} = e^{-5}$$
$$-\alpha(10) = -5$$
$$\alpha = 0.5$$

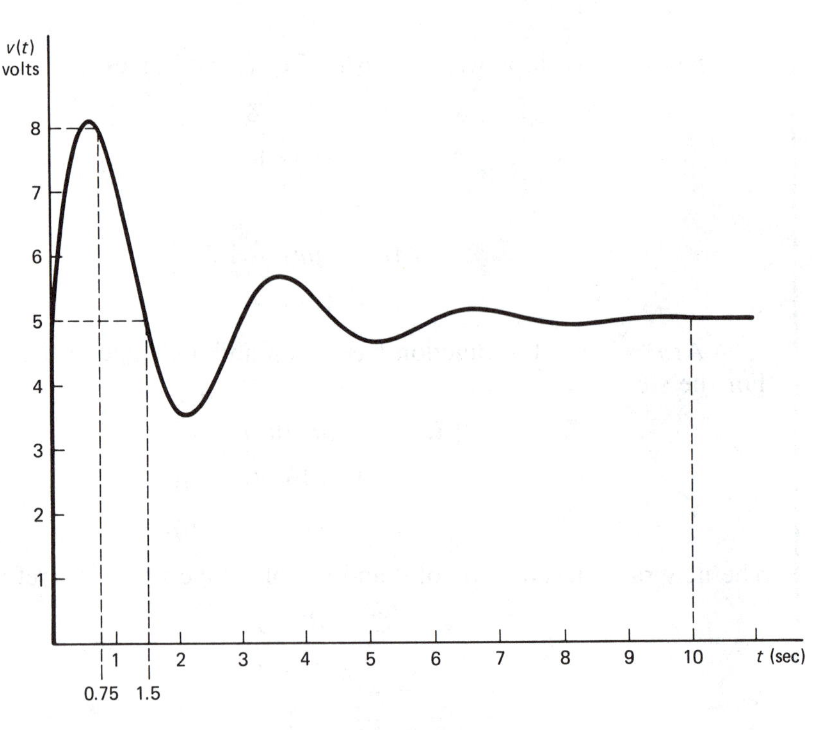

Figure 2-26

To find ω we find a convenient location where we know ωt and the time it occurs. At $t = 1.5$ sec, ωt is π radians.

$$\omega t = \pi \text{ at } 1.5 \text{ sec}$$

$$\omega = \frac{\pi}{1.5} = \frac{2\pi}{3}$$

At this point we have the following information about the function:

$$v(t) = 5u(t) + K_2 e^{-0.5t} \sin\left(\frac{2\pi}{3}t + \theta\right)u(t)$$

At $t = 0^+$ the function has a value of 5; therefore,

$$5 = 5 + K_2 e^0 \sin(0 + \theta)$$

which means that either K_2 or θ is equal to zero. K_2 must not equal zero since the function must have a sinusoidal function in it. This means that θ must equal zero.

We may not always be able to solve θ in this equation at $t = 0$. We may have to solve it simultaneously with the equation we will use to find K_2. To find K_2 we need to find a location where:

1. The sine part is not equal to zero.
2. The exact value of the function, with the time it occurs, is known.

We cannot evaluate K_2 at $t = 1.5$ sec since the sine function is zero at this time. At $t = 0.75$ sec the sine part of the function is not zero, which makes it a suitable point to use.

$$8 = 5 + K_2 e^{-0.5t} \sin\left(\frac{2\pi}{3} t\right)$$

$$3 = K_2 e^{-0.5(0.75)} \sin\left[\frac{2\pi}{3}(0.75)\right]$$

$$K_2 = 4.3650$$

The complete answer is

$$v(t) = 5u(t) + 4.3650e^{-0.5t} \sin\left(\frac{2\pi}{3} t\right) u(t)$$

When writing mathematical expressions from a given graph, we often find a waveform that repeats. This type of waveform has an infinite number of terms, but we can express it as a summation. Summations are not difficult if a few simple rules are followed.

1. Write out the terms that complete approximately two cycles. The first term commonly will not be in the summation, and we need enough terms to see what is repeating and how it is repeating.
2. Starting with the second or third term, write down everything that does not change from term to term, leaving space for the items that are changing.
3. Fill in the blanks with some form of the counting variable, n. Start n at any convenient value since we can change the starting value of n later. Table 2-5 shows some common forms of the counting variable and the sequence that each form follows.
4. After you supply a form of n for all the blanks that were left from step 2, try a value of n one increment before where you are currently starting n. If this gives you the beginning terms you left out, the entire function can be expressed by the summation alone. If this does not produce the beginning terms, they must be expressed separately before the summation.
5. Change the summation to the shortest form. This is done by examining the forms used for n. If there are mostly $n + 1$ terms, substitute $n = n$

TABLE 2-5

n	$n + 1$	$(-1)^n$	$(-1)^{n+1}$	$2n$	$2(n + 1)$
0	1	1	−1	0	2
1	2	−1	1	2	4
2	3	1	−1	4	6
3	4	−1	1	6	8

-1 into all forms of n. If there are mostly $n - 1$ terms, substitute $n = n + 1$ for all n's. If there are mostly n terms, leave it alone.

Following these steps will produce the best form of a summation in the shortest amount of time.

EXAMPLE 2-17

Write the equation for Fig. 2-27 in a summation form.

SOLUTION

Step 1. First write the terms for the first couple of cycles. Each cycle is 2 sec.

$$2u(t) - 3u(t-1) + 3u(t-2) - 3u(t-3) + 3u(t-4) + \cdots$$

Step 2. Starting with the second term, we write down everything that stays constant. We will also arbitrarily start n at zero.

$$\sum_{n=0}^{\infty} \quad 3u(t - \quad)$$

Notice that we have left a blank for the sign of the term.

Step 3. Find the form of the counting variable, n, to use in the blanks. From Table 2-5 we can find the forms required to fill in the blanks.

$$\sum_{n=0}^{\infty} (-1)^{n+1} 3u[t - (n+1)]$$

Step 4. When we let $n = -1$ it does not give us the first term of the equation, and therefore we will have to write it as a separate term.

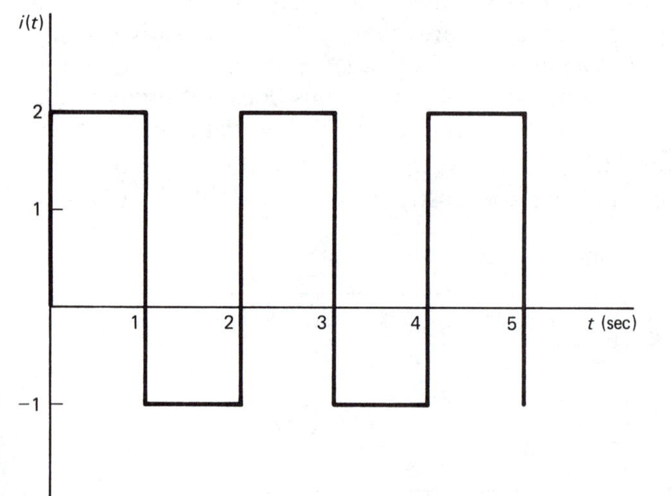

Figure 2-27

Step 5. There are mostly $n + 1$ terms in the summation. We will let $n = n - 1$ to shorten the expression.

$$\sum_{n-1=0}^{\infty} (-1)^{n-1+1} 3u[t - (n - 1 + 1)]$$

$$\sum_{n=1}^{\infty} (-1)^{n} 3u(t - n)$$

The equation in its shortest form is

$$i(t) = 2u(t) + \sum_{n=1}^{\infty} (-1)^{n} 3u(t - n)$$

EXAMPLE 2-18

Write the equation for the graph shown in Fig. 2-28 in a summation form.

SOLUTION

Step 1. Write the terms of the first couple of cycles. First we must find the ω for the sine wave on which this function is based. At $t = 2$ sec, ωt is equal to one-half of a complete cycle.

$$\omega t = \frac{2\pi}{2} \qquad \text{at } t = 2 \text{ sec}$$

$$\omega(2) = \pi$$

$$\omega = \frac{\pi}{2}$$

The first term of the waveform is then

$$\sin\left(\frac{\pi}{2} t\right) u(t)$$

At $t = 2$ sec, we need to start another sine function to cancel the first

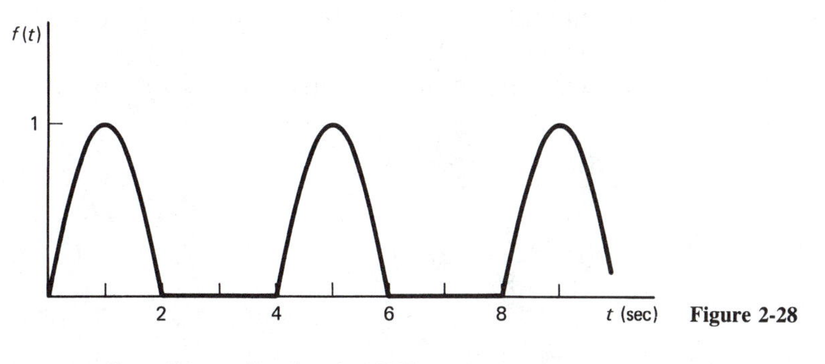

Figure 2-28

term. (There are a couple of ways to do this, but we must always write our functions in a shifted form.)

$$\sin\left(\frac{\pi}{2}t\right)u(t) + \sin\left[\frac{\pi}{2}(t-2)\right]u(t-2)$$

Notice that at $t = 2$ sec the second term will be 180° out of phase with the first term, and the two terms will cancel each other from $t = 0$ to ∞.

At $t = 4$ sec we need to start a new sine wave and then cancel it at $t = 6$ sec. The third and fourth terms will therefore be

$$\sin\left[\frac{\pi}{2}(t-4)\right]u(t-4) + \sin\left[\frac{\pi}{2}(t-6)\right]u(t-6)$$

Step 2. Starting with the second term, we write everything that stays constant.

$$\sum_{n=1}^{\infty} \sin\left[\frac{\pi}{2}(t - \quad)\right]u(t - \quad)$$

Step 3. Using Table 2-5, we can fill in the blanks.

$$\sum_{n=1}^{\infty} \sin\left[\frac{\pi}{2}(t-2n)\right]u(t-2n)$$

Step 4. In this particular case when we start n at 0 we get the first term. So we just start at $n = 0$ instead of at 1, and the completed function is

$$f(t) = \sum_{n=0}^{\infty} \sin\left[\frac{\pi}{2}(t-2n)\right]u(t-2n)$$

Step 5. Since all the terms above do not have $n + 1$ or $n - 1$ form, we will leave the summation as it is.

2-9 DERIVATIVES AND INTEGRALS OF WAVEFORMS

There are a number of circuits that integrate and differentiate an input signal. The simplest circuits are the inductor and capacitor. Assuming the initial conditions of zero, the voltage across these devices given the current is

$$v(t) = L\frac{di(t)}{dt} \tag{2-22}$$

$$v(t) = \frac{1}{C}\int_0^t i(t)\,dt \tag{2-23}$$

and the current through these devices given the voltage is

$$i(t) = \frac{1}{L}\int_0^t v(t)\,dt \tag{2-24}$$

$$i(t) = C\frac{dv(t)}{dt} \tag{2-25}$$

Since even the simplest components integrate and differentiate, we need to know what to expect when a function is integrated or differentiated. We may integrate or differentiate functions either mathematically or graphically. Either method may be superior, depending on the particular case. First we will look at the mathematical way of integrating and differentiating, then the graphic method.

There are three areas that may not have been covered in beginning calculus that are common to complex functions in electronics. First are graphs with discontinuities in them. The discontinuity will be either a step or an impulse. These will be handled by evaluating the function on the left of the discontinuity, the right of the discontinuity, and then at the discontinuity. On either side of a discontinuity, normal calculus procedures are used.

We have already covered, in concept, the integral and derivative of the step and impulse function in Sections 2-3 and 2-4. Section 2-3 showed the integral of a step is the ramp function.

$$\int_0^t u(t)\,dt = tu(t) \tag{2-26}$$

In Section 2-4 we defined the impulse by finding the slope (the derivative) at the step function's discontinuity.

$$\frac{d}{dt}u(t) = \delta(t) \tag{2-27}$$

Since the derivative of the step is the impulse, the integral of the impulse must be the step function.

$$\int_0^t \delta(t)\,dt = u(t) \tag{2-28}$$

Notice that we instantaneously sum up the area contained in the impulse at $t = 0$. The derivative of the impulse normally does not show up in practical circuits. This is more theoretical and will not be covered here.

Second is removing $u(t)$ from an integral. We will always be integrating from $t = 0$ to t. Since we are concerned with the function to the right of $t = 0$ we may use $t = 0^+$ for the lower limit of integration. In this interval, $t = 0^+$ to t, $u(t)$ has a constant value of 1 and may be factored out of the integral like any other constant.

Third is finding the integral and derivative of a shifted function. Since we always write our shifted function in the form of $f(t - a)u(t - a)$, we may integrate the function in its unshifted form, $f(t)u(t)$, and then shift the result back to time $t = a$.

EXAMPLE 2-19

Show that the preceding paragraph is true for the following function.

$$\int_0^t 4(t - 2)u(t - 2)\, dt$$

SOLUTION The unshifted form of $4(t - 2)u(t - 2)$ is $4tu(t)$, which we will integrate. Since $u(t)$ has a constant value of 1 from $t = 0^+$ to ∞, we may remove $u(t)$ from the integral as we could any other constant.

$$\int_0^t 4t\, dt\, u(t) = \frac{4t^2}{2}\bigg|_0^t u(t)$$

$$= 2t^2u(t)$$

Then we shift the function to start at $t = 2$ sec.

$$2(t - 2)^2u(t - 2)$$

We will now work the problem without shifting back to the origin, and we should get the same result.

$$\int_0^t 4(t - 2)u(t - 2)\, dt$$

Since the function is zero until $t = 2$ sec, we are actually integrating from $t = 2^+$ to t.

$$\int_2^t 4(t - 2)u(t - 2)\, dt$$

Over these limits $u(t - 2)$ has a constant value of 1 and may therefore be removed from under the integral.

$$\left[\int_2^t 4(t - 2)\, dt\right]u(t - 2)$$

$$\left(\int_2^t 4t\, dt + \int_2^t -8\, dt\right)u(t - 2)$$

$$\left[\frac{4t^2}{2}\bigg|_2^t + (-8t)\bigg|_2^t\right]u(t - 2)$$

$$[2t^2 - 2(2)^2 - 8t + 8(2)]u(t - 2)$$

$$2(t^2 - 4t + 4)u(t - 2)$$

$$2(t - 2)^2u(t - 2)$$

The derivative of a shifted function is handled in a similar fashion. Shift the function to the origin, take the derivative, and shift the result back.

When taking the derivative of a function, we must consider each term of the function as the product of two functions of t. One of the functions will typically be $u(t)$. Therefore, the derivative of each term will be

$$\frac{d}{dt} f(t)u(t) = f'(t)u(t) + f(t)\delta(t)$$

Since in the second term $\delta(t)$ is zero at any time except at $t = 0$, the only time $f(t)$ makes any difference is at $t = 0$. We may, therefore, use $f(0)$ in the second term.

$$\frac{d}{dt} f(t)u(t) = f'(t)u(t) + f(0)\delta(t) \tag{2-29}$$

EXAMPLE 2-20

Find the derivative of

$$4u(t) + 2\cos(4t)\, u(t)$$

SOLUTION Take the derivative of each term.

$$\frac{d}{dt} 4u(t) = (0)u(t) + 4\delta(t) = 4\delta(t)$$

$$\frac{d}{dt} 2\cos(4t)\, u(t) = -(2)(4)\sin(4t)\, u(t) + 2\cos[4(0)]\, \delta(t)$$

$$= -8\sin(4t)\, u(t) + 2\delta(t)$$

The derivative of the function is the sum of the two previous results.

$$= 6\delta(t) - 8\sin(4t)\, u(t)$$

EXAMPLE 2-21

Mathematically find the integral of the function in Fig. 2-29. Graph the result.

SOLUTION First we find the equation of the function.

$$g(t) = -10u(t) + 20u(t-4) - 5(t-8)u(t-8)$$
$$+ 5(t-10)u(t-10) + 5\delta(t-12)$$

Next we shift each term back to the origin, integrate it, and shift the result back to the proper time.

$$\int_0^t -10u(t)\, dt = -10tu(t) \quad \Rightarrow \quad -10tu(t)$$

$$\int_0^t 20u(t)\, dt = 20tu(t) \quad \Rightarrow \quad 20(t-4)u(t-4)$$

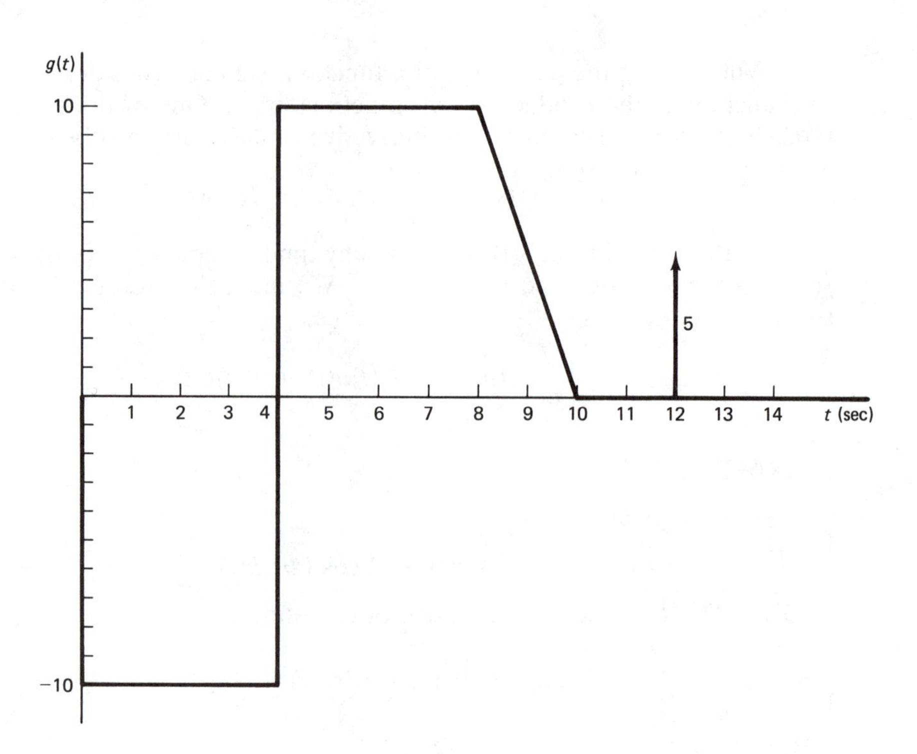

Figure 2-29

$$\int_0^t -5tu(t)\,dt = \frac{-5}{2}t^2u(t) \quad \Rightarrow \quad \frac{-5}{2}(t-8)^2u(t-8)$$

$$\int_0^t 5tu(t)\,dt = \frac{-5}{2}t^2u(t) \quad \Rightarrow \quad \frac{5}{2}(t-10)^2u(t-10)$$

$$\int_0^t 5\delta(t)\,dt = 5u(t) \quad \Rightarrow \quad 5u(t-12)$$

Combining all terms gives

$$\int_0^t g(t)\,dt = -10tu(t) + 20(t-4)u(t-4)$$
$$- \frac{5}{2}(t-8)^2u(t-8) + \frac{5}{2}(t-10)^2u(t-10) + 5u(t-12)$$

Figure 2-30 shows the graph of the function. Notice that the fourth term of the result does not cancel the third term the way that ramp functions cancel. Instead, the fourth term cancels the decreasing effect the third term has on the function. Therefore, the third and fourth terms combine to have the effect of a ramp with a slope of -10.

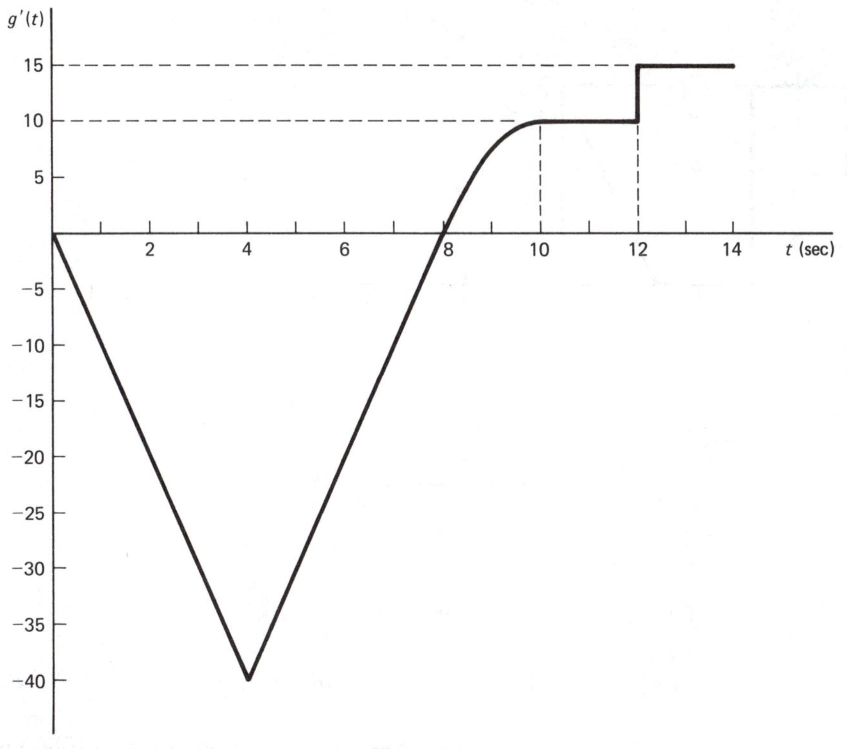

Figure 2-30

EXAMPLE 2-22

Mathematically find the derivative of the function shown in Fig. 2-31. Graph the result.

SOLUTION Again we write the equation of the function first.

$$g(t) = 5u(t) - 5u(t-1) + 5(t-2)u(t-2) - 10u(t-3) - 5(t-3)u(t-3)$$

Next we shift each term back to the origin, find the derivative, and then shift each term back to the proper time delay.

$$\frac{d}{dt}5u(t) = (0)u(t) + 5\delta(t) \qquad \Rightarrow \quad 5\delta(t)$$

$$\frac{d}{dt}-5u(t) = -(0)u(t) - 5\delta(t) \qquad \Rightarrow \quad -5\delta(t-1)$$

$$\frac{d}{dt}5tu(t) = 5u(t) + 5(0)\delta(t) \qquad \Rightarrow \quad 5u(t-2)$$

$$\frac{d}{dt}-10u(t) = -(0)u(t) - 10\delta(t) \quad \Rightarrow \quad -10\delta(t-3)$$

$$\frac{d}{dt}-5tu(t) = -5u(t) - 5(0)\delta(t) \quad \Rightarrow \quad -5u(t-3)$$

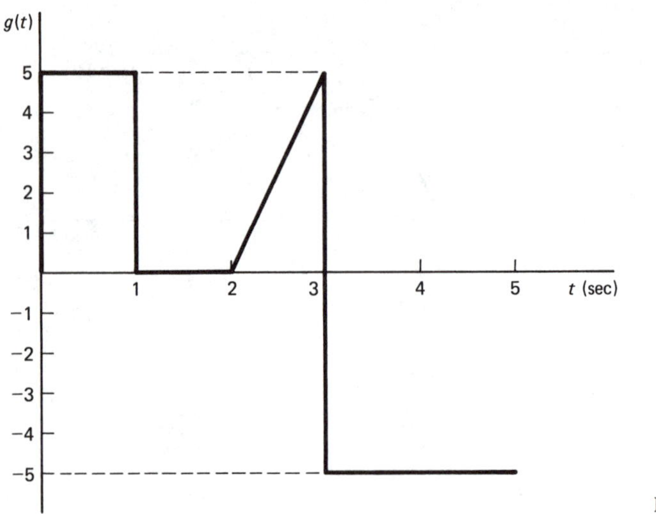

Figure 2-31

The resulting equation is

$$\frac{d}{dt}g(t) = 5\delta(t) - 5\delta(t-1) + 5u(t-2) - 10\delta(t-3) - 5u(t-3)$$

Figure 2-32 shows the graph of the result. Notice the $-10\delta(t)$ is still drawn from the x-axis even though the end of it could be considered starting at a level of 5. This is because the impulse actually has an infinite height and we stop it at -10 only to note its area, not its height.

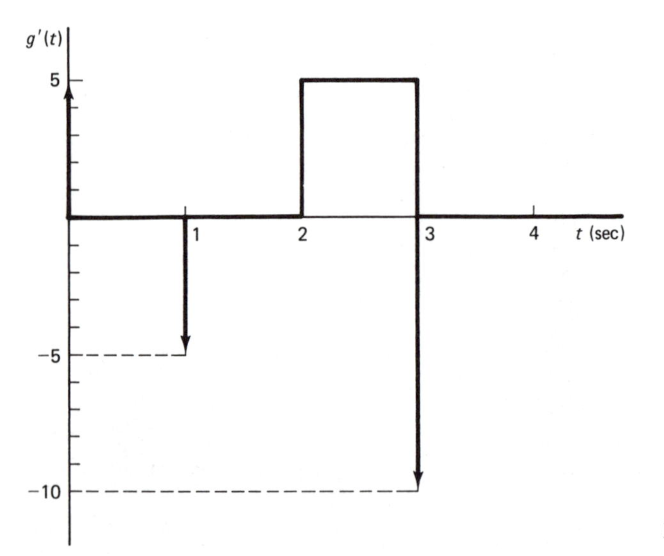

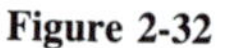

Figure 2-32

The graphic method of integration is simply a matter of calculating the area under a curve. Sliding a sheet of paper from left to right, we can observe, to the left of the paper, the rate of area being summed and if the area is increasing, decreasing, or remaining constant.

EXAMPLE 2-23

Graphically find the integral of the function in Fig. 2-33. (Note that it is the same graph used in Example 2-21.)

SOLUTION From $t = 0$ to $t = 4$ sec the area is

$$-10 \times 4 = -40$$

Therefore, we have summed -40 V-sec when we get to $t = 4$ sec. If we slide a piece of paper from left to right across $g(t)$, we see that a constant amount of area is being added as we move the paper. Since this is a constant area being summed, we will have a ramp starting at zero area at $t = 0$ and ending with an area of -40 at $t = 4$ sec. This is shown in Fig. 2-34a.

From $t = 4$ to $t = 8$ will also add a constant amount of area as we slide our paper across the function, but this will be a positive area. The total area added when we reach $t = 8$ sec is

$$(8 - 4) \times 10 = 40$$

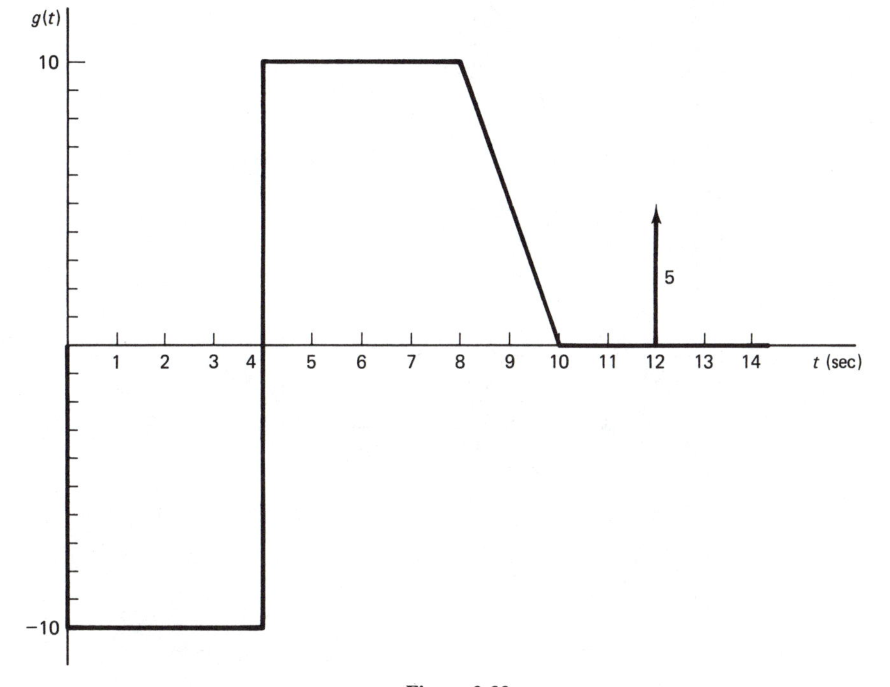

Figure 2-33

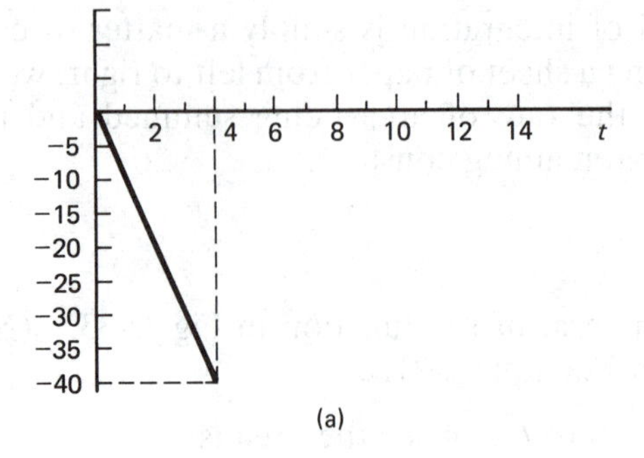

(a)

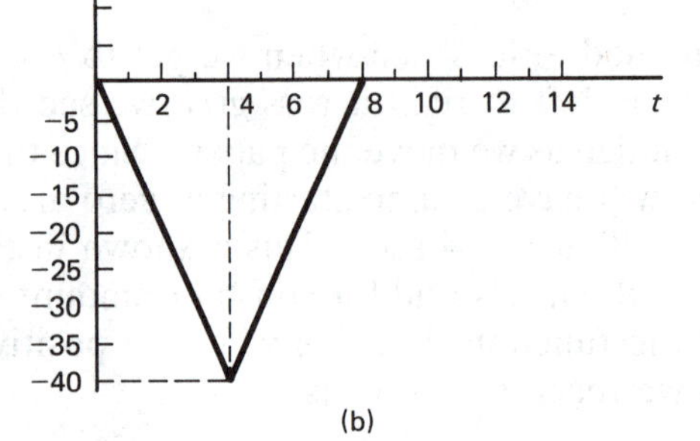

(b)

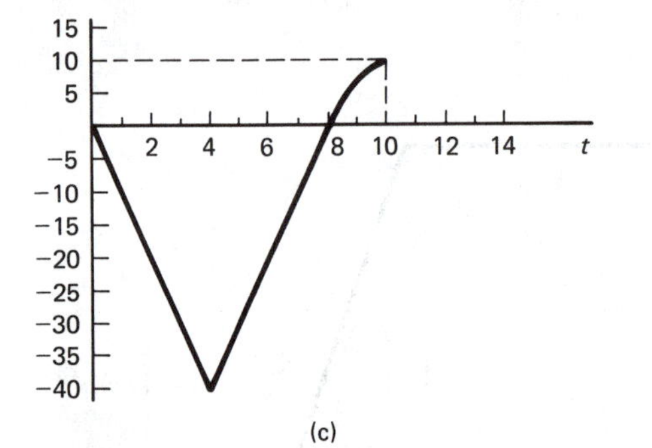

(c)

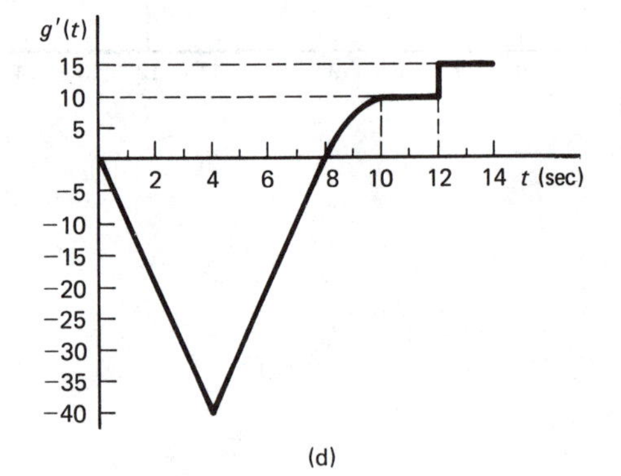

(d)

Figure 2-34

This is added to our present area already summed up to $t = 4$ sec shown in Fig. 2-34b. Since the area is positive, we will have a positive ramp. This ramp will take us back to zero at $t = 8$ sec because the negative and positive areas summed up to $t = 8$ sec are equal.

From $t = 8$ to $t = 10$ the area under the curve is

$$\frac{1}{2}(10 - 8) \times 10 = 10$$

As we move our paper across the function in this interval, less and less positive area is being summed. Therefore, the integral is not a ramp. The area is mostly added closer to $t = 8$ sec, and at $t = 10$ sec, zero is being added. Figure 2-34c shows this being added to our graph.

From $t = 10$ to $t = 12$ no area is being added; therefore, the integral value stays at a constant value of 10 in this interval. At $t = 12$ the impulse instantaneously adds an area of 5, and after $t = 12$ no more area is added. The final graph of the integral is shown in Fig. 2-34d.

The graphic method of finding the derivative is just finding the slope. The only tricky part is the derivative of a step; however, from the mathematical method, we found that this generates an impulse with an area equal to the level change of the step. The direction of the impulse is positive if the right side of the discontinuity is a higher level than the left side.

EXAMPLE 2-24

Graphically find the derivative of Fig. 2-35a. (Note it is the same graph used in Example 2-22.)

SOLUTION The solution is shown in Fig. 2-35b. Let's look at what happens at each time. At $t = 0$ there is a step going up as we go from the left to the right side of $t = 0$. The change is 5, so we get an impulse at $t = 0$ with an area of 5.

From $t = 0^+$ to $t = 1^-$ the slope is zero, and therefore the derivative is zero.

Since at $t = 1$ there is a step change going down with a difference of 5, we get another impulse of area 5 but in the negative direction.

From $t = 2$ to $t = 3$ sec there is a ramp with a slope of 5. The derivative will then be 5.

At $t = 3$ there is a step down. The magnitude of this step is 10. This will yield a negative impulse going down with an area of 10.

After $t = 3$ the slope of the function is zero. The derivative will then also be zero.

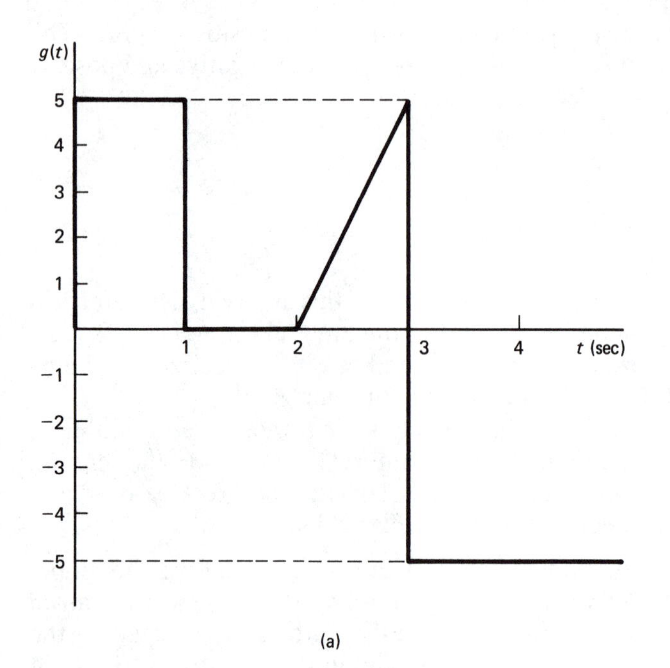

(a)

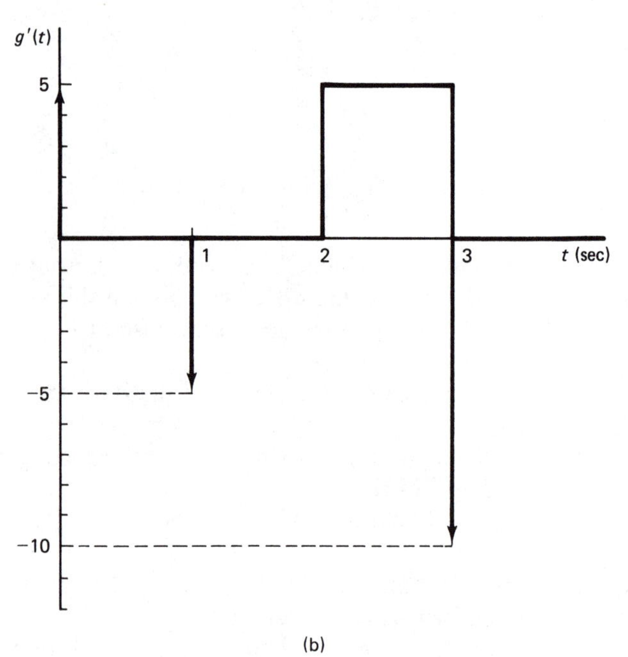

(b)

Figure 2-35

PROBLEMS

Section 2-1

1. Draw a graph of the function.

$$v(t) = 5u(t)$$

2. Draw a graph of the function.

$$f(t) = 4u(t - 3)$$

3. Write the equation for Fig. 2-36.

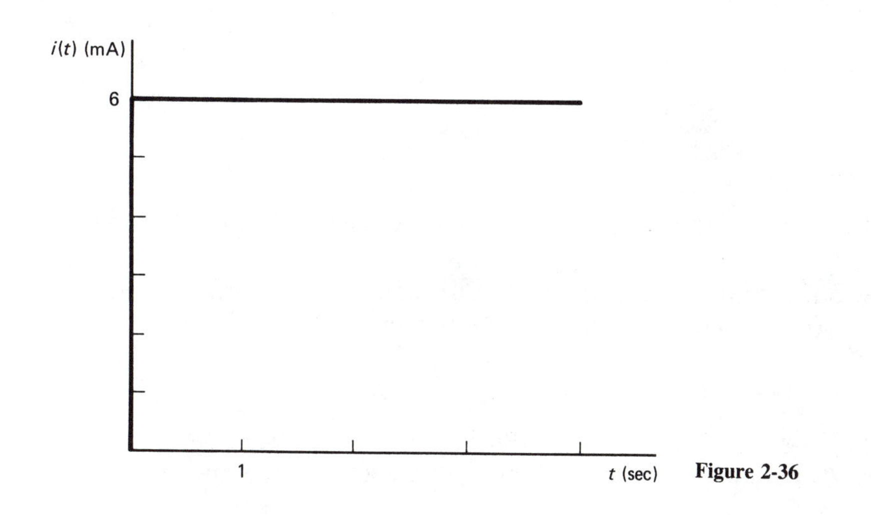

Figure 2-36

Section 2-2

4. Draw a graph of the function.

$$f(t) = \frac{3}{5}tu(t)$$

5. Draw a graph of the function.

$$i(t) = (t - 2)u(t - 2)$$

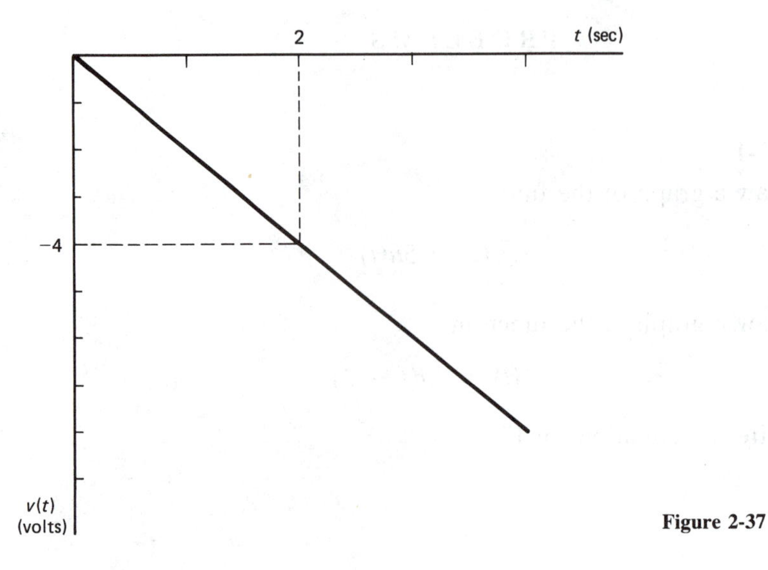

Figure 2-37

6. Write the equation of Fig. 2-37.

Section 2-3

7. Show that the derivative of $f_3(t)$ as defined by Eq. (2-7) is equal to $f_2(t)$.

8. Show that the integral of $f_2(t)$ as defined by Eq. (2-7) is equal to $f_3(t)$. Assume zero for the initial condition.

Section 2-4

9. Draw the graph of $f(t) = 4\delta(t)$.

10. Approximate the function as an impulse and draw the graph.

$$f(t) = 10u(t) - 10u(t - 0.01)$$

Section 2-5

11. For the function shown:
- **(a)** Find the time constant.
- **(b)** Find the damping constant.
- **(c)** Make a table of values for the function from $t = 0$ to when the function is zero (5τ) in steps of $\frac{1}{2}$ time constant.
- **(d)** Using the table in part (c), draw the function.

$$v(t) = 5e^{-2\times 10^3 t}u(t)$$

12. Write the equation for Fig. 2-38.

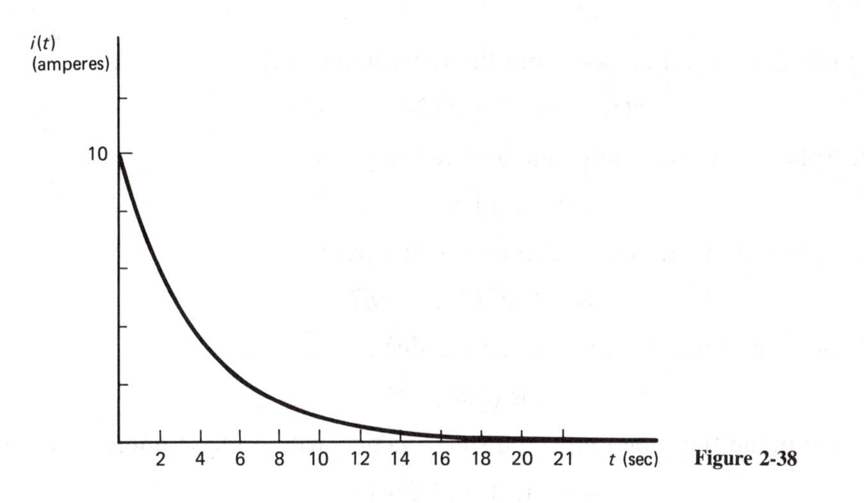

Figure 2-38

Section 2-6

13. Convert the following functions into a form suitable for evaluating the functions on a calculator in radians and then in degrees.
  (a) $14 \sin (6t + 30°)$
  (b) $9 \cos (5t + 15°)$

14. For the function shown:
  (a) Find the frequency in rad/sec and herz.
  (b) Find the voltage at $t = 0$.
  (c) Find the time where the peaks and zero crossings occur for the first cycle.

$$v(t) = 4 \sin (30t + 40°)\, u(t)$$

15. For the function shown:
  (a) Find the time the function decays to zero.
  (b) Find the value of the function at $t = 0$.
  (c) Make a table showing the time and value of the positive sinusoidal peaks, negative sinusoidal peaks, and zero crossings from $t = 0$ until the function is zero.
  (d) Draw the function from the table in part (c).

$$i(t) = 4e^{-2t} \sin (5t)\, u(t)$$

16. For the function shown:
  (a) Find the time the function decays to zero.
  (b) Find the value of the function at $t = 0$.
  (c) Make a table showing the time and value of the positive sinusoidal peaks, negative sinusoidal peaks, and zero crossings for the first cycle.

(d) Draw the function from the table in part (c).

$$v(t) = 6e^{-20t} \sin (24\pi t + 35°)$$

17. Split the function into sine and cosine parts.

$$10 \sin (4\pi t + 57°)$$

18. Split the function into sine and cosine parts.

$$8e^{-2t} \sin (7\pi t + \pi/7)$$

19. Split the function into sine and cosine parts.

$$4 \cos (3\pi t + 40°)$$

20. Combine the function into a sine function with a phase angle.

$$4 \sin (5t) + 3 \cos (5t)$$

21. Combine the function into a sine function with a phase angle.

$$6e^{-7t} \sin (4\pi t) + 5e^{-7t} \cos (4\pi t)$$

22. Combine the function into a sine function with a phase angle.

$$4 \sin (2\pi t) + 3e^{-4t} \cos (5\pi t)$$

Section 2-7

23. Shift the function so that it will start at $t = 2$ sec.

$$4tu(t)$$

24. Shift the function so that it will start at $t = 5$ sec.

$$5 \sin (4\pi t)\, u(t)$$

25. Shift the function so that it will start at $t = 3$ μsec.

$$7e^{-2t} \sin (3t + 40°)\, u(t)$$

26. Shift the function back to the origin.

$$4\delta(t - 2)$$

27. Shift the function back to the origin.

$$4e^{(-2t-6)} \sin [4(t - 3) + 10°]\, u(t - 3)$$

28. Shift the function back to the origin.

$$(4t + 25)u(t - 5)$$

29. Draw a graph of the function.

$$v(t) = 5(t - 4)u(t - 4)$$

30. Draw a graph of the function.

$$i(t) = 10\delta(t - 1)$$

31. Draw a graph of the function.

$$v(t) = 5e^{-2(t-3)}u(t - 3)$$

32. Draw a graph of the function showing two complete cycles.

$$i(t) = 4 \sin [5(t - 4)]\, u(t - 4)$$

Section 2-8

33. Draw a graph of the function.

$$f(t) = 2u(t) + 3u(t - 1) - u(t - 2) - 4u(t - 3)$$

34. Draw a graph of the function.

$$v(t) = 3\delta(t) + 10tu(t) - 30u(t - 3)$$

35. Draw a graph of the function.

$$i(t) = 8u(t) - 4tu(t) - 5\delta(t) + 8(t - 2)u(t - 2)$$

36. Draw a graph of the function.

$$f(t) = 5tu(t) - 5(t - 2)u(t - 2) - 10u(t - 4)$$

37. Draw a graph of the function.

$$v(t) = 4u(t) + 4e^{-0.5t} \sin (0.2\pi t + 30°)$$

38. Write the mathematical expression for Fig. 2-39.

39. Write the mathematical expression for Fig. 2-40.

40. Write the mathematical expression for Fig. 2-41.

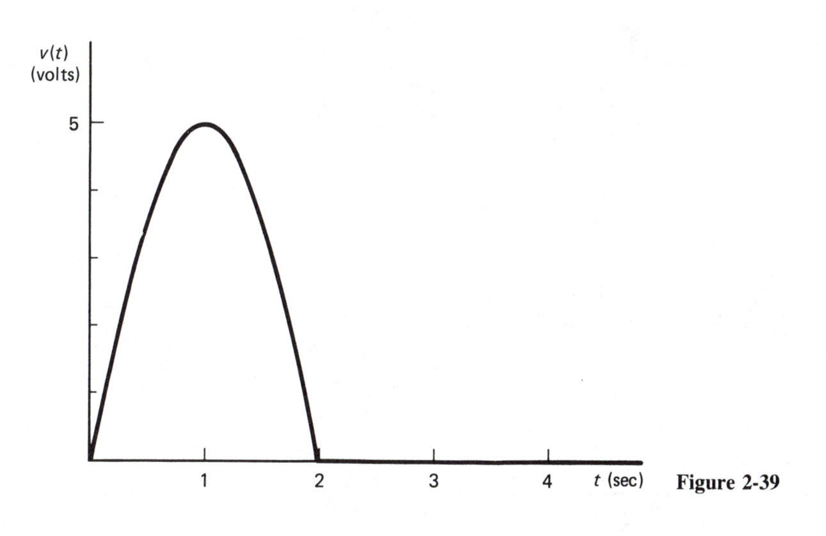

Figure 2-39

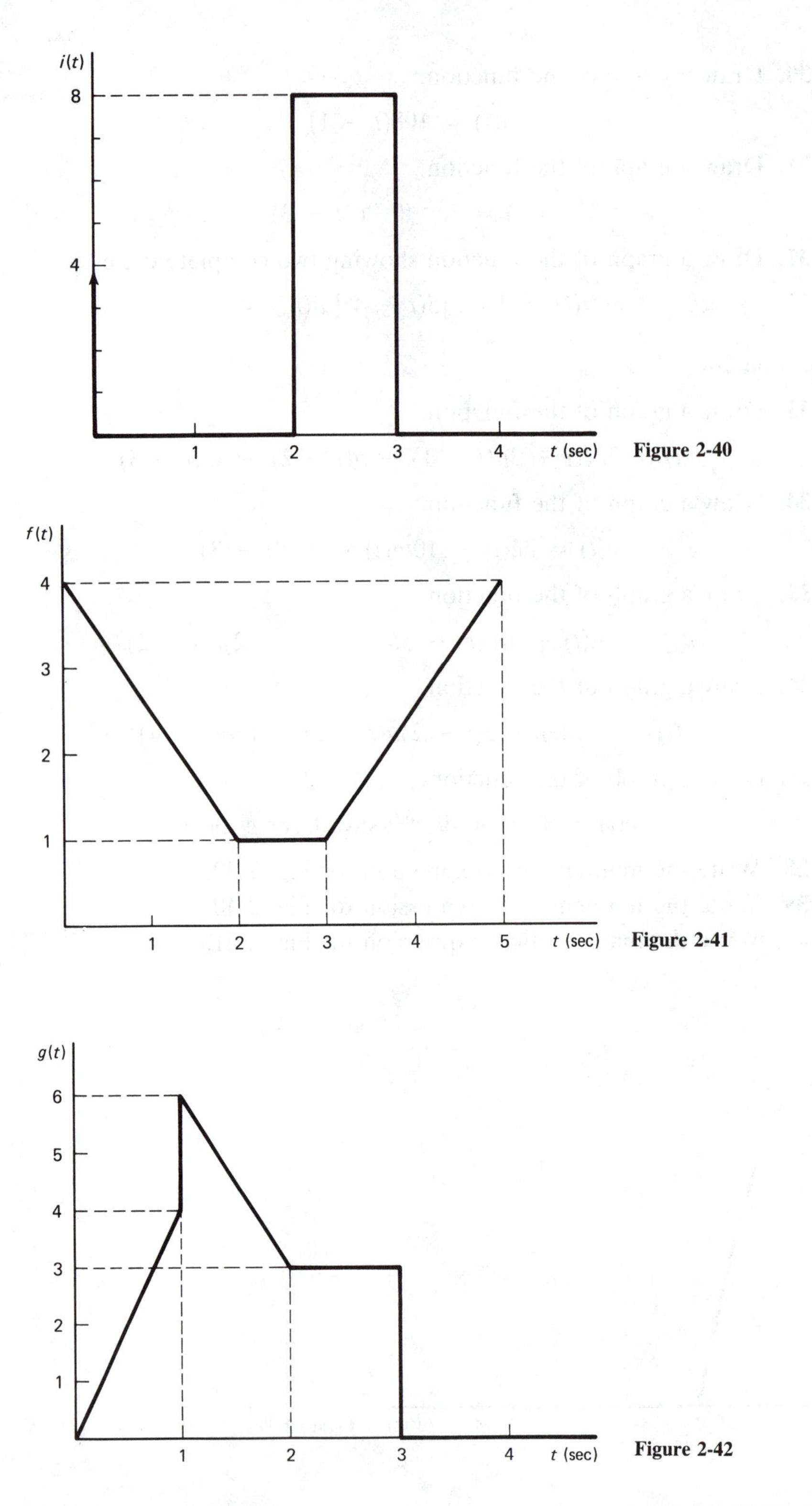

Figure 2-40

Figure 2-41

Figure 2-42

41. Write the mathematical expression for Fig. 2-42.

42. Draw a graph of the function.

$$f(t) = \sum_{n=0}^{\infty} 4(-1)^n u(t - 2n)$$

43. Draw a graph of the function.

$$v(t) = -5tu(t) + \sum_{n=0}^{\infty} 5u(t - n)$$

44. Draw a graph of the function.

$$\begin{aligned} i(t) = \sum_{n=0}^{\infty} & 4(t - 5n)u(t - 5n) - u[t - (2 + 5n)] \\ & - 4[t - (2 + 5n)]u[t - (2 + 5n)] + u[t - (3 + 5n)] \\ & - 4[t - (3 + 5n)]u[t - (3 + 5n)] \\ & + 4[t - (5 + 5n)]u[t - (5 + 5n)] \end{aligned}$$

45. A function generator has a square-wave output of 1 kHz. It has a positive swing of 20 V and a negative swing of 20 V. Draw a graph of the waveform and write the equation in a summation form. Assume that a positive swing starts at $t = 0$.

46. A function generator has a digital square-wave ouput with a variable duty cycle. Write the equation in a summation form for the output when it has a duty cycle of 70%. Assume that the signal goes to 5 V at $t = 0$, and assume that the signal is at 10 kHz.

47. The full-wave rectifier shown in Fig. 2-43 has a 60-Hz 110-V rms signal applied to the input of the transformer. Assuming no voltage drop across the diodes, write the equation in a summation form for the output of the rectifier.

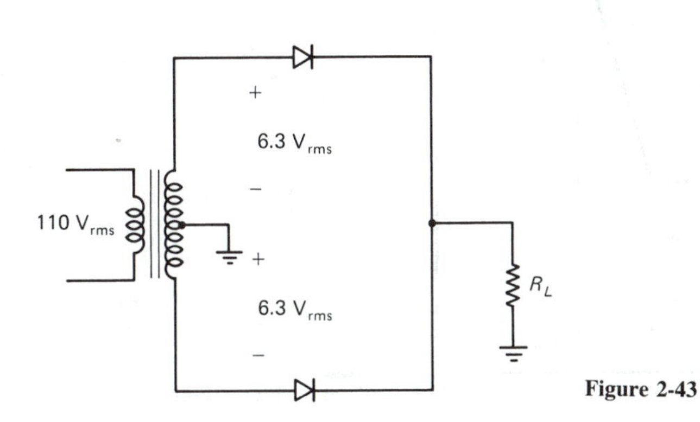

Figure 2-43

Section 2-9

48. Mathematically find the integral of the function.

$$f(t) = 4u(t) + 5\delta(t - 2) - 4u(t - 3)$$

49. Mathematically find the integral of the function.

$$f(t) = \delta(t) + 7u(t) - 8(t - 1)u(t - 1)$$

50. Mathematically find the derivative of the function.

$$f(t) = 3tu(t) - 3(t - 4)u(t - 4) - 12u(t - 4)$$

51. Mathematically find the derivative of the function.

$$f(t) = u(t) + 5(t - 2)^2u(t - 2)$$

52. Mathematically find the integral of the function.

$$f(t) = e^{-4t}u(t) + \sin [3(t - 1) + 30°]\, u(t - 1)$$

53. Mathematically find the integral of the function.

$$f(t) = 5u(t - 1) + e^{-2(t-2)} \sin [4(t - 2)]\, u(t - 2)$$

54. Mathematically find the voltage across a 1-μF capacitor that has the current $i(t)$ applied to it.

$$i(t) = \delta(t) + tu(t) - (t - 2)u(t - 2)$$

55. Mathematically find the derivative of the function.

$$f(t) = 6e^{-3t}u(t) + \sin [4(t - 3) + 75°]\, u(t - 3)$$

56. Mathematically find the derivative of the function.

$$f(t) = e^{-4(t-2)} \sin [3(t - 2) + 40°]\, u(t - 2)$$

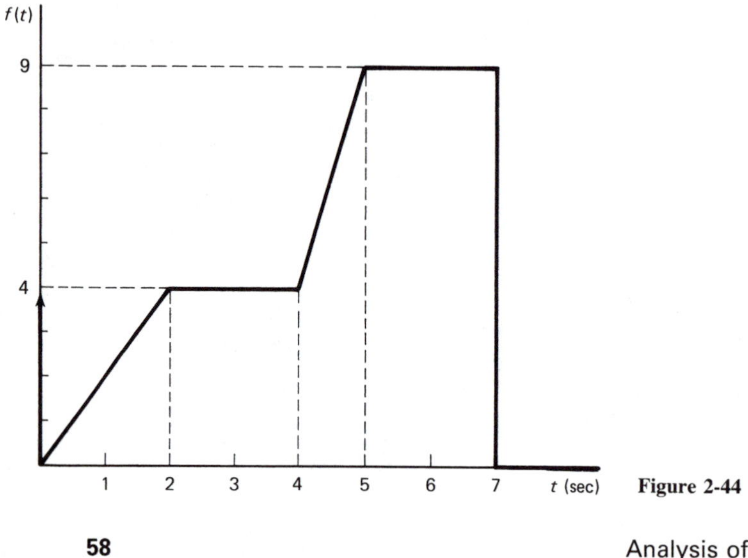

Figure 2-44

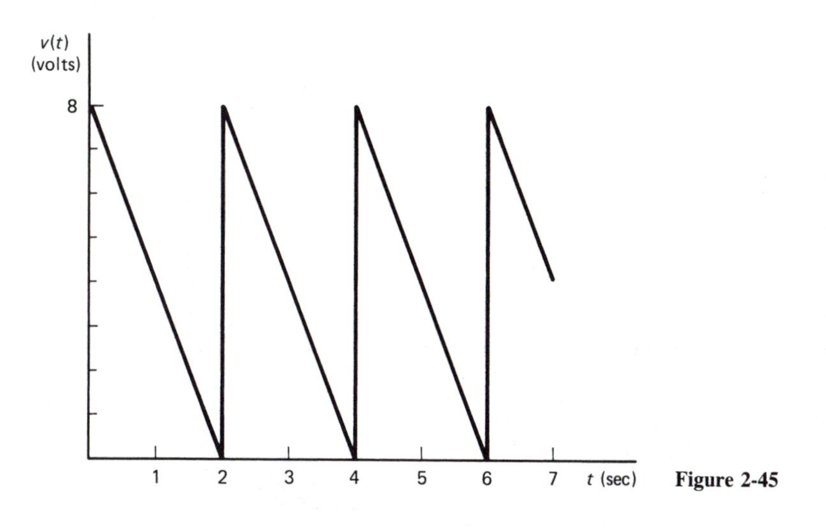

Figure 2-45

57. Mathematically find the voltage across a 5-mH inductor that has the current $i(t)$ going through it.

$$i(t) = u(t) + 4(t - 3)u(t - 3)$$

58. Graphically find the integral of Fig. 2-44.
59. A 2-H inductor has across it the voltage shown in Fig. 2-45. Find the current through the inductor graphically and mathematically.
60. Graphically find the derivative of Fig. 2-46.
61. Figure 2-47 shows the current applied to a $\frac{1}{2}$-F capacitor. Write the equation for the voltage across the capacitor.

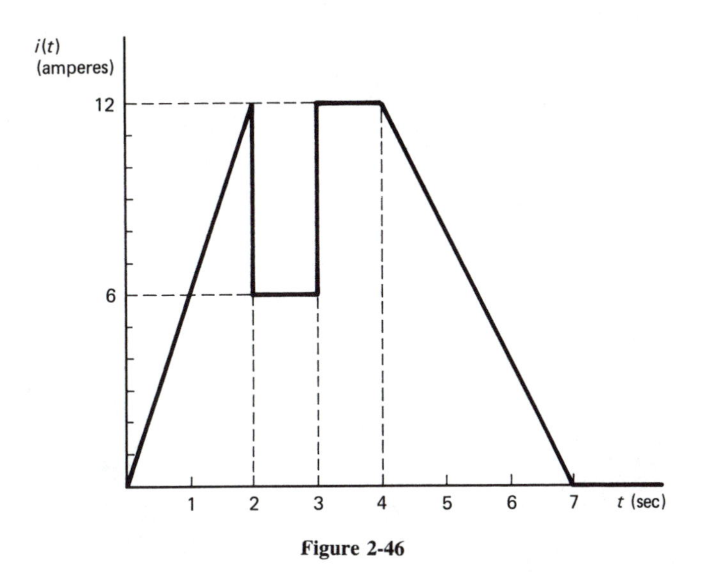

Figure 2-46

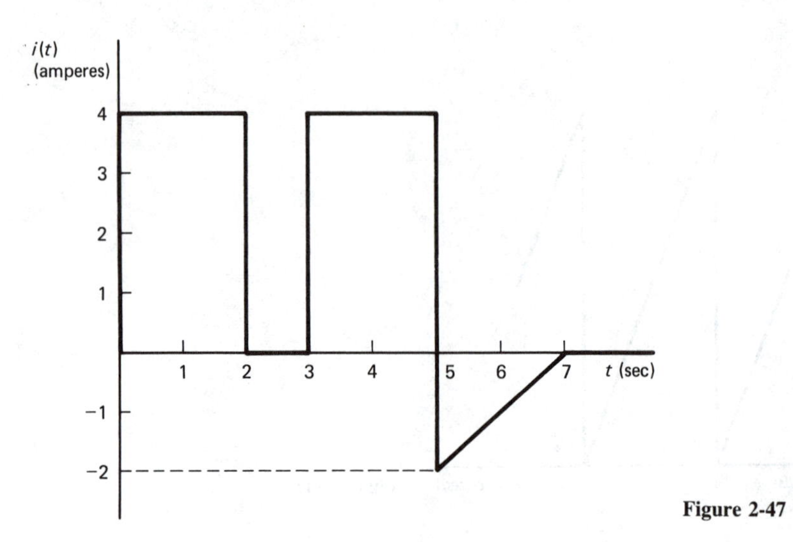

Figure 2-47

CHAPTER 3

Laplace Transform

3-0 INTRODUCTION

In Chapter 1 we explained what Laplace transforms are. This chapter covers finding the Laplace and inverse Laplace transform function. After learning these techniques we may apply Laplace to electronic circuits. Because of the nature of practical electronic circuits, we require only two equations to find the inverse Laplace transform. We will make extensive use of Laplace operations, which makes a long list of Laplace transform tables unnecessary. Strive to memorize the minimum number of equations in this chapter and Laplace transforms will be easy to use.

This chapter is like learning how to ride a bicycle. You can watch someone ride for a long time but you will never learn until you get on the bicycle and try. You will not learn Laplace transforms by reading examples and watching someone work problems—you must work the problems.

Since this chapter will have quadratic equations in a special form, we need to define the form before starting. The quadratic equation for the roots of

$$as^2 + bs + c$$

is usually expressed as

$$\text{roots} = \frac{-b \pm \sqrt{b^2 - 4ac}}{2a}$$

In our case the coefficient of the squared variable is typically unity ($a = 1$). Therefore, if we let $a = 1$, the quadratic function becomes

$$s^2 + bs + c$$

and the quadratic equation becomes

$$\text{roots} = \frac{-b}{2} \pm \sqrt{\left(\frac{b}{2}\right)^2 - c}$$

If $(b/2)^2$ is $<c$, the function is complex and the roots will have a real part and an imaginary part:

$$\text{roots} = -\alpha \pm j\omega$$

When this is the case we can use a factored form that shows us the roots:

$$(s + \alpha)^2 + \omega^2$$

3-1 LAPLACE TRANSFORMS BY TABLE

3-1-1 The Laplace Integral

The Laplace transform is defined as

$$F(s) = \int_0^\infty f(t)e^{-st}\,dt \tag{3-1}$$

This can also be stated in a shorthand notation that we will use most of the time.

$$\mathscr{L}[f(t)] = F(s) \tag{3-2}$$

The time-domain function and its corresponding Laplace function are called a transform pair.

This is a single-sided Laplace transform since it goes from zero to infinity instead of from minus infinity to plus infinity. This means that all of our time-domain functions must exist at or after $t = 0$. This implies that all time functions are multiplied by $u(t - a)$, where a must be equal to or greater than zero.

Note that functions in the time domain are shown with lowercase characters and functions in the Laplace domain are shown with uppercase characters.

Our purpose is to learn Laplace transforms as a tool, but just to take some of the mystery out of it, we will derive a transform pair.

EXAMPLE 3-1

Find $\mathscr{L}[u(t)]$ using the Laplace integral.

SOLUTION Using Eq. (3-1) and given

$$f(t) = u(t)$$

$$F(s) = \int_0^\infty f(t)e^{-st}\,dt$$

$$= \int_0^\infty 1e^{-st}\,dt$$

We recognize the integral form

$$\int_0^\infty e^u du = e^u \Big|_0^\infty$$

Letting

$$u = -st$$

we get

$$du = -s\,dt$$

Therefore,

$$\frac{-1}{s}\int_0^\infty e^{-st}(-s\,dt)$$

$$\frac{-e^{-st}}{s}\Big|_0^\infty = 0 - \frac{-1}{s}$$

The result is

$$\mathcal{L}[u(t)] = \frac{1}{s}$$

Appendix A shows the Laplace transform of several functions that we will need. Also, the table of operations and the table of identities are listed. We will need to refer to these tables as we learn Laplace transforms.

Let's first look at the identities listed in Section A-3. The first entry (I-1) shows the proper way to write the Laplace transform of a function $f(t)$ as stated in Eq. (3-2), and the inverse Laplace of a function $F(s)$. The second entry (I-2) states that $f(t)$ is not equal to $F(s)$ but that

$$f(t) \Leftrightarrow F(s)$$

which is read $f(t)$ transforms to $F(s)$.

Entry (I-3) show that a constant can be moved inside or outside the Laplace and inverse Laplace notation. Entry (I-4) shows that the Laplace transform of a sum of functions is equal to the sum of the Laplace transform of each function, and the inverse Laplace transform of a sum of functions is equal to the sum of the inverse Laplace transform of each function. Both of these identities are true because the Laplace transform is an integral function.

3-1-2 Laplace Transform by Table

The Laplace transform of a function will be found by using the transform pairs and the transform operations in Appendix A. We will first begin by using the table of transform pairs, and later learn how to use the transform operations. The best way to learn is by example.

EXAMPLE 3-2

Find the Laplace transform of the voltage

$$v(t) = 2\delta(t) + 10u(t) + 12tu(t)$$

SOLUTION

$$V(s) = \mathcal{L}[2\delta(t) + 10u(t) + 12tu(t)]$$

Using identity (I-4) gives

$$= \mathcal{L}[2\delta(t)] + \mathcal{L}[10u(t)] + \mathcal{L}[12tu(t)]$$

Using identity (I-3) yields

$$= 2\mathcal{L}[\delta(t)] + 10\mathcal{L}[u(t)] + 12\mathcal{L}[tu(t)]$$

Using transform pairs (P-1), (P-2), and (P-3), respectively, leads to

$$= 2 + \frac{10}{s} + \frac{12}{s^2}$$

Then we put the answer in a proper form that is a factored numerator and factored denominator.

$$V(s) = \frac{2(s + 2)(s + 3)}{s^2}$$

EXAMPLE 3-3

Find $\mathcal{L}[\sin(2\pi t + 35°)]$.

SOLUTION Since we are using single-sided Laplace transforms, this expression (or any other function) must be assumed to be multiplied by $u(t)$. First we must split the sine function into a sine and cosine.

$$1 \angle 35° = 0.81915 + j0.57358$$

Therefore,

$$\sin(2\pi t + 35°) = 0.81915 \sin(2\pi t) + 0.57358 \cos(2\pi t)$$

Using (P-5) and (P-6) gives

$$\frac{(0.81915)(2)}{s^2 + 4} + \frac{0.57358s}{s^2 + 4}$$

$$\mathcal{L}[\sin(2\pi t + 35°)] = \frac{0.57358(s + 2.8563)}{s^2 + 4}$$

3-2 INVERSE LAPLACE TRANSFORM BY TABLE

Finding the inverse by tables is simply a matter of putting the function into a form that appears in a table and finding the inverse. We will keep this simple since we will find other techniques more powerful, and this method can get quite complicated.

EXAMPLE 3-4

Find

$$\mathcal{L}^{-1}\left[\frac{5}{s}\right]$$

SOLUTION This is just a case of factoring out the 5 so that we can use (P-2).

$$5\mathcal{L}^{-1}\left[\frac{1}{s}\right] = 5u(t)$$

EXAMPLE 3-5

Find

$$\mathcal{L}^{-1}\left[\frac{s + 6}{s^2 + 9}\right]$$

SOLUTION From the denominator we see that this is a sinusoidal type of function. We can just split the numerator to see that this is a sine function and a cosine function.

$$\mathcal{L}^{-1}\left[\frac{s}{s^2 + 3^2}\right] + \mathcal{L}^{-1}\left[\frac{(2)(3)}{s^2 + 3^2}\right]$$

Using (P-6) and (P-5), we have

$$\cos(3t) + 2\sin(3t)$$

and combining into a single sine function gives

$$2 + j = 2.2361 \underline{/26.565^\circ}$$

$$2.2361 \sin(3t + 26.565^\circ)$$

3-3 LAPLACE OPERATIONS

By using only a few transform pairs and the transform operations we will be able to find the Laplace transform for a large variety of functions. The transform operations are very powerful and we will go through an example of each. The transform operations are shown in Section A-2.

3-3-1 Shifted Function Operation

Once you know how to find the Laplace and the inverse Laplace transformation of unshifted functions, this operation allows you to find the Laplace and the inverse Laplace of shifted functions using the same transform pairs. Transform operation (O-1) is

$$h(t-a)u(t-a) \Leftrightarrow e^{-as}H(s)$$

This means that you shift the function back to the origin to get $h(t)u(t)$, find the Laplace transform $H(s)$, and then multiply by e^{-as}. We may think of $(t-a)$ as a time shift and e^{-as} as a "Laplace time shift."

It is very important that the time shifting is always the first thing to do, and the Laplace time shifting is always the last thing to do. You may have to use other transform operations to find $H(s)$, but the time shifting first and last rule always takes precedent.

EXAMPLE 3-6

Find

$$\mathcal{L}[4\delta(t-1) + 2u(t-2) - 2u(t-3) - 3(t-3)u(t-3)]$$

SOLUTION

Step 1. Shift each term back to the origin.

$$h_1(t-a)u(t-a) = 4\delta(t-1) \Rightarrow h_1(t)u(t) = 4\delta(t)$$

With the impulse function, $u(t)$ is not required since $\delta(t)$ is zero for $t < 0$. The function $h(t)$ in this case is $4\delta(t)$ with $u(t)$ understood.

$$h_2(t-a)u(t-a) = 2u(t-2) \quad \Rightarrow h_2(t)u(t) = 2u(t)$$

$$h_3(t-a)u(t-a) = -2u(t-3) \quad \Rightarrow h_3(t)u(t) = -2u(t)$$

$$h_4(t-a)u(t-a) = -3(t-3)u(t-3) \quad \Rightarrow h_4(t)u(t) = -3tu(t)$$

Step 2. Find the unshifted Laplace transform.

$$H_1(s) = \mathcal{L}[4\delta(t)] = 4$$

$$H_2(s) = \mathcal{L}[2u(t)] = \frac{2}{s}$$

$$H_3(s) = \mathcal{L}[-2u(t)] = \frac{-2}{s}$$

Notice that the last two functions transform into the same thing when seen as an unshifted function.

$$H_4(s) = \mathscr{L}[-3tu(t)] = \frac{-3}{s^2}$$

Step 3. Shift the functions back to the correct time by using (O-1) and sum the terms.

$$4e^{-s} + \frac{2e^{-2s}}{s} - \frac{2e^{-3s}}{s} - \frac{3e^{-3s}}{s^2}$$

Sometimes a function is not in the proper form of a shifted function. That is, the function does not have the same form as the step function's argument. There are two ways to change a function into a shifted form. Using an algebraic method you multiply out, add and subtract terms, and collect terms appropriately so that you get a proper shifted form. Using a graphic method you draw the function and rewrite the equation from the graph.

EXAMPLE 3-7

Find the Laplace transform of the function $v(t)$. Show both methods to force the function into a proper shifted form.

$$v(t) = (2t - 1)u(t - 2)$$

SOLUTION This function is not in the proper form of a shifted function because $(2t - 1)$ is not in the form of the step function's argument, $(t - 2)$.

Algebraic method. First multiply out.

$$v(t) = 2tu(t - 2) - u(t - 2)$$

Then you add terms to the function to create the form you need, and subtract the same terms so that the function's value will not change.

$$v(t) = 2tu(t - 2) - [4u(t - 2)] - u(t - 2) + [4u(t - 2)]$$

$$v(t) = (2t - 4)u(t - 2) + (-1 + 4)u(t - 2)$$

$$v(t) = 3u(t - 2) + 2(t - 2)u(t - 2)$$

Graphic method. Figure 3-1a shows the function $(2t - 1)$, and Fig. 3-1b shows this function multiplied by $u(t - 2)$. We write the equation from Fig. 3-1b using methods shown in Chapter 2.

$$v(t) = 3u(t - 2) + 2(t - 2)u(t - 2)$$

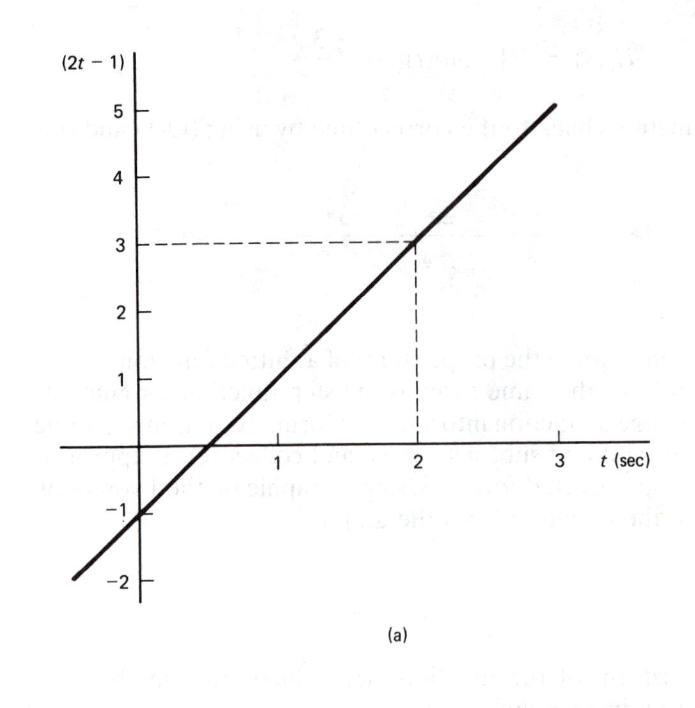

(a)

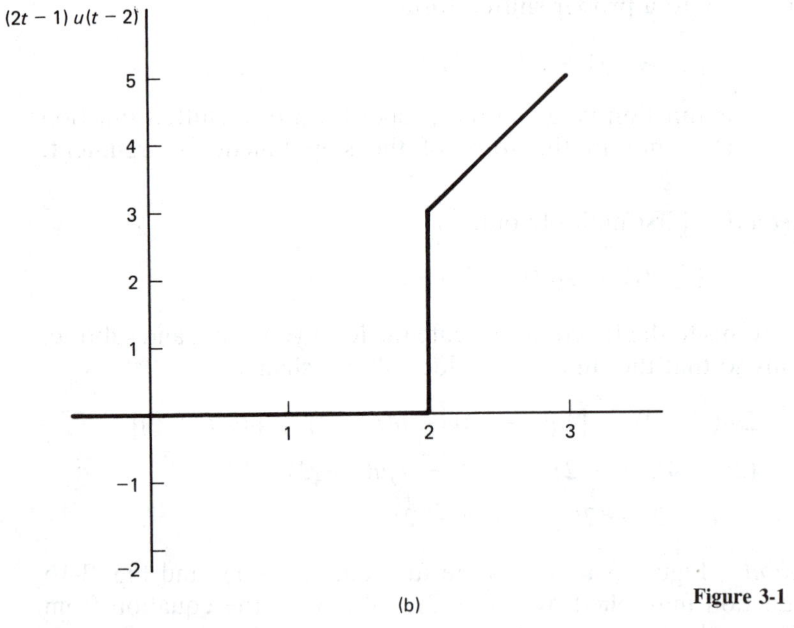

(b)

Figure 3-1

Now that we have $v(t)$ in a proper form we can write the Laplace transform. First we shift each term back to the origin.

$$h_1(t - a)u(t - a) = 3u(t - 2) \qquad \Rightarrow h_1(t)u(t) = 3u(t)$$

$$h_2(t - a)u(t - a) = 2(t - 2)u(t - 2) \Rightarrow h_2(t)u(t) = 2tu(t)$$

Find the unshifted Laplace transform.

$$H_1(s) = \mathcal{L}[3u(t)] = \frac{3}{s}$$

$$H_2(s) = \mathcal{L}[2tu(t)] = \frac{2}{s^2}$$

Shift each term back in Laplace, and sum the terms.

$$V(s) = \frac{3e^{-2s}}{s} + \frac{2e^{-2s}}{s^2}$$

3-3-2 Exponential Multiplier Operation

In electronics a lot of functions decay and therefore have an exponential multiplier. If you know the Laplace transform of the function the exponential is multiplied by, you can find the Laplace transform using this operation.

Transform operation (O-2) is

$$e^{-bt}h(t)u(t) \Leftrightarrow H(s + b)$$

EXAMPLE 3-8

Find $\mathcal{L}[e^{-2t}tu(t)]$.

SOLUTION

$$e^{-bt}h(t)u(t) = e^{-2t}tu(t) \Rightarrow h(t)u(t) = tu(t)$$

From (P-3),

$$H(s) = \mathcal{L}[tu(t)] = \frac{1}{s^2}$$

Now we replace s with $s + 2$.

$$H(s + b) = \mathcal{L}[e^{-2t}tu(t)] = \frac{1}{(s + 2)^2}$$

The next example shows the use of the exponential multiplier with a shifted function. Remember the first and last rule for a shifted function.

EXAMPLE 3-9

Find the Laplace transform.

$$i(t) = e^{-3(t-4)} \sin [7(t - 4)]\, u(t - 4)$$

SOLUTION First we must shift this function back to the origin.

$$h_1(t - a)u(t - a) = e^{-3(t-4)} \sin [7(t - 4)]\, u(t - 4)$$

$$\Rightarrow h_1(t)u(t) = e^{-3t} \sin (7t)u(t)$$

Notice this is in the form of an exponential times a function for which we know the Laplace transform.

$$e^{-bt}h_2(t)u(t) = e^{-3t} \sin (7t)\, u(t) \Rightarrow h_2(t)u(t) = \sin (7t)\, u(t)$$

Using (P-5) yields

$$H_2(s) = \mathscr{L}[\sin (7t)\, u(t)] = \frac{7}{s^2 + 49}$$

Using (O-2), we add in the effect of the exponential multiplier. Put parentheses around each s so that the only thing inside the parentheses is s, and replace the s with the $s + b$ term.

$$H_1(s + b) = \mathscr{L}[e^{-3t} \sin (7t)\, u(t)] = \frac{7}{(s + 3)^2 + 49}$$

Finally, we shift the completed transform equation back to the proper time.

$$I(s) = \mathscr{L}[e^{-3(t-4)} \sin [7(t - 4)]\, u(t - 4)] = \frac{7e^{-4s}}{(s + 3)^2 + 49}$$

3-3-3 *t* Multiplier Operation

A function mulitiplied by t will appear occasionally, and if we know the Laplace transform of the function, this operation will allow us to find the transform of the function when multiplied by t. Successive use of the operation will allow us to find the Laplace transform of a function multiplied by t raised to any positive power.

Transform operation (O-3) is

$$th(t)u(t) \Leftrightarrow -\frac{d}{ds}[H(s)]$$

This involves taking the derivative of a function which, in most cases, is a fraction. The derivative of a fraction is given by

$$\frac{d}{ds}\left[\frac{F(s)}{G(s)}\right] = \frac{F'(s)G(s) - G'(s)F(s)}{G^2(s)} \tag{3-3}$$

EXAMPLE 3-10

Find $\mathscr{L}[(t-2)^2u(t-2)]$.

SOLUTION First unshift the function and find the Laplace transform.

$$h_1(t-a)u(t-a) = (t-2)^2u(t-2) \Rightarrow h_1(t)u(t) = t^2u(t)$$

Next we split this into a function multiplied by t. This is just a matter of how you look at it.

$$[t][h_2(t)u(t)] = [t][tu(t)] \Rightarrow h_2(t)u(t) = tu(t)$$

Then, using (P-3), we obtain

$$H_2(s) = \mathscr{L}[tu(t)] = \frac{1}{s^2}$$

Next we take the negative derivative of the function $H_2(s)$ to get $H_1(s)$.

$$H_1(s) = -\frac{d}{ds}[H_2(s)] = -\frac{d}{ds}\left(\frac{1}{s^2}\right)$$

$$= -\frac{(0)(s^2) - (2s)(1)}{s^4}$$

Simplifying, we have

$$H_1(s) = \mathscr{L}[t^2u(t)] = \frac{2}{s^3}$$

Finally, we shift the function back to the correct time.

$$\mathscr{L}[(t-2)^2u(t-2)] = \frac{2e^{-2s}}{s^3}$$

EXAMPLE 3-11

Find the Laplace transform of the voltage.

$$v(t) = t^2e^{-5t}u(t)$$

SOLUTION Using (O-2), we can write

$$e^{-bt}h_1(t)u(t) = e^{-5t}t^2u(t) \Rightarrow h_1(t) = t^2u(t)$$

and then find the Laplace transform of t^2 using (O-3), as we did in Example 3-10.

$$H_1(s) = \mathscr{L}[t^2u(t)] = \frac{2}{s^3}$$

Finally, using (O-2), we replace s in $H_1(s)$ with $(s+b)$.

$$H_1(s+b) = \frac{2}{(s+5)^3}$$

3-3-4 Derivative Operation

The Laplace transform can be used to find derivatives of functions that may be complex to find in the time domain. Transform operation (O-4a) applies to the derivative of a function switched on at $t = 0$, and transform operation (O-4b) applies to a function switched on at $t = 0$ and then applied to a differentiating circuit.

Transform operation (O-4a) is

$$\left\{\frac{d}{dt}[h(t)]\right\} u(t) \Leftrightarrow sH(s) - h(0)$$

Transform operation (O-4b) is

$$\frac{d}{dt}[h(t)u(t)] \Leftrightarrow sH(s)$$

EXAMPLE 3-12

Find the derivative of 2 cos $(3t)$ that is switched on at $t = 0$.

SOLUTION Since the derivative is switched on at $t = 0$, and this is not a switched function applied to a differentiator, we use (O-4a).

$$\left\{\frac{d}{dt}[h(t)]\right\} u(t) = \frac{d}{dt}[2\cos(3t)]\,u(t)$$

$$\Rightarrow h(t) = 2\cos(3t)$$

$$\Rightarrow h(0) = 2\cos[3(0)] = 2$$

$$\Rightarrow H(s) = \frac{2s}{s^2 + 9}$$

Substituting into $sH(s) - h(0)$, we have

$$\mathcal{L}\left[\left\{\frac{d}{dt}\, 2\cos(3t)\right\} u(t)\right] = \frac{2s^2}{s^2 + 9} - 2$$

Adding the terms together, we get

$$\frac{-18}{s^2 + 9}$$

Taking the inverse Laplace transform using the Laplace pairs (P-5), we have

$$\left\{\frac{d}{dt}[2\cos(3t)]\right\} u(t) = -6\sin(3t)\,u(t)$$

Notice in Example 3-12 that an impulse does not appear in the answer. If we had a switched cosine function there would be a discontinuity at $t = 0$ and therefore there would be an impulse appearing in the derivative. In this case we switched on the output of a differentiator that was differentiating a cosine function.

3-3-5 Integral Operation

When using this operation we do not have the same problem as we did with the derivative operation. Since the lower integral limit is zero, it does not matter whether the function started before zero or not. We will be finding the area under the curve for $t > 0$.

The integral operation (O-5) is

$$\int_0^t h(t)u(t)\,dt \Leftrightarrow \frac{H(s)}{s}$$

EXAMPLE 3-13

Assume, for this problem, that we do not know the Laplace pair for $tu(t)$, but we do know the Laplace pair for $u(t)$. Find the Laplace transform for $tu(t)$ using (O-5).

SOLUTION We know that the integral of $u(t)$ is $tu(t)$ or

$$\int_0^t h(t)u(t)\,dt = \int_0^t 1u(t)\,dt = tu(t)$$

and we know

$$h(t)u(t) = 1u(t)$$

$$\Leftrightarrow H(s) = \frac{1}{s}$$

Substituting into (O-5) gives

$$\frac{H(s)}{s} = \frac{1}{s^2}$$

$$\mathcal{L}\left[\int_0^t u(t)\,dt\right] = \mathcal{L}[tu(t)] = \frac{1}{s^2}$$

3-4 PRACTICAL INVERSE TECHNIQUES

Practical electronic circuits restrict the type of Laplace functions we will encounter. Since the roots of the denominator of these Laplace functions will either be real or complex conjugate, we need only two techniques to find the inverse of practical Laplace circuits. There are a few definitions that are essential in order to understand these two methods of inverse Laplace.

3-4-1 Inverse Equations Definitions

$F(s)$ is the function to find the inverse Laplace transform of. It can be represented as a numerator polynomial, $N(s)$, divided by a denominator polynomial, $D(s)$.

$$F(s) = \frac{N(s)}{D(S)} \tag{3-4}$$

where the degree of the $N(s)$ polynomial is less than the $D(s)$ degree. The roots of $D(s)$ are called poles since the function $F(s)$ will approach infinity when s is equal to any of the poles. (When graphed, this looks like a flagpole.) The roots of $N(s)$ are called zeros since the function $F(s)$ is zero when s is equal to one of these roots.

The numerator polynomial must be at least one degree less than the denominator polynomial. If it is not, we divide the numerator polynomial by the denominator polynomial. The result will be a constant plus a new $N(s)/D(s)$. The constant will invert into an impulse, and then we will work on the resulting $N(s)/D(s)$.

It is easiest to work with the function $F(s)$ in a certain form. First we make the coefficient of the highest power of s in the numerator 1. Then we do the same to the denominator. We factor the numerator and denominator, and cancel any factors we can.

EXAMPLE 3-14

Put the following function in the proper form of $N(s)/D(s)$.

$$F(s) = \frac{6(s+2)(s+3)(s+5)}{(s+1)(3s^2+18s+24)}$$

SOLUTION The numerator is already in a properly factored form; however, the denominator needs to be factored. We also see that this function needs to be divided out since both numerator and denominator polynomials are of the same degree. First let's factor out the 3 in the denominator's quadratic.

$$F(s) = \frac{6(s+2)(s+3)(s+5)}{3(s+1)(s^2+6s+8)} = \frac{2(s+2)(s+3)(s+5)}{(s+1)(s^2+6s+8)}$$

Next we could factor the rest of the denominator and look for terms to cancel, or we could multiply out and do the division first. If we take the first choice and if there is a cancellation, the division will be easier. Let's factor first.

$$F(s) = \frac{6(s+2)(s+3)(s+5)}{(s+1)(s+2)(s+4)} = \frac{6(s+3)(s+5)}{(s+1)(s+4)}$$

Next we multiply out the numerator and the denominator to get the constant plus the new $N(s)/D(s)$.

$$\begin{array}{r|l} & \quad\quad\quad 1 \\ \hline s^2 + 5s + 4 & s^2 + 8s + 15 \\ & s^2 + 5s + \ \ 4 \\ \hline & \quad\quad 3s + 11 \end{array}$$

The result is

$$F(s) = 6\left[1 + \frac{3(s + 3.6667)}{(s + 1)(s + 4)}\right]$$

$$= 6 + \frac{18(s + 3.6667)}{(s + 1)(s + 4)}$$

To find the inverse Laplace transform we will split up the function $F(s)$ into a summation of fractions and find the inverse Laplace transform of each term. Each real pole will have a fraction. With multiple-order real poles there will be multiple fractions equal in number to the order of the pole. Therefore, the poles will determine the form of the inverse Laplace transform. Knowing how to split up the function is not important other than to realize what is going on mathematically. The sum of fractions in Laplace will not be stressed, and usually we will go directly to the inverse.

EXAMPLE 3-15

What will the summation of fractions look like for the following function?

$$\frac{N(s)}{(s + 2)(s + 7)^3}$$

SOLUTION The numerator of each function will be a function of s.

$$\frac{N_1(s)}{s + 2} + \frac{N_2(s)}{(s + 7)^3} + \frac{N_3(s)}{(s + 7)^2} + \frac{N_4(s)}{s + 7}$$

When $F(s)$ is split up into terms (as shown in Example 3-15), the numerators will have some constants. Since this book is intended to show how to use the Laplace transforms, we will not go through the mathematical theory of "why we do what we do" to isolate constants. To evaluate these constants we will remove each pole from the function $F(s)$ by multiplying $F(s)$ by the factor containing the pole of interest. A shorthand notation will be used to simplify writing these functions. For real poles,

$$R(s)_{-a} = F(s)(s + a)^r \tag{3-5a}$$

$$R(-a) = F(s)(s + a)^r \big|_{s=-a} \tag{3-5b}$$

where r is the order of the pole. Notice that $R(-a)$ means to substitute the pole value into the specific function $R(s)_{-a}$. We must keep this in mind since normal function notation does not imply a specific function. The value in this case, $-a$, implies that we substitute into the $R(s)_{-a}$ function only.

For complex-conjugate poles,

$$R(s)_{-\alpha+j\omega} = F(s)[(s + \alpha)^2 + \omega^2]^r \tag{3-6a}$$

$$R(-\alpha + j\omega) = F(s[(s + \alpha)^2 + \omega^2]^r \,|_{s=-\alpha+j\omega} \tag{3-6b}$$

Once again the value we substitute in, $-\alpha + j\omega$, implies that we substitute in to $R(s)_{-\alpha+j\omega}$.

EXAMPLE 3-16

For the function, find $R(s)_{-3}$, $R(s)_{-3+j6}$, and $R(-3)$.

$$F(s) = \frac{(s + 2)(s + 4)}{(s + 3)[(s + 3)^2 + 36]}$$

SOLUTION Substituting into Eq. (3-5a) yields

$$R(s)_{-a} = F(s)(s + a)^r$$

$$R(s)_{-3} = \frac{(s + 2)(s + 4)}{(s + 3)[(s + 3)^2 + 36]}(s + 3)$$

$$R(s)_{-3} = \frac{(s + 2)(s + 4)}{(s + 3)^2 + 36}$$

Substituting into Eq. (3-6a) gives

$$R(s)_{-\alpha+j\omega} = F(s)[(s + \alpha)^2 + \omega^2]$$

$$R(s)_{-3+j6} = \frac{(s + 2)(s + 4)}{(s + 3)[(s + 3)^2 + 36]}[(s + 3)^2 + 36]$$

$$R(s)_{-3+j6} = \frac{(s + 2)(s + 4)}{(s + 3)}$$

$R(-3)$ implies that we substitute the -3 root into the $R(s)_{-3}$ function.

$$R(-3) = \frac{(-3 + 2)(-3 + 4)}{(-3 + 3)^2 + 36}$$

$$= -0.027778$$

3-4-2 Inverse Transform of Real Poles

The equations used to find the inverse of real poles are

$$M_n = \frac{d^{n-1}}{ds^{n-1}} R(s)_{-a}\Big|_{s=-a} \tag{3-7a}$$

$$R(s)_{-a} = F(s)(s + a)^r \tag{3-7b}$$

$$F(s) = \sum_{n=1}^{r} F_n(s) + \text{rest of function} \tag{3-8a}$$

$$F_n(s) = \frac{M_n/(n-1)!}{(s+a)^{r-n+1}} + \text{rest of function} \tag{3-8b}$$

$$f(t) = \sum_{n=1}^{r} f_n(t) + \text{rest of inverse} \tag{3-9a}$$

$$f_n(t) = \frac{M_n}{(n-1)!\,(r-n)!} t^{r-n} e^{-at} u(t) \tag{3-9b}$$

where in Eqs. (3-7) through (3-9), $n = 1, 2, 3, \ldots, r$.

Equation (3-7a) will give us a magnitude, M_n, which will be a real number. Equations (3-8) show how the function, $F(s)$, splits up into a summation of fractions. Equations (3-8) are not important other than to show mathematically what is going on, and generally we will skip this step. Equations (3-9) show us what to do with the magnitude, M_n, to find the inverse due to the poles $-a$.

We will first look at first-order poles where $r = 1$. This is called partial fraction expansion. This is also called the cover-up method since the magnitude equation, Eq. (3-7), becomes $F(s)$ missing one denominator factor, or one factor is covered up. When $r = 1$ is substituted into Eq. (3-7) through (3-9), n will only go to 1. The equations we need to find the inverse, (3-7a), (3-7b), and (3-9b), will become

$$M_1 = R(s)_{-a} \tag{3-10a}$$

$$R(s)_{-a} = F(s)(s + a) \tag{3-10b}$$

$$f_1(t) = M_1 e^{-at} u(t) + \text{rest of inverse} \tag{3-10c}$$

EXAMPLE 3-17

Find the inverse of

$$F(s) = \frac{(s + 10)(s + 5)}{(s + 2)(s + 1)}$$

SOLUTION Since the order of the numerator polynomial is the same as the denominator polynomial, we must first divide out.

$$\begin{array}{r|l} & \quad\quad\quad\quad\quad\ \ 1 \\ s^2 + 3s + 2 & s^2 + 15s + 50 \\ & \underline{s^2 + \ \ 3s + \ \ 2} \\ & \quad\quad 12s + 48 \end{array}$$

Thus the function in a different form is

$$F(s) = 1 + \frac{12s + 48}{(s + 2)(s + 1)}$$

The first term will become an impulse in the time domain. We will now work with the second term. Since both poles are first order, we may use Eq. (3-10). [We could also use Eq. (3-7) and (3-9).]

Due to the pole -2,

$$M_1 = R(s)_{-a} = \left.\frac{12s + 48}{s + 1}\right|_{s=-2} = \frac{12(-2) + 48}{-2 + 1} = -24$$

Then finding the inverse due to this pole using Eq. (3-9) or Eq. (3-10c), we have

$$f_1(t) = M_1 e^{-at}u(t) + \text{rest of inverse}$$
$$= -24e^{-2t}u(t) + \text{rest of inverse}$$

Due to the pole -1,

$$M_1 = R(s)_{-a} = \left.\frac{12s + 48}{s + 2}\right|_{s=-1}$$
$$= \frac{12(-1) + 48}{-1 + 2} = 36$$

Then finding the inverse due to this pole,

$$f_1(t) = M_1 e^{-at}u(t) + \text{rest of inverse}$$
$$= 36e^{-t}u(t) + \text{rest of inverse}$$

The complete inverse is then the sum of each inverse we found.

$$f(t) = \delta(t) - 24e^{-2t}u(t) + 36e^{-t}u(t)$$

With multiple-order poles, $r > 1$, the variable n goes from 1 to the value of r, and we must use the full equations in Eqs. (3-7) and (3-9).

EXAMPLE 3-18

Find $v(t)$.

$$V(s) = \frac{s^2 + 2s + 26}{(s + 2)(s + 3)^3}$$

SOLUTION First we will factor the numerator to see if we can cancel any factors.

$$\frac{(s+1)^2+25}{(s+2)(s+3)^3}$$

Next we will find the inverse due to pole -2. Using Eq. (3-10a), we find that

$$M_1 = R(s)_{-a} = \left.\frac{(s+1)^2+25}{(s+3)^3}\right|_{s=-2} = 26$$

The inverse due to pole -2 using Eq. (3-10c) is

$$f_1(t) = M_1 e^{-at}u(t) + \text{rest of inverse}$$

$$v(t) = 26e^{-2t}u(t) + \text{rest of inverse}$$

Next we find the inverse due to the multiple-order pole -3. Since we will be taking multiple derivatives of fractional functions, it is sometimes easier to make the numerator a polynomial. This way we can combine terms before taking the next derivative. From Eq. (3-7a),

$$M_n = \left.\frac{d^{n-1}}{ds^{n-1}}R(s)_{-a}\right|_{s=-a}$$

Since $r = 3$, then $n = 1, 2, 3$.

$$M_1 = \left.\frac{s^2+2s+26}{s+2}\right|_{s=-3} = -29$$

$$M_2 = \left.\frac{d}{ds}\frac{s^2+2s+26}{s+2}\right|_{s=-3}$$

$$= \left.\frac{s^2+4s-22}{(s+2)^2}\right|_{s=-3} = -25$$

$$M_3 = \left.\frac{d^2}{ds^2}\frac{s^2+2s+26}{s+2}\right|_{s=-3}$$

$$= \left.\frac{d}{ds}\frac{s^2+4s-22}{(s+2)^2}\right|_{s=-3}$$

Since we do not need the next derivative, we do not need to simplify the result of this derivative.

$$\left.\frac{(2s + 4)(s + 2)^2 - [(s^2 + 4s - 22)(2)(s + 2)(1)]}{(s + 2)^4}\right|_{s=-3} = -52$$

Using Eqs. (3-9) gives

$$f_n(t) = \frac{M_n}{(n - 1)!\,(r - n)!} t^{r-n} e^{-at} u(t)$$

$$f_1(t) = \frac{-29}{(1 - 1)!\,(3 - 1)!} t^{3-1} e^{-3t} u(t)$$

$$= -14.5t^2 e^{-3t} u(t)$$

$$f_2(t) = \frac{-25}{(2 - 1)!\,(3 - 2)!} t^{3-2} e^{-3t} u(t)$$

$$= -25te^{-3t} u(t)$$

$$f_3(t) = \frac{-52}{(3 - 1)!\,(3 - 3)!} t^{3-3} e^{-3t} u(t)$$

$$= -26e^{-3t} u(t)$$

The inverse due to the pole -3 is

$$f(t) = f_1(t) + f_2(t) + f_3(t) + \text{rest of inverse}$$

$$= -14.5t^2 e^{-3t} u(t) - 25te^{-3t} - 26e^{-3t} + \text{rest}$$

The complete solution is

$$v(t) = 26e^{-2t}\, u(t) - (14.5t^2 + 25t + 26)e^{-3t} u(t)$$

When the pole is zero, the inverse of multiple-order poles looks like a special case; however, it is not.

EXAMPLE 3-19

Find

$$\mathcal{L}^{-1}\left[\frac{(s + 2)^2}{s^4}\right]$$

SOLUTION The denominator looks very different from what we have seen, but if we put in the pole, 0, it will look the same as previous denominators we have worked with.

$$\frac{(s + 2)^2}{(s + 0)^4}$$

Using Eq. (3-7a) yields

$$M_1 = (s + 2)^2 \big|_{s=0} = 4$$

$$M_2 = \frac{d}{ds}(s + 2)^2 \Bigg|_{s=0} = 2(s + 2) \Bigg|_{s=0} = 4$$

$$M_3 = \frac{d}{ds} 2(s + 2) \Bigg|_{s=0} = 2 \Bigg|_{s=0} = 2$$

$$M_4 = \frac{d}{ds} 2 \Bigg|_{s=0} = 0$$

Using Eqs. (3-9b) gives

$$f_1(t) = \frac{4}{0!\,3!} t^3 e^{-0t} = 0.66667t^3u(t)$$

$$f_2(t) = \frac{4}{1!\,2!} t^2 e^{-0t} = 2t^2u(t)$$

$$f_3(t) = \frac{2}{2!\,1!} t^1 e^{-0t} = t^1u(t)$$

$$f_4(t) = \frac{0}{3!\,1!} te^{-0t} = 0$$

The complete solution is

$$(0.66667t^3 + 2t^2 + t)u(t)$$

This may look like a lot of writing, but most of this can be done mentally.

3-4-3 Inverse Transform of Complex Poles

Equations (3-7) through (3-10) can be used for complex poles. However, each complex root will generate an exponential time-domain inverse. When this is combined with the conjugate root inverse, the result will be a sinusoidal function by using Euler identities. Since in practical circuits any complex root will have the conjugate with it, we can solve for the sinusoidal function directly.

The equations used to find the inverse of complex poles are:

$$M_n \angle\theta_n \quad \text{is defined as} \tag{3-11a}$$

$$M_1 \angle\theta_1 = R(-\alpha + j\omega)$$

$$M_2 \angle\theta_2 = R'(-\alpha + j\omega) - \frac{r}{2j\omega}R(-\alpha + j\omega)$$

$$M_3 \angle\theta_3 = R''(-\alpha + j\omega) - \frac{2r}{2j\omega}R'(-\alpha + j\omega) + \frac{r(r+1)}{(2j\omega)^2}R(-\alpha + j\omega)$$

$$M_4 \angle\theta_4 = R'''(-\alpha + j\omega) - \frac{3r}{2j\omega}R''(-\alpha + j\omega)$$

$$+ \frac{3r(r+1)}{(2j\omega)^2}R'(-\alpha + j\omega) - \frac{r(r+1)(r+2)}{(2j\omega)^3}R(-\alpha + j\omega)$$

$$R(s)_{-\alpha + j\omega} = F(s)[(s+\alpha)^2 + \omega^2]^r \tag{3-11b}$$

$$f(t) = \sum_{n=1}^{r} f_n(t) + \text{rest of inverse} \tag{3-12a}$$

$$f_n(t) = \frac{2(-1)^{\text{int}(r/2)}M_n}{(2\omega)^r(n-1)!\,(r-n)!}t^{r-n}e^{-\alpha t}$$

$$\times \sin\left[\omega t + \theta_n + \frac{1+(-1)^r}{2}(90°)\right]u(t) \tag{3-12b}$$

where in Eqs. (3-11) and (3-12):

1. $n = 1, 2, 3, \ldots, r$
2. $\text{int}(r/2)$ means the integer portion. Therefore,

$$\text{int}\left(\frac{5}{2}\right) = \text{int}\left(\frac{4}{2}\right) = 2$$

Equations (3-11a) are the magnitude equations. Enough magnitude equations are listed to see the pattern being followed. Also notice that the numerical constants can be found by using Pascal's triangle.

Fortunately, it is rare to see r equal to or greater than 2 in complex poles. We will first look at single-order complex poles ($r = 1$). When $r = 1$, (3-11) and (3-12) reduce to

$$M_1 \angle\theta_1 = R(-\alpha + j\omega) \tag{3-13a}$$

$$R(s)_{-\alpha + j\omega} = F(s)[(s+\alpha)^2 + \omega^2] \tag{3-13b}$$

$$f_1(t) = \frac{M_1}{\omega}e^{-\alpha t}\sin(\omega t + \theta_1)\,u(t) \tag{3-13c}$$

EXAMPLE 3-20

Find the inverse transform of

$$\frac{(s+2)(s+4)}{(s+1)[(s+2)^2+9]}$$

SOLUTION First we will find the inverse of the single-order real pole using Eq. (3-10).

$$M_1 = \left.\frac{(s+2)(s+4)}{(s+2)^2+9}\right|_{s=-1} = 0.3$$

$$f_1(t) = 0.3e^{-t}u(t) + \text{rest of inverse}$$

Next we will find the inverse due to the complex poles $s = -2 + j3$. Using Eq. (3-13a) gives

$$M_1 \angle\theta_1 = R(-\alpha + j\omega)$$

$$M_1 \angle\theta_1 = \left.\frac{(s+2)(s+4)}{s+1}\right|_{s=-2+j3}$$

$$= \frac{(-2+j3+2)(-2+j3+4)}{-2+j3+1} = 3.4205 \angle 37.875^\circ$$

Using Eq. (3-13c) yields

$$f_1(t) = \frac{M_1}{\omega}e^{-\alpha t}\sin(\omega t + \theta_1)\,u(t)$$

$$f_1(t) = \frac{3.4205}{3}e^{-2t}\sin(3t + 37.875^\circ)\,u(t) + \text{rest}$$

$$= 1.1402e^{-2t}\sin(3t + 37.875^\circ)\,u(t) + \text{rest}$$

Combining all the parts, the total inverse is

$$0.3e^{-t}u(t) + 1.1402e^{-2t}\sin(3t + 37.875^\circ)\,u(t)$$

Example 3-20 had complex poles that contained both a real part and an imaginary part. It is common to find complex poles containing only an imaginary part: For example, $s^2 + 25$. This type of complex pole looks completely different; however, if we put in the 0 for the real part, it will look the same $[(s+0)^2 + 25]$. This is similar to the case with real poles, as shown in Example 3-19.

The next example shows the inverse of a multiple-order complex pole. This is a simple example so that the order of what to do can be easily seen.

EXAMPLE 3-21

Find

$$\mathcal{L}^{-1}\left[\frac{(s+2)(s+3)}{[(s+1)^2+25]^2}\right]$$

SOLUTION The magnitude equation (3-11a) will require $R(-1+j5)$ and $R'(-1+j5)$.

$$R(-1+j5) = s^2+5s+6\Big|_{s=-1+j5}$$

$$= -23+j15$$

$$R'(-1+j5) = \frac{d}{ds}R(s)_{-1+j5}\Big|_{s=-1+j5}$$

$$= 2s+5\Big|_{s=-1+j5}$$

$$= 3+j10$$

Next we find $M_1 \angle\theta_1$ and $M_2 \angle\theta_2$ using Eq. (3-11a).

$$M_1 \angle\theta_1 = R(-\alpha+j\omega)$$

$$= -23+j15 = 27.459 \angle 146.89°$$

$$M_2 \angle\theta_2 = R'(-\alpha+j\omega) - \frac{r}{2j\omega}R(-\alpha+j\omega)$$

$$= (3+j10) - \frac{2}{2j5}(-23+j15) = 5.4000 \angle 90°$$

Using Eq. (3-12b), we find $f_1(t)$ and $f_2(t)$.

$$f_n(t) = \frac{2(-1)^{\text{int}(r/2)}M_n}{(2\omega)^r(n-1)!\,(r-n)!}t^{r-n}e^{-\alpha t}$$

$$\times \sin\left[\omega t + \theta_n + \frac{1+(-1)^r}{2}(90°)\right]u(t)$$

To simplify this equation we evaluate two terms.

$$(-1)^{\text{int}(r/2)} = (-1)^{\text{int}(2/2)} = -1$$

$$\frac{1+(-1)^r}{2}(90°) = \frac{1+(-1)^2}{2}(90°) = 90°$$

$$f_1(t) = \frac{2(-1)(27.459)}{[2(5)]^2(1-1)!\,(2-1)!}t^{2-1}e^{-1t}\sin(5t + 146.89° + 90°)$$
$$= -0.54918te^{-t}\sin(5t + 236.89°)\,u(t)$$
$$= 0.54918te^{-t}\sin(5t + 56.889°)\,u(t)$$
$$f_2(t) = \frac{2(-1)(5.4)}{[2(5)]^2(2-1)!\,(2-2)!}t^{2-2}e^{-1t}\sin(5t + 90° + 90°)$$
$$= -0.108e^{-t}\sin(5t + 180°)\,u(t)$$
$$= 0.108e^{-t}\sin(5t)\,u(t)$$

Since the second-order complex pole is all of the poles, the complete solution is the sum of these two inverse parts.

$$0.54918te^{-t}\sin(5t + 56.889°)\,u(t) + 0.108e^{-t}\sin(5t)\,u(t)$$

PROBLEMS

Section 3-1

1. Find the Laplace transform of $tu(t)$ using the Laplace integral, Eq. (3-1).
2. Find $V(s)$.
$$v(t) = \delta(t) + 5tu(t)$$
3. Find $\mathcal{L}[i(t)]$.
$$i(t) = \cos(3t + 30°)\,u(t)$$

Section 3-2

4. Find $f(t)$.
$$F(s) = \frac{s^2 + 2s}{s^3}$$
5. Find
$$\mathcal{L}^{-1}\left[\frac{4(s + 2)}{s^2 + 4}\right]$$

Section 3-3

6. Find the Laplace transform of the following functions using the shifted function operation presented in Section 3-3-1.
   (a) $3u(t - 2) - 2\delta(t - 4)$
   (b) $e^{-3(t-1)}u(t - 1) + \sin[3(t - 2)]\,u(t - 2)$

7. Find the Laplace transform of the following functions using the methods presented in Section 3-3-1.
   (a) $(t - 3)u(t - 2)$
   (b) $\cos(2t - 60°)\, u(t - 1)$
8. Find the Laplace transform of the following functions using the exponential multiplier operation presented in Section 3-3-2.
   (a) $e^{-4t}u(t)$
   (b) $e^{-2t}\cos(3t - 30°)\, u(t)$
9. Find the Laplace transform of the following functions using the exponential multiplier operation and the shifted function operation.
   (a) $e^{-2(t-1)}(t - 1)u(t - 1) + e^{-3(t-2)}u(t - 2)$
   (b) $e^{-(t-3)}\sin(4t - 12)\, u(t - 3)$
10. Find the Laplace transform of the following functions using the t multiplier operation.
    (a) $e^{-4t}tu(t)$
    (b) $t\cos(3t)\, u(t)$
    (c) $(t - 2)e^{-(t-2)}\sin[4(t - 2)]\, u(t - 2)$
11. Using the derivative operations, find the Laplace domain derivative of the following functions.
    (a) $\left\{\dfrac{d}{dt}[t]\right\} u(t)$
    (b) $\left\{\dfrac{d}{dt}[\sin(4t + 45°)]\right\} u(t)$
    (c) $\left\{\dfrac{d}{dt}[e^{-6t}]\right\} u(t)$
    (d) $\left\{\dfrac{d}{dt}[\cos(2t)]\right\} u(t)$
12. Using the integral operation, find the Laplace domain integral of the following functions.
    (a) $\int_0^t \delta(t) + u(t - 2)\, dt$
    (b) $\int_0^t \cos(2t)\, u(t)\, dt$
    (c) $\int_0^t (t - 2)\, u(t - 2)\, dt$

Section 3-4

13. Find $R(s)_{-3}$, $R(s)_{-2}$, and $R(s)_{-4+j3}$.

$$\frac{s^2(s + 1)}{(s + 3)^2(s + 2)[(s + 4)^2 + 9]}$$

14. Find R(−7) and R(−4).

$$\frac{(s + 2)^2[(s + 2)^2 + 25]}{(s + 4)^2(s + 7)[(s + 2)^2 + 16]^3}$$

15. Find the inverse Laplace transform of the single-order real poles.

(a) $\dfrac{s + 2}{(s + 4)(s + 5)(s + 16)}$

(b) $\dfrac{s[(s + 2)^2 + 4]}{(s + 3)(s + 2)(s + 4)}$

16. Find the inverse Laplace transform of the multiple-order real poles.

(a) $\dfrac{s + 2}{(s + 4)^3}$

(b) $\dfrac{(s + 4)^2(s + 5)}{s^3}$

(c) $\dfrac{2(s + 3)e^{-7s}}{(s + 2)^2 s^2}$

17. Find the inverse Laplace transform of the real poles.

(a) $\dfrac{(s^2 + 2s + 3)(s^2 + 7s + 12)}{s^2(s + 1)(s + 3)}$

(b) $\dfrac{(s + 3)(s + 4)}{(s + 1)(s + 2)^3}$

18. Find the inverse Laplace transform of the single-order complex poles.

(a) $\dfrac{s^2}{(s + 3)^2 + 49}$

(b) $\dfrac{s + 2}{[(s + 2)^2 + 9][(s + 1)^2 + 16]}$

19. Find the inverse Laplace transform of the multiple-order complex poles.

(a) $\dfrac{s^2}{[(s + 2)^2 + 1]^3}$

(b) $\dfrac{s^2(s + 4)}{[(s + 1)^2 + 4]^2}$

20. Find the inverse Laplace transform of the following functions.

(a) $\dfrac{s^2(s + 3)}{s^2(s + 2)[(s + 1)^2 + 25]}$

(b) $\dfrac{s^2 + 9s + 25}{(s + 1)[(s + 2)^2 + 1]^2}$

CHAPTER 4

Circuit Analysis Using Laplace Transforms

4-0 INTRODUCTION

Knowing Laplace transforms, we can solve circuits for the complete solution of voltages and currents. First we convert the time-domain circuit to the Laplace domain circuit. Then the analysis proceeds the same as dc analysis would. Once the Laplace voltage or current is found, we find the inverse Laplace for the time-domain value.

Since we have not solved circuits in the time domain, it may seem that Laplace is the long method. This is not true. When Laplace is not used, simultaneous differential equations may have to be solved.

4-1 INITIAL VOLTAGES AND CURRENTS

A circuit may have switches that may be physical switches or more typically electronic switches (transistor switches). These will cause reactive components to have initial conditions starting at the time we want to analyze the circuit.

We can analyze a circuit starting at any arbitrary time, which we will consider to be $t = 0$. This is done by drawing an equivalent circuit that is mathematically identical to the physical circuit. We must know the initial conditions of the circuit elements at our selected $t = 0$ time in order to draw the equivalent circuit. Laplace could be used to find the initial conditions, but it is more practical to find the initial conditions in the time domain since frequently the circuit is in a dc steady-state condition.

Since reactive components store energy and later release the energy into the circuit, they may have an initial condition associated with them. Resistors do not store energy, and therefore will never have an initial condition caused by them.

To find the initial condition of a reactive component, we consider its condition at our arbitrary time ($t = 0$). The condition of the component will be the same at $t = 0^-$, 0, and 0^+. Since in this time period the current through the inductor stays the same, we will use the current to describe its initial

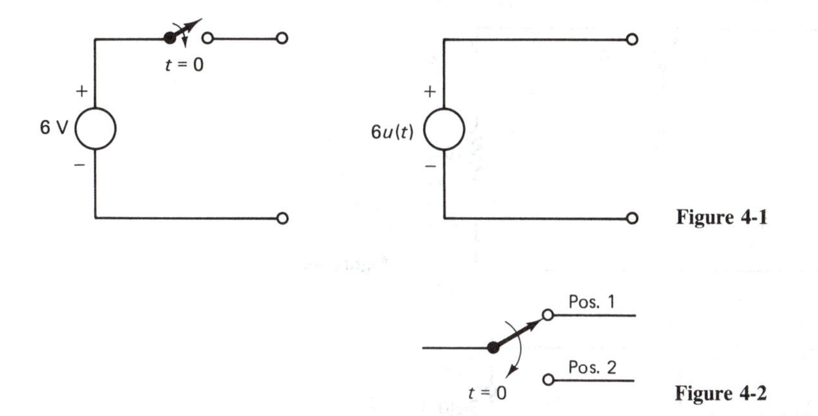

Figure 4-1

Figure 4-2

condition. Since the voltage across a capacitor also stays the same in this time period, we will use the voltage to describe the capacitor's initial condition.

This allows us to redraw the circuit using initial conditions and $u(t)$ instead of switches. Let's first examine switches. Figure 4-1 shows a 6-V source switched on at $t = 0$. This can be written as $6u(t)$ without the switch. This is mathematically the same as the physical circuit.

Figure 4-2 shows a two-position switch. This book assumes that a switch is simultaneously disconnected from one terminal and connected to the other terminal. The switch is always in contact with one terminal or the other. As shown in Fig. 4-2, the switch is in position 1 from $-\infty$ to $t = 0^-$ and is connected to position 2 at $t = 0^+$.

In order to have an equivalent circuit when we change the switches over to a $u(t)$ form, we must show the initial conditions of the reactive components. The following examples will show this technique.

EXAMPLE 4-1

At $t = 0$ the switch in Fig. 4-3 is thrown. Find the initial condition on the inductor. Draw the equivalent circuit.

SOLUTION When we look at the circuit from $t = -\infty$ to 0^-, we have the circuit shown in Fig. 4-4a. The circuit is at steady state, and the inductor therefore looks like a short as shown in Fig. 4-4b. The current through the inductor at $t = 0^-$ (just before the switch is thrown) will be the same at $t = 0^+$ since current will not change instantaneously through an inductor.

$$i(0^-) = i(0^+) = \frac{5}{2} = 2.5 \text{ A}$$

The circuit after $t = 0$ looks like a 5-V source switched on at $t = 0$ to a circuit containing an initially fluxed inductor and series resistor. Figure 4-5 shows this circuit.

The circuits of Figs. 4-3 and 4-5 are mathematically identical. Note that $u(t)$ is associated with the initial condition so that everything becomes active at $t = 0$.

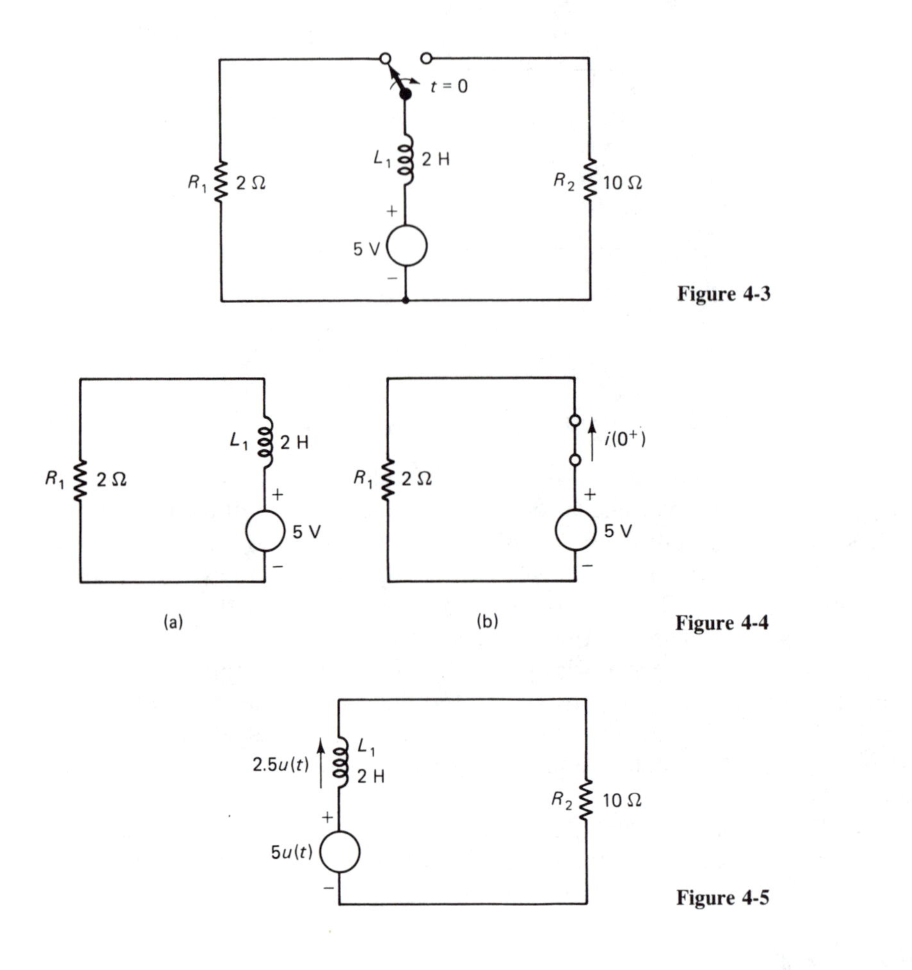

Figure 4-3

Figure 4-4

Figure 4-5

EXAMPLE 4-2

Draw an equivalent circuit for Fig. 4-6 without switches.

SOLUTION Figure 4-7a shows the circuit just before the switch is opened. Notice that the capacitor has been replaced with an open circuit since this is what the capacitor looks like in dc steady state. The circuit is a series circuit with the voltage across the R_2 being the same voltage as across the capacitor. For convenience we measure this voltage (Fig. 4-7a), so we get a positive voltage. The voltage measured at $t = 0^-$ is

$$V = \frac{ER_2}{R_1 + R_2} = \frac{10(6)}{4 + 6} = 6 \text{ V}$$

Since the voltage across the capacitor cannot change instantaneously, it will have the same value at $t = 0^+$.

Figure 4-7b shows the circuit without switches and with everything starting at $t = 0$.

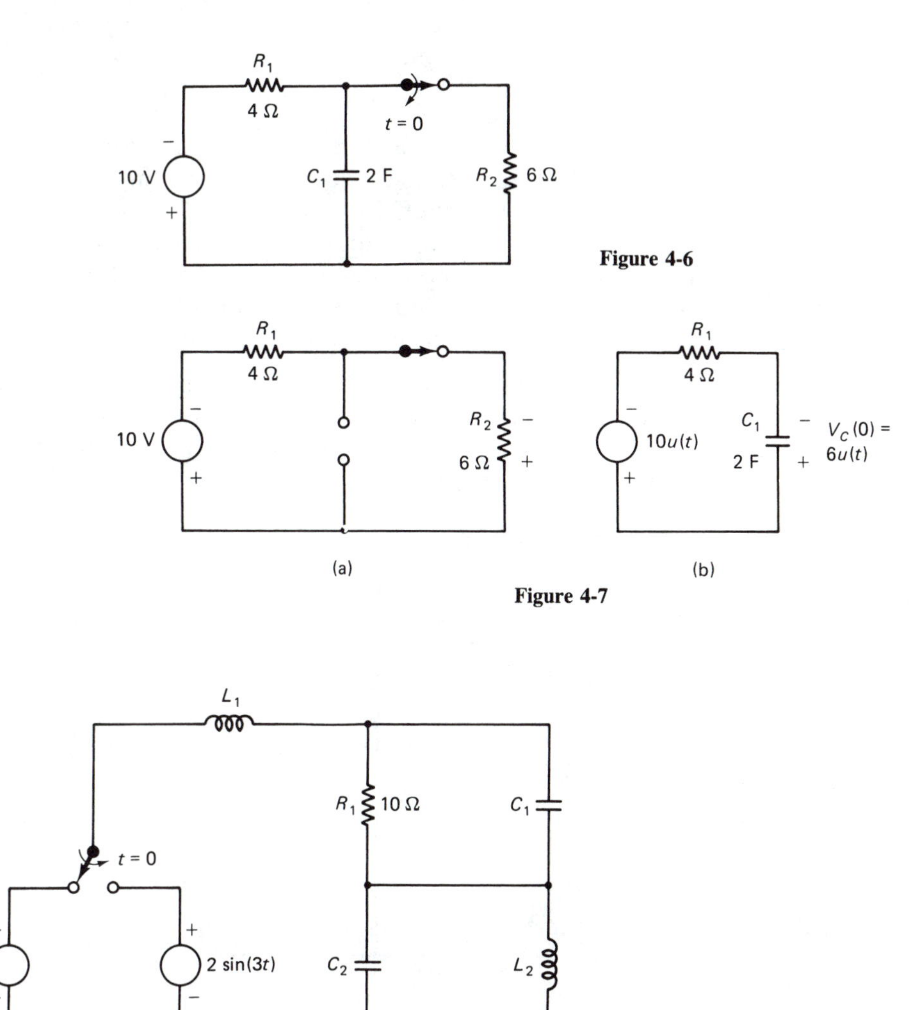

Figure 4-6

Figure 4-7

Figure 4-8

EXAMPLE 4-3

Draw an equivalent circuit for Fig. 4-8 without switches.

SOLUTION Figure 4-9a shows the circuit at $t = 0^-$, which is just before the switch is changed. The capacitors have been replaced with an open and the inductors have been replaced with a short since the circuit is in dc steady state. Notice that the inductor L_2 shorts any voltage across C_2. This circuit is essentially a voltage source across a 10-Ω resistor, and the current through the series is

$$I = \frac{E}{R} = \frac{5}{10} = 0.5 \text{ A}$$

This current must go through both inductors, which makes both their currents 0.5 A at $t = 0^-$. The voltage across capacitor C_1 at this time is the voltage across R_1, as seen by using Kirchhoff's voltage law.

Since the current through an inductor and the voltage across a capacitor cannot change instantaneously, the values at $t = 0^-$ will be the same at $t = 0^+$. Figure 4-9b shows the circuit equivalent circuit. Notice that since the switch has been removed, the sinusoidal source must have $u(t)$ added to it.

The circuit may not be in a steady-state condition just before a switch is changed, but we use the same concept. We find the voltage across the capacitors and the current through the inductors at $t = 0^-$, which are the initial conditions at $t = 0^+$.

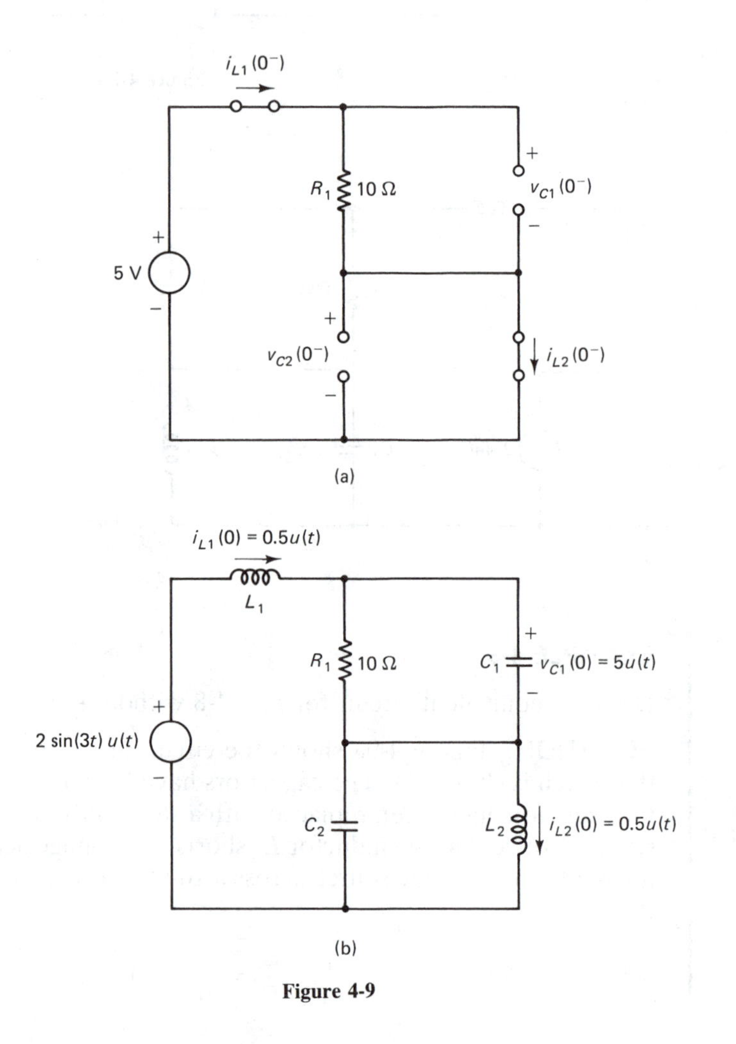

Figure 4-9

4-2 LAPLACE IMPEDANCE

Each circuit element, R, L, and C, has a value in the Laplace domain. Their value in the Laplace domain will be called the Laplace impedance. The Laplace impedance is the ratio of the Laplace voltage to the Laplace current.

$$Z(s) = \frac{E(s)}{I(s)} \tag{4-1}$$

The mathematically correct units for the voltage, current, and impedance are unimportant to us, but we may call them Laplace volts, Laplace amperes, and Laplace ohms.

From the preceding section we found that inductors and capacitors may have an initial condition associated with them. When there is an initial value it may be expressed as either a voltage or a current. We found that the natural way of expressing an initial condition for an inductor was with a current, and the natural way of expressing an initial condition for a capacitor was with a voltage. This form will be called the natural equivalent circuit.

4-2-1 Laplace Resistance

Figure 4-10 shows a resistor with an arbitrary voltage applied to it. By using Ohm's law, the current through it is

$$i(t) = \frac{e(t)}{R}$$

Since these are functions of time, we may take the Laplace transform of both sides of the equation. Since R is a constant multiplier it will not be affected by this process. Taking the Laplace transform gives

$$I(s) = \frac{E(s)}{R}$$

or

$$R = \frac{E(s)}{I(s)}$$

Therefore, by Eq. (4-1) the Laplace impedance for resistance is

$$Z_R(s) = R \tag{4-2}$$

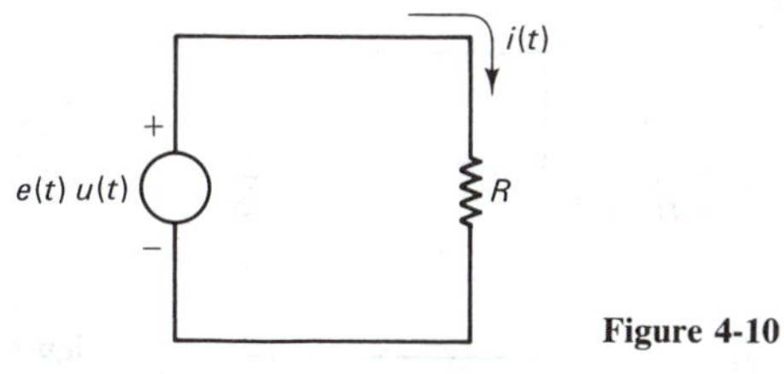

Figure 4-10

For convenience, the function notation may be dropped, and the notation of Z_R may be used. The subscript is not necessary other than to indicate to which component in a circuit the impedance refers.

Since resistors cannot store energy, there will never be an initial condition caused by a resistor. A resistor may have a current at $t = 0^-$, but it does not store this energy to be released later. The capacitor and inductor will store energy and release it at a later time.

4-2-2 Laplace Inductance

Figure 4-11 shows an inductor with an arbitrary voltage applied to it. The current through the inductor is

$$i(t) = \frac{1}{L}\int_0^t e(t)u(t)\,dt$$

Taking the Laplace transform of both sides using (0-5) yields

$$I(s) = \frac{1}{L}\left[\frac{E(s)}{s}\right] = \frac{E(s)}{sL}$$

and solving for $E(s)/I(s)$ gives

$$sL = \frac{E(s)}{I(s)}$$

Therefore, by Eq. (4-1) the Laplace impedance for inductance is

$$Z_L(s) = sL \tag{4-3}$$

Here again this may be expressed as Z_L.

Inductors store energy and therefore may have an initial condition. The initial condition can be represented as a current, $i(0)$, which will always be a constant. In the time domain this adds a current to the integral.

$$i(t) = \frac{1}{L}\int_0^t e(t)\,dt + i(0)u(t)$$

Taking the Laplace of the equation reveals the same Laplace impedance except that a step function current is added.

$$I(s) = \frac{E(s)}{sL} + \frac{i(0)}{s}$$

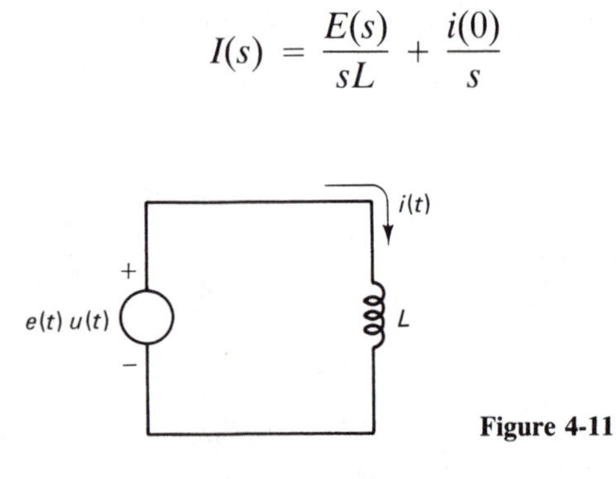

Figure 4-11

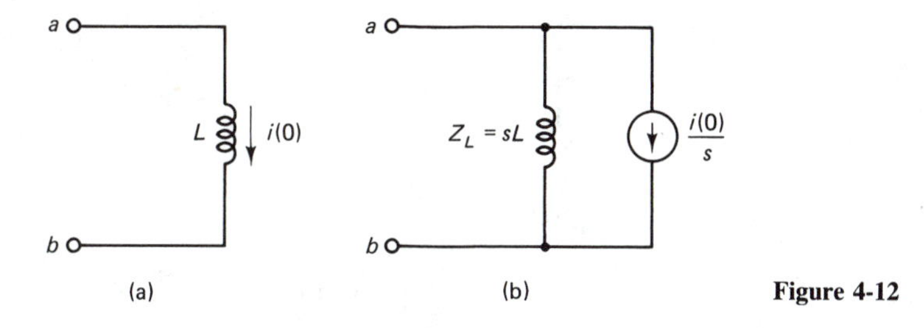

Figure 4-12

This is a sum of currents, and can be represented mathematically by an initially unfluxed inductor in parallel with a dc current source. Figure 4-12a shows an inductor with an initial current. Figure 4-12b shows the equivalent Laplace circuit, which is called the natural equivalent circuit for the inductor.

We must remember that Fig. 4-12b is an equivalent circuit. The two circuits, Figs. 4-12a and 4-12b, are equivalent only as seen from the terminals a and b. When we want to find the current through or voltage across the inductor, it must be done at the terminals, which will include the current source. It is essential to understand this in order to use equivalent circuits.

4-2-3 Laplace Capacitance

Figure 4-13 shows a capacitor with an arbitrary voltage applied to it. The current through the capacitor is

$$i(t) = C\frac{d}{dt}[e(t)u(t)]$$

Taking the Laplace transform using (0-4b) gives

$$I(s) = CsE(s)$$

and solving for $E(s)/I(s)$, we obtain

$$\frac{1}{sC} = \frac{E(s)}{I(s)}$$

Therefore, by Eq. (4-1) the Laplace impedance for capacitance is

$$Z_C(s) = \frac{1}{sC} \tag{4-4}$$

As with the resistor and inductor notation, we may drop the reference to s and simply write Z_C.

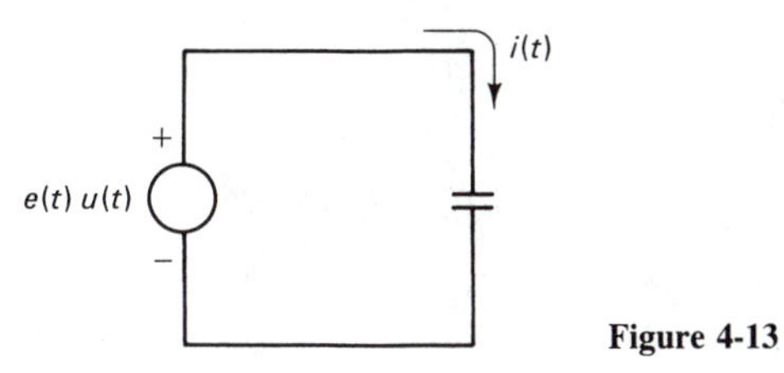

Figure 4-13

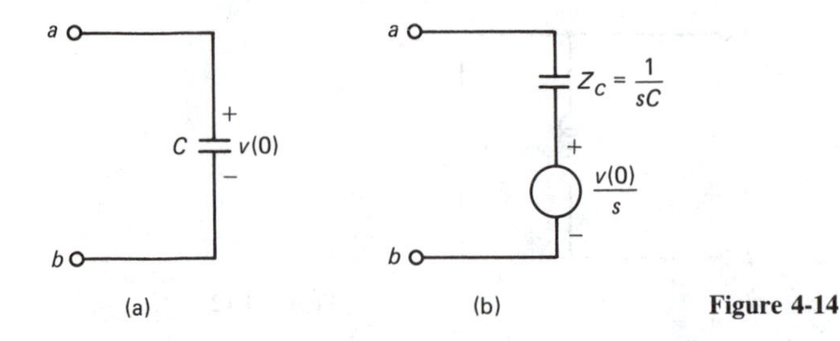

Figure 4-14

Capacitors store energy and therefore may have an initial condition. The initial condition can be represented as a voltage, $v(0)$, which will always be a constant. In the time domain this adds a voltage to the derivative.

$$i(t) = C\frac{d}{dt}[e(t)u(t)] + v(0)u(t)$$

Taking the Laplace of the equation reveals the same Laplace impedance except that a step voltage is added.

$$I(s) = CsE(s) + \frac{v(0)}{s}$$

This is a sum of voltages and can be represented mathematically by an initially uncharged capacitor in series with a dc voltage source. Figure 4-14a shows a capacitor with an initial voltage. Figure 4-14b shows the equivalent Laplace circuit, which is called the natural equivalent circuit for the capacitor.

This is an equivalent circuit and is equivalent only at terminals *a* and *b*. When the voltage across or current through the capacitor is found, it must be found at the terminals, which will include the voltage source.

4-2-4 Laplace Impedance and Source Conversion

In some cases we need the fluxed inductor represented as an impedance with a series voltage source. We may also require a charged capacitor represented as an impedance with a parallel current source. This is accomplished by using source conversion.

Source conversion is done exactly as it is done in dc circuit analysis. Keep in mind that the circuit, as with all equivalent circuits, is equivalent only at the terminals where the conversion was done.

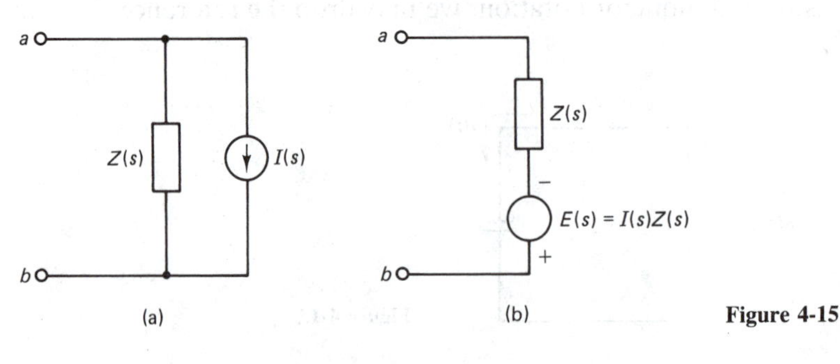

Figure 4-15

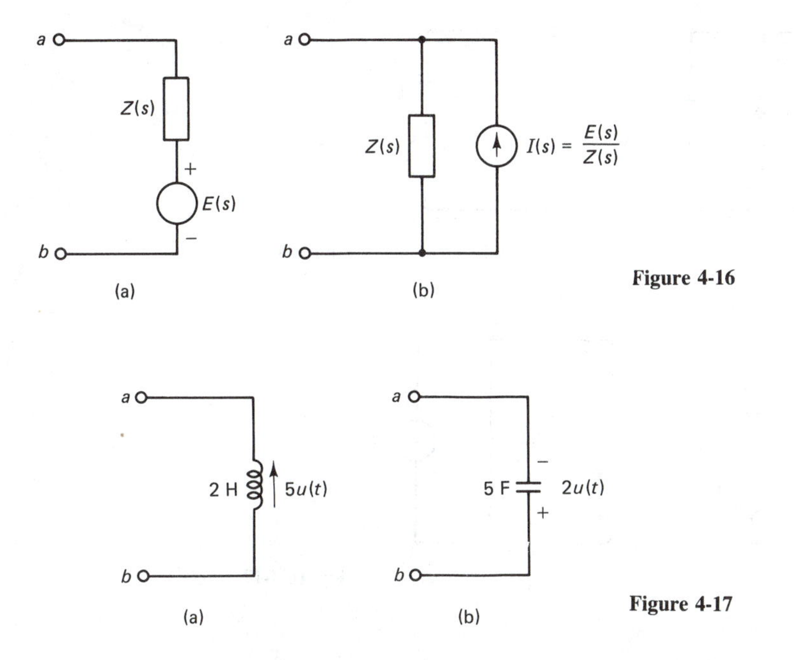

Figure 4-16

Figure 4-17

Figure 4-15 shows conversion of a parallel current source and impedance (Norton equivalent) to a series voltage source and impedance (Thévenin equivalent). Figure 4-16 shows a series voltage source and impedance conversion to a parallel current source and impedance. The value of the impedance is the same in both equivalent circuits; however, the voltage and current for this impedance may be different. Only the voltage and current at terminals *a* and *b* are the same.

Notice the similarity between source conversion and Ohm's law. To get a voltage you multiply a current times an impedance, and to get a current you divide a voltage by an impedance.

An easy way to determine polarity or current direction is to remove everything from terminals *a* and *b* and assume that a pure resistance is attached. Then put the source in the equivalent circuit so that the current direction through this resistance is the same in both circuits.

EXAMPLE 4-4

Draw the Laplace natural equivalent circuit for the circuits in Fig. 4-17. Use source conversion and draw the resulting Laplace equivalent circuit.

SOLUTION Figure 4-18 shows the Laplace natural equivalent circuits.

Converting the parallel current source to a series voltage source, we have

$$E(s) = Z(s)I(s) = 2s\,\frac{5}{s} = 10$$

The equivalent circuit is shown in Fig. 4-19a.

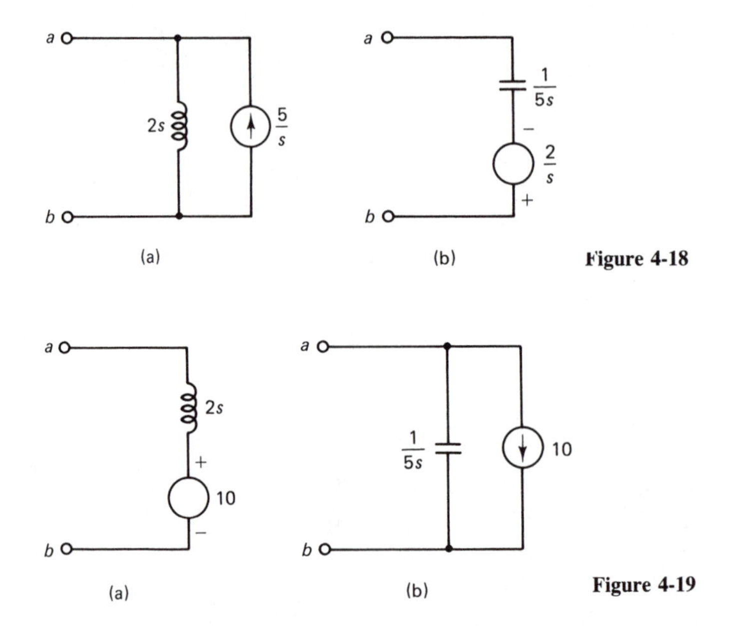

Figure 4-18

Figure 4-19

Converting the series voltage source to a parallel current source, we

$$I(s) = \frac{E(s)}{Z(s)} = \frac{\dfrac{2}{s}}{\dfrac{1}{5s}} = 10$$

The resulting circuit is shown in Fig. 4-19b.

From Example 4-4, we notice in the converted circuits of Fig. 4-19 that we have a constant for the source values. This corresponds to an impulse in the time domain. Therefore, in the time domain we could represent a fluxed inductor as an unfluxed inductor in series with an impulse voltage source and a charged capacitor as a parallel uncharged capacitor with an impulse current source.

4-3 LAPLACE CIRCUIT

To find the Laplace circuit we find the initial conditions and convert each element to their natural equivalent circuit. Depending on the circuit analysis method used, we may have to convert to all voltage sources or all current sources. A few examples will demonstrate this.

EXAMPLE 4-5

Convert the time-domain circuit shown in Fig. 4-20 to the Laplace domain circuit. Convert all sources to voltage sources.

SOLUTION Using the methods described in Section 4-2, we replace the components that have an initial condition with their natural equivalent circuit as shown in Fig. 4-21a. Z_{L1} is the only component with a current source, and the only component on which we need to use source conversion. The resulting circuit is shown in Fig. 4-21b.

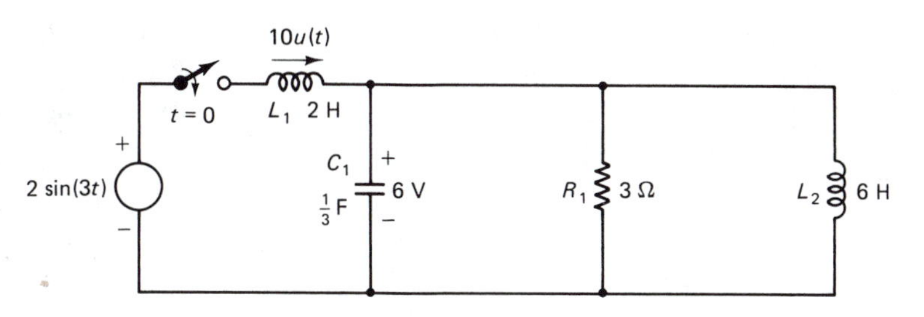

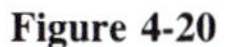

Figure 4-20

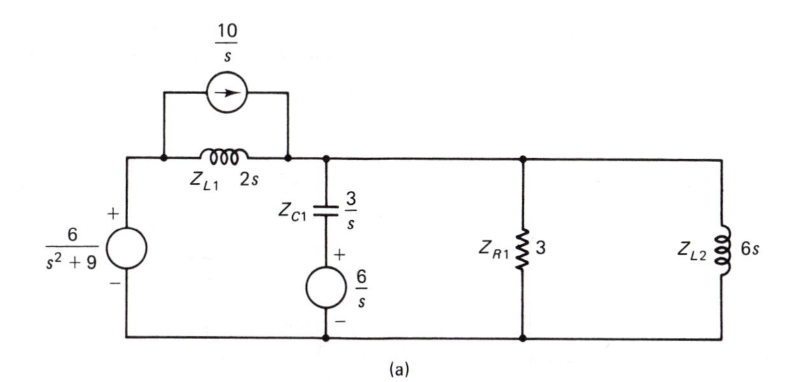

(a)

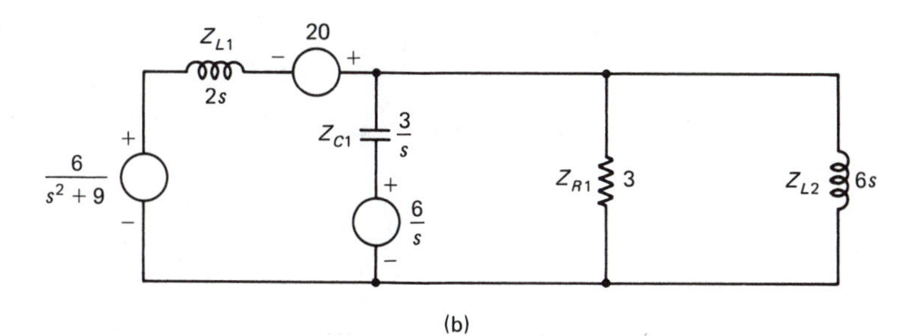

(b)

Figure 4-21

EXAMPLE 4-6

For the circuit shown in Fig. 4-22, draw the Laplace circuit with the reactive components in their natural equivalent circuit.

SOLUTION At $t = 0^-$ the resistor R_3 and capacitor C_1 are shorted. The circuit, just before the switch is thrown, is shown in Fig. 4-23a. The only reactive component with an initial condition is L_1, which has an initial current of

$$I(0^-) = I(0^+) = \frac{E}{R_1 \parallel R_2} = \frac{10}{2} = 5 \text{ A}$$

We can now remove the switch of Fig. 4-22, replace the voltage source with $10u(t)$, and indicate the initial current through L_1. This is shown in Fig. 4-23b. From Fig. 4-23b we can convert the circuit to the Laplace domain as shown in Fig. 4-23c.

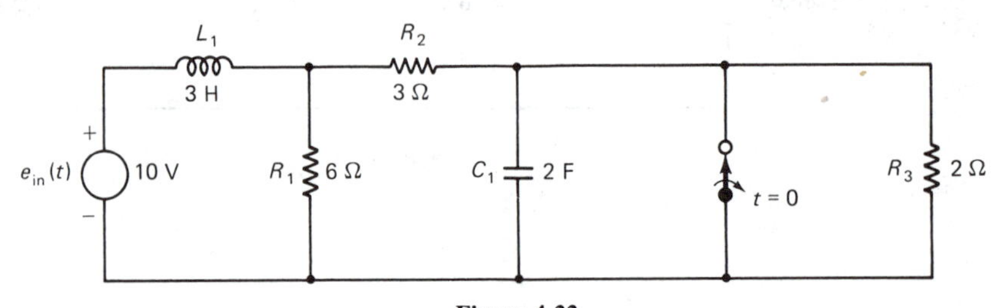

Figure 4-22

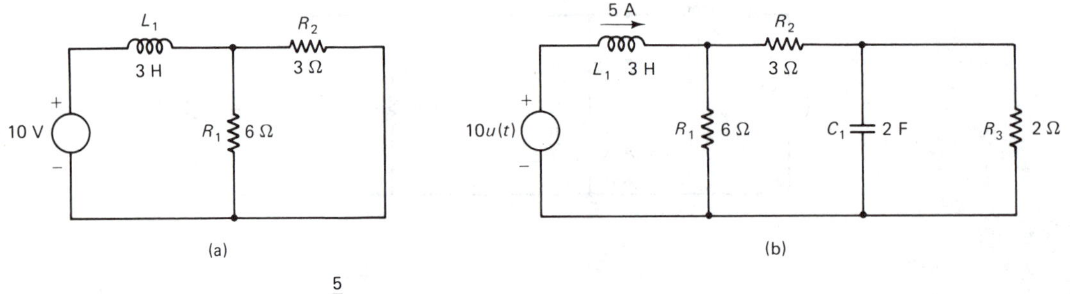

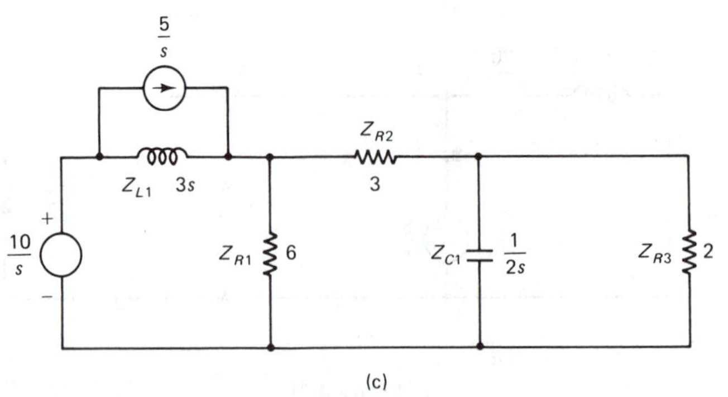

(c)

Figure 4-23

4-4 SOLVING SIMPLE CIRCUITS

Once we have the Laplace circuit, we solve for the Laplace voltage, $V(s)$, and/or Laplace currents, $I(s)$, as if the circuit was a dc circuit involving only resistors. The difference is that the values in Laplace will be functions of s instead of numerical values. We then find the inverse of $V(s)$ and/or $I(s)$, which will give us the complete response from $t = 0$ to ∞.

The response of a voltage or current can be divided into two components, the transient response and the steady-state response. When we solve a circuit using Laplace, we find both of these responses at the same time. The transient response is composed of the transient and the steady-state terms of the equation. The transient response is from $t = 0$ until when the transient terms of the equation become insignificant. A term is transient if it becomes insignificant in a finite amount of time which may be as small as 1 msec or as long as days. The steady-state response is composed of only the steady-state terms.

One of the major advantages of using Laplace is finding the transient response. Transient responses can have large values that do not appear in the steady-state response, and therefore go undetected when only steady-state analysis is used. These large values can cause fuses to blow or circuit components to be destroyed.

A few examples will demonstrate how to use Laplace to analyze a circuit. Once the basic concept is understood, it is just a matter of reviewing dc circuit analysis. A review of basic dc circuit analysis equations is given in Appendix C.

EXAMPLE 4-7

Find the current through R_1 in Fig. 4-24.

SOLUTION We must first convert the time-domain circuit to a Laplace domain circuit. This is shown in Fig. 4-25.

Since this is a current split between parallel branches, we should use the current-divider rule.

$$I_R(s) = \frac{I_{\text{in}}(s)Z_{C1,C2}}{Z_{R1} + Z_{C1,C2}}$$

$$= \frac{\dfrac{3}{s}\left(\dfrac{1}{2s} + \dfrac{1}{4s}\right)}{5 + \left(\dfrac{1}{2s} + \dfrac{1}{4s}\right)}$$

$$= \frac{0.45}{s(s + 0.15)}$$

Notice that the series capacitors are added rather than being the product over the sum as in time-domain circuit analysis. When we use Laplace, all impedances and sources are used as if they were resistors or dc sources.

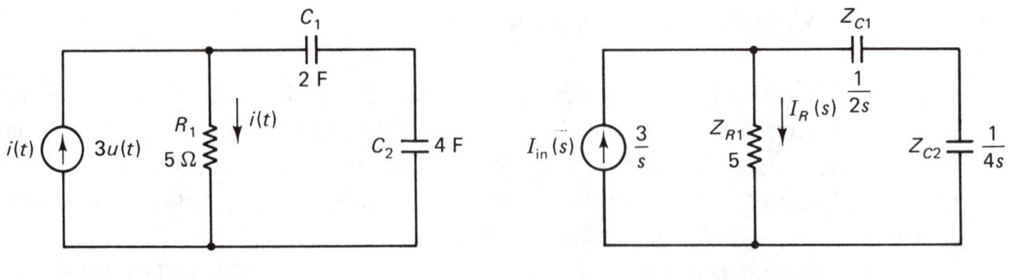

Figure 4-24

Figure 4-25

Finding the inverse Laplace, we have the time-domain current.

$$i_R(t) = 3u(t) - 3e^{-0.15t}u(t)$$

EXAMPLE 4-8

The circuit shown in Fig. 4-26 shows an inductor that was switched from one circuit path to the one shown. Previous to the switching, 1 A of current was established. Find $i_R(t)$ and $v_R(t)$. Find the steady-state value of the current.

SOLUTION First we convert the time-domain circuit to the Laplace circuit as shown in Fig. 4-27a. The initial condition current source is then converted to a series voltage source in Fig. 4-27b. Since this initial voltage source is in series with the supply voltage source, we can combine them as shown in Fig. 4-27c.

We have two approaches that could easily be applied. We could divide the resulting voltage source by the total circuit impedance, since it is a series circuit, and find the voltage across the impedance $Z_R(s)$ using Ohm's law. The other approach would be to use the voltage-divider rule to find the voltage across the impedance $Z_R(s)$ and Ohm's law to find the current. Let's use the second approach.

$$V_R(s) = \frac{E(s)Z_R}{Z_R(s) + Z_L(s)}$$

$$= \frac{\left(\dfrac{6}{s^2 + 9} - 3\right)12}{12 + 3s}$$

$$= \frac{-12(s^2 + 7)}{(s^2 + 9)(s + 4)}$$

The inverse Laplace is then

$$v_R(t) = -11.04e^{-4t}u(t) + 1.6 \sin (3t - 36.870°)$$

Since the component to find the current across is a resistor, we could divide by the resistance value in the time domain. If the component were

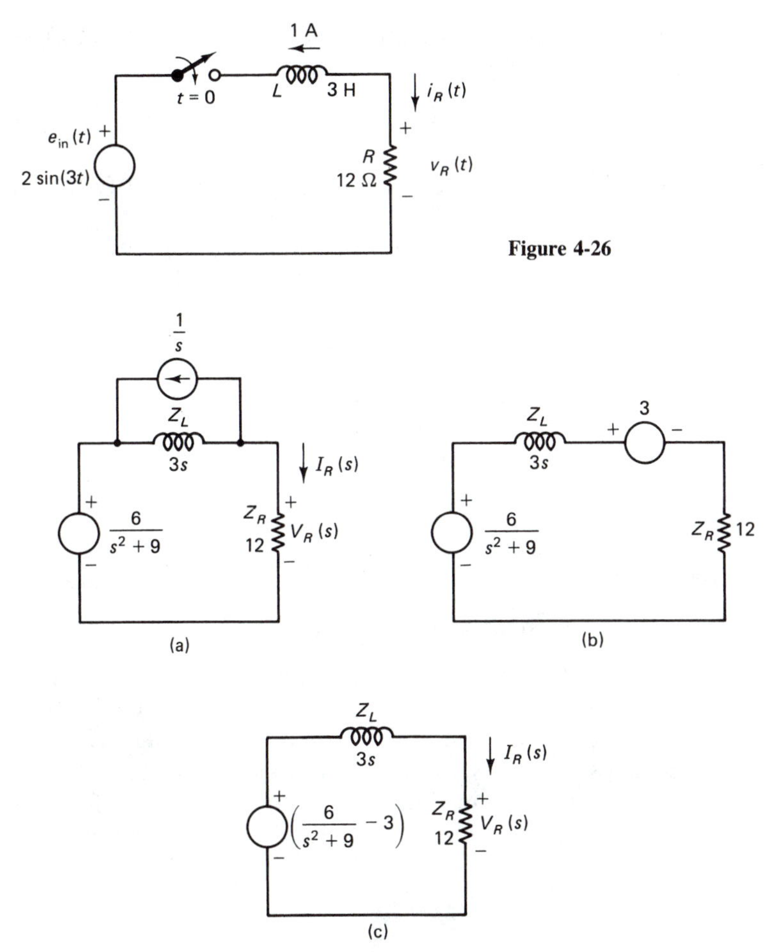

Figure 4-26

Figure 4-27

a capacitor, we could take the derivative of the voltage and multiply by the capacitance value. An inductor would involve the integral. In the Laplace domain, however, all cases are done by using $I(s) = E(s)/Z(s)$.

In this case it would be simpler to divide the time-domain voltage by the resistance, but we will use the Laplace approach to demonstrate the Laplace method.

$$I_R(s) = \frac{V_R(s)}{R}$$

$$= \frac{\dfrac{-12(s^2 + 7)}{(s^2 + 9)(s + 4)}}{12}$$

$$= \frac{-(s^2 + 7)}{(s^2 + 9)(s + 4)}$$

Notice that the only quantity changed is the constant multiplier. The inverse Laplace is therefore

$$i_R(t) = -0.92e^{-4t}u(t) + 0.13333 \sin(2t - 36.870°)$$

To find the steady-state current, we must identify the transient terms in the current equation. The first term has e to a negative power. This term will be insignificant after a period of time, and is therefore transient. The second term does not have a factor like this and will always be a significant value. The steady-state value will then be the equation with the transient terms removed. In this case it is

$$i_{ss}(t) = 0.13333 \sin(3t - 36.870°)$$

This is the answer that would be found using steady-state ac circuit analysis.

EXAMPLE 4-9

For the Laplace circuit shown in Fig. 4-28, find the voltage V_{ab}.

SOLUTION The only way to solve for a voltage across an open circuit is to use Kirchhoff's voltage law. We must first find all the other voltages around a loop. Since we know the current through Z_{R2}, we can find the voltage across Z_{R2}. By using Kirchhoff's current law at the top node and using Ohm's law, we can find the current and voltage for Z_{R1}.

Using Fig. 4-29, we can apply Kirchhoff's current law at the top node.

$$I_{in} - I_1 - I_2 = 0$$

$$\begin{aligned} I_1 &= I_{in} - I_2 \\ &= 4 - \frac{s + 1}{s + 0.25} \\ &= \frac{3s}{s + 0.25} \end{aligned}$$

Using Ohm's law, we can find the voltage across Z_{R1} and Z_{R2}.

$$V_1 = (1)\frac{3s}{s + 0.25} = \frac{3s}{s + 0.25}$$

$$V_2 = (1)\frac{s + 1}{s + 0.25} = \frac{s + 1}{s + 0.25}$$

Since we are looking for V_{ab}, we measure a with respect to b as shown in Fig. 4-29. (The first subscript is always the point measured, and the

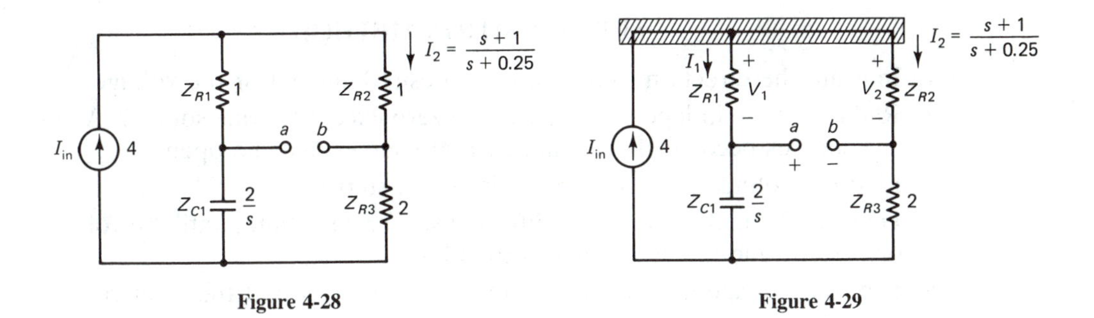

Figure 4-28

Figure 4-29

second subscript is always used as reference.) Writing Kirchhoff's voltage law gives

$$V_1 - V_2 + V_{ab} = 0$$

$$V_{ab} = V_2 - V_1$$

$$= \frac{s + 1}{s + 0.25} - \frac{3s}{s + 0.25}$$

$$= \frac{-2s + 1}{s + 0.25}$$

The inverse Laplace transform is

$$V_{ab}(t) = -2\delta(t) + 1.5e^{-0.25t u(t)}$$

4-5 SOLVING MULTIPLE-SOURCE CIRCUITS

A network with multiple sources is nearly impossible to solve in the time domain when inductors and capacitors are present. Using Laplace will make the process much easier. Since Laplace makes the analysis a dc process, this section is written as a review of some dc network techniques. There are a number of methods in dc circuit analysis used to solve multiple-source circuits which are all valid to use in Laplace. We will concentrate on three methods.

4-5-1 Superposition

This method is best suited for finding one or a few voltages and currents when a multiple-source network is being solved. It also allows us to see the effect that each independent source has on a network or component since we work with each source separately using single-source techniques.

SUPERPOSITION METHOD

1. Indicate the direction to measure the desired current or the voltage.
2. Reduce all the independent sources to zero except for one source. Voltage sources become shorts, and current sources become opens.
3. Find the voltage or current as indicated in step 1.
4. Do steps 2 and 3, leaving in a different source each time, until the voltage or current due to each source is found.
5. Sum the voltage or current due to each source to find the total voltage or current.

EXAMPLE 4-10

Find the Laplace current through the inductor of Fig. 4-30 using the superposition method.

SOLUTION We first convert the circuit to a Laplace circuit as shown in Fig. 4-31. An arbitrary current direction is chosen since a direction was not indicated in the problem.

We will find the current due to the initial voltage from the capacitor first. We remove all sources except the source we are working with, as shown in Fig. 4-32a. The circuit can be simplified to the circuit shown in Fig. 4-32b. We will call the current I'. Using Ohm's law, the current is

$$I' = \frac{\frac{4}{s}}{s + 10 + \frac{2}{s}} = \frac{4}{s^2 + 10s + 2}$$

Next we will find the current due to the current source. We will call this current I''. Figure 4-33a shows all the sources removed except the current source, and Fig. 4-33b shows the circuit simplified. Notice that the

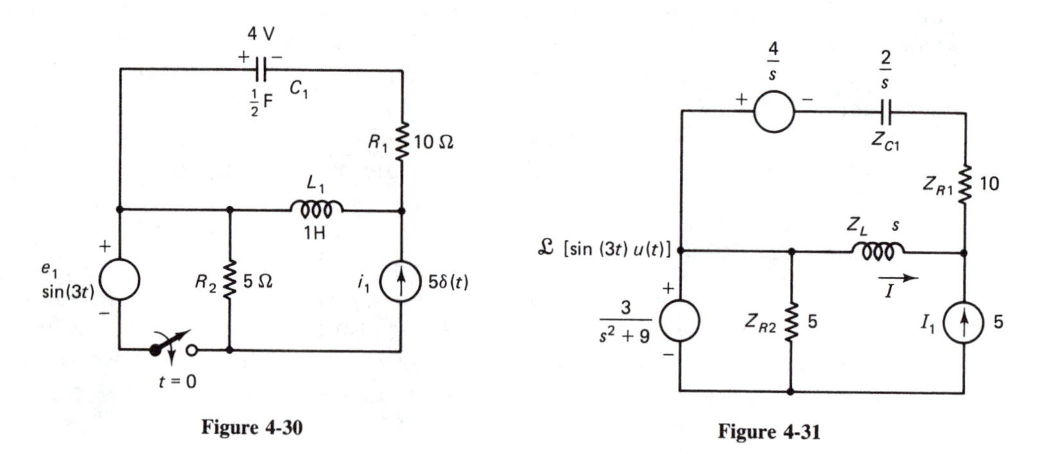

Figure 4-30

Figure 4-31

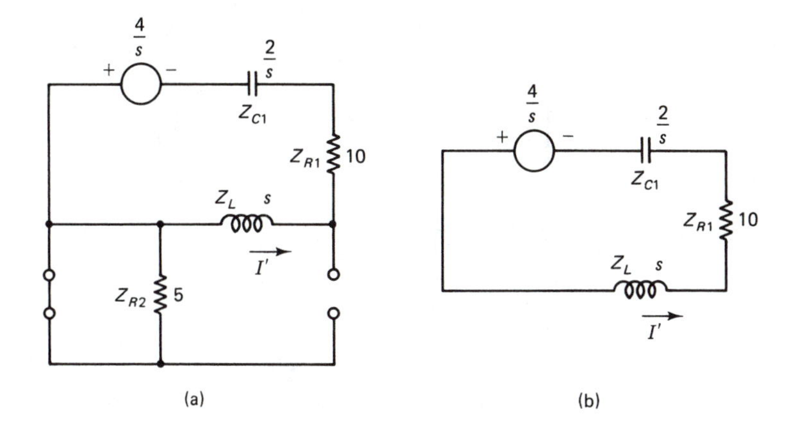

Figure 4-32

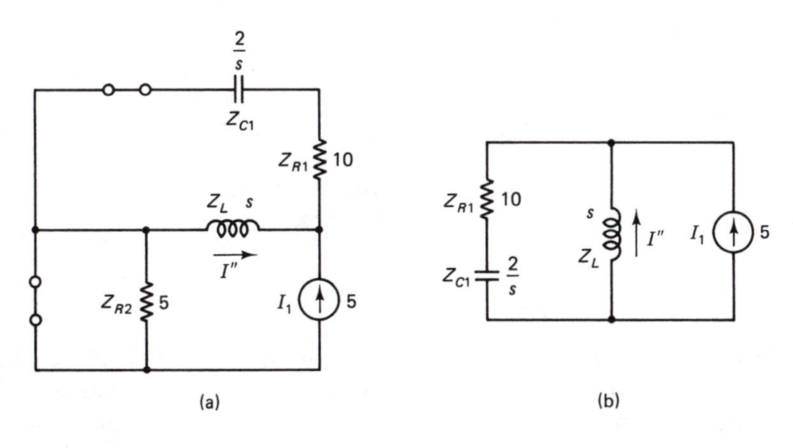

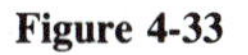
Figure 4-33

current I'' is opposite to the direction of current flow we choose. Therefore, I'' will be a negative quantity. Using the current divider rule yields

$$I'' = \frac{-5\left(10 + \dfrac{2}{s}\right)}{\left(10 + \dfrac{2}{s}\right) + s} = \frac{-50(s + 0.2)}{s^2 + 10s + 2}$$

Finally, we will find the current due to the sinusoidal voltage source. Figure 4-34 shows this current, I'''. Notice that there is no complete path

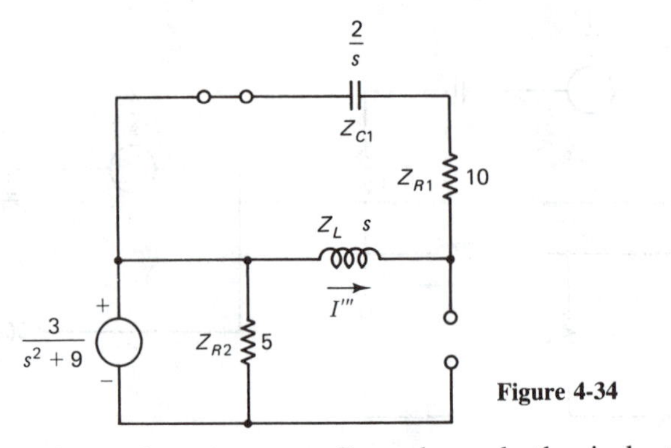

Figure 4-34

for current from the source to flow through the inductor. The current, therefore, is

$$I''' = 0$$

Since we measured the currents due to each source in the same direction, we sum the results.

$$I_T = I' + I'' + I'''$$

$$= \frac{4}{s^2 + 10s + 2} + \frac{-50(s + 0.2)}{s^2 + 10s + 2} + 0$$

$$= \frac{-50(s + 0.12)}{(s + 0.20417)(s + 9.7958)}$$

4-5-2 Mesh Current

When there are several quantities to find in a network, it is easier to solve for all of them at the same time. The mesh current method solves for the loop currents. Once these are known, any current or voltage in the circuit can be solved. There are a few variations of this method. We will use the following form:

MESH CURRENT METHOD

1. Draw the circuit so that none of the wires cross over each other.
2. Convert all independent and dependent current sources to voltage sources.
3. Draw a clockwise loop current inside each enclosed area (called a window) of the network to indicate the direction the loop currents are measured.
4. For each loop write an equation inside each window as follows.
   (a) Sum all the components, except sources, in the window and multiply by the loop current.
   (b) Locate any component, other than sources, that has another loop current passing through it. For each component, subtract the component value times this other loop current.

(c) Set the terms found above equal to the clockwise algebraic summation of the rises (added) and drops (subtracted) of the voltage sources.

5. Solve the simultaneous equations for the currents. This is usually done with determinants.
6. The current through a component showing two loop currents is equal to the difference of the two loop currents. This value of the current is measured in the direction of the current you subtracted from. Therefore, a component with I_3 and I_4 through it could be either: (a) $I_3 - I_4$ measured in the direction of I_3, or (b) $I_4 - I_3$ measured in the direction of I_4.

EXAMPLE 4-11

Write the mesh equations for Fig. 4-35.

SOLUTION Figure 4-36a shows the circuit in Laplace. The initial condition for the inductor must be converted to a voltage source. Figure 4-36b shows this conversion, and the loop currents have been indicated.

The first equation is found in the top window or the I_1 window. There are two components and one voltage source in this window. The capacitor is the only component that has another current going through it. Voltage sources are treated separately and not considered as components in this context. The first equation is

$$\left(10 + \frac{1}{2s}\right)I_1 - 0I_2 - \frac{1}{2s}I_3 = \frac{-5}{s^2 + 25}$$

Experience will show that it is best to remove fractions from these equations rather than removing them when working with the determinants. Multiplying by $[2s(s^2 + 25)]$, the equation becomes

$$(s^2 + 25)(20s + 1)I_1 - 0I_2 - (s^2 + 25)I_3 = -10s$$

For the second window or I_2 loop, the 4-Ω resistor is the only component with two currents through it. The equation is

$$-0I_1 + (4s + 4)I_2 - 4I_3 = \frac{5}{s^2 + 25} - 8$$

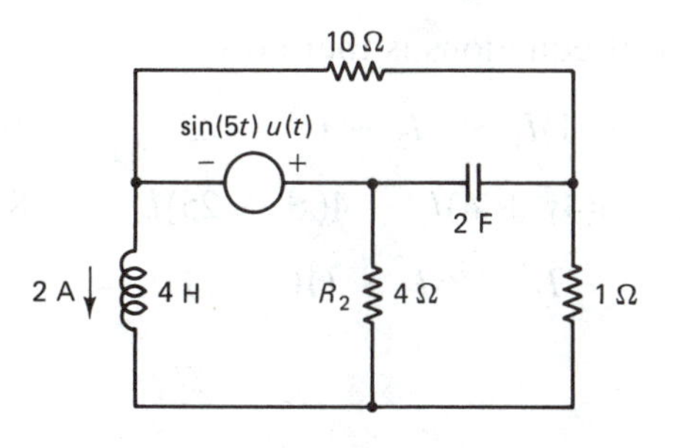

Figure 4-35

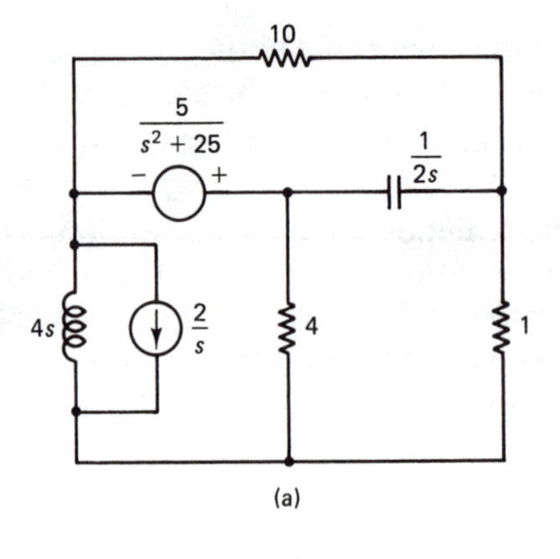

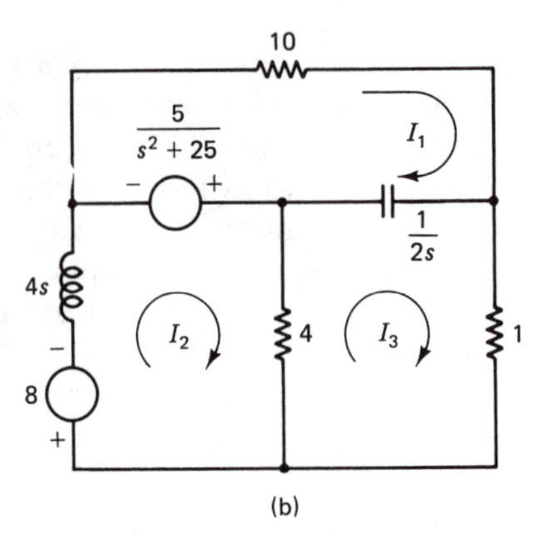

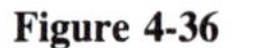
Figure 4-36

Notice how the voltage sources are added or subtracted depending on the way they are measured and using a clockwise summation. Multiply by $(s^2 + 25)$ to remove the fractions.

$$-0I_1 + (s^2 + 25)(4s + 4)I_2 - 4(s^2 + 25)I_3 = -8s^2 - 195$$

The last equation from the third loop is

$$-I_1 - 8sI_2 + (10s + 1)I_3 = 0$$

The complete set of equations is therefore

$$(s^2 + 25)(20s + 1)I_1 - 0I_2 - (s^2 + 25)I_3 = -10s$$

$$-0I_1 + (s^2 + 25)(4s + 4)I_2 - 4(s^2 + 25)I_3 = -8s^2 - 195$$

$$-I_1 - 8sI_2 + (10s + 1)I_3 = 0$$

EXAMPLE 4-12

Using mesh analysis, find the voltage across the capacitor in Fig. 4-37.

SOLUTION Figure 4-38 shows the Laplace circuit with all sources in a voltage source form. The loop equations are

$$\left(2s + \frac{2}{s}\right)I_1 - \frac{2}{s}I_2 = 4 - \frac{3}{s}$$

$$-\frac{2}{s}I_1 + \left(\frac{2}{s} + 3\right)I_2 = \frac{3}{s} - \frac{2}{s}$$

Removing the fractions gives

$$(2s^2 + 2)I_1 - 2I_2 = 4s - 3$$

$$-2I_1 + (3s + 2)I_2 = 1$$

Solving the denominator determinant, we obtain

$$D = \begin{vmatrix} 2s^2 + 2 & -2 \\ -2 & 3s + 2 \end{vmatrix} = (2s^2 + 2)(3s + 2) - 4$$

$$= 6s^3 + 4s^2 + 6s$$

Solving for the currents I_1 and I_2 yields

$$I_1 = \frac{\begin{vmatrix} 4s - 3 & -2 \\ 1 & 3s + 2 \end{vmatrix}}{D} = \frac{12s^2 - s - 4}{D}$$

$$I_2 = \frac{\begin{vmatrix} 2s^2 + 2 & 4s - 3 \\ -2 & 1 \end{vmatrix}}{D} = \frac{2s^2 + 8s - 4}{D}$$

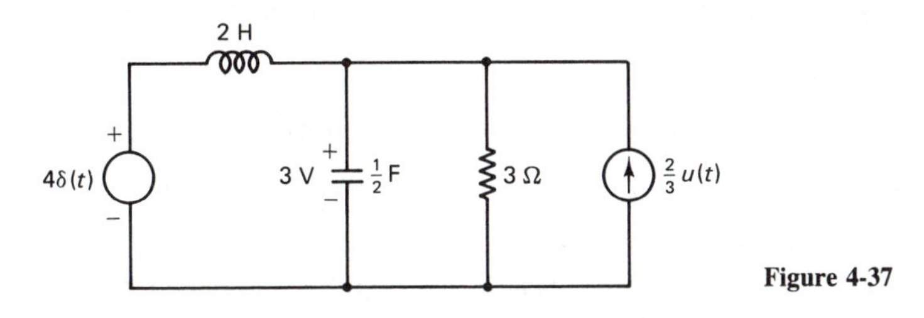

Figure 4-37

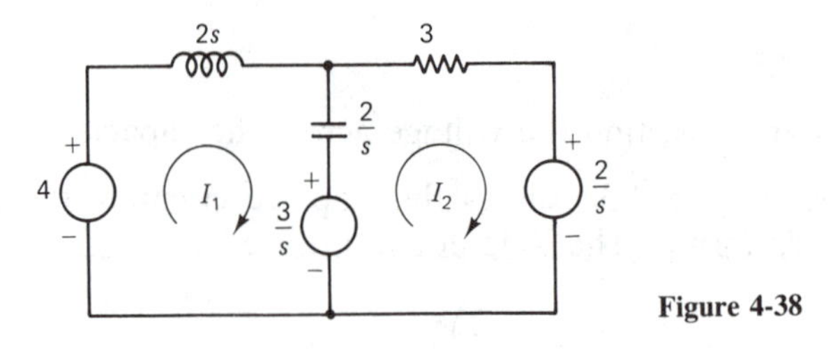

Figure 4-38

Assuming that we want to find the current through the capacitor measured in the downward or I_1 direction,

$$I_C = I_1 - I_2 = \frac{[12s^2 - s - 4] - [2s^2 + 8s - 4]}{D}$$

$$= \frac{5(s - 0.9)}{3[(s + 0.3333)^2 + 0.88889]}$$

The voltage across the capacitor is the Laplace current times the Laplace impedance plus the initial condition source.

$$V_C(s) = \frac{1}{sC} I_C + \frac{3}{s}$$

$$= \frac{10(s - 0.9)}{3s[(s + 0.3333)^2 + 0.88889]} + \frac{3}{s}$$

Since we are interested in finding the inverse Laplace transform, we should not combine the $3/s$ with the longer term. Instead, we may find the inverse of the longer term and add the result of the inverse of $3/s$. The complete inverse is

$$v(t) = 5.4886e^{-t/3} \sin (0.94281t + 33.133°)\, u(t)$$

A good test for this result is to let $t = 0$ and see if we get $v(0) = 3$.

4-5-3 Node Voltage

This method is similar to mesh except that we solve for the node voltages with respect to a chosen reference node. This method is frequently used for solving op-amp filters.

NODE VOLTAGE METHOD

1. Draw the circuit so that none of the wires cross over each other.
2. Convert all independent and dependent voltage sources to current sources.
3. Choose a node to reference voltages to. Number the remaining nodes. Each node should include as many components as possible.

4. For each node except the reference node, write an equation as follows.
   (a) Sum all the admittances (reciprocal of impedances) attached to the node and multiply by the node voltage.
   (b) Locate any component, other than sources, which has its other electrode attached to a node other than the reference node. Subtract this component value times the other node voltage.
   (c) Set the terms found above equal to the algebraic summation of the current sources entering (added) and leaving (subtracted) the node.
5. Solve the simultaneous equations for the node voltages. This is usually done with determinants.
6. The voltage across a component is equal to the difference of the two voltages where its electrodes are attached. This value of the voltage is measured at the voltage node that is subtracted from, using the voltage location of the subtracting node as reference. Therefore, a component between V_3 and V_4 could be either: (a) $V_3 - V_4$ measured at V_3 with respect to V_4, or (b) $V_4 - V_3$ measured at V_4 with respect to V_3.

EXAMPLE 4-13

Write the node equation for the circuit in Fig. 4-39.

SOLUTION Figure 4-40 shows the Laplace circuit with all current sources. From the circuit we write the equations

$$\left(\frac{1}{2s} + \frac{1}{4} + 1\right)V_1 - V_2 + 0V_3 = \frac{4}{s(s^2 + 16)}$$

$$-V_1 + \left(1 + 2s + \frac{1}{3}\right)V_2 - \frac{1}{3}V_3 = 20$$

$$-0V_1 - \frac{1}{3}V_2 + \left(\frac{1}{3} + \frac{1}{s}\right)V_3 = \frac{-2}{s}$$

Multiplying the three equations by $4s(s^2 + 16)$, 3, and $3s$, respectively, and combining terms, we have

$$(s^2 + 16)(5s + 2)V_1 - 4s(s^2 + 16)V_2 + 0V_3 = 16$$

$$-3V_1 + (6s + 4)V_2 - V_3 = 60$$

$$0V_1 - sV_2 + (s + 3)V_3 = -6$$

These equations are ready to be solved by determinants. Notice that if we replaced the V's with I's, the equations would look just like a set of mesh equations.

With one modification, the node voltage method is typically used to analyze op-amp filters. We can modify this method, so we may leave voltage sources in the network instead of converting to current sources. If a voltage

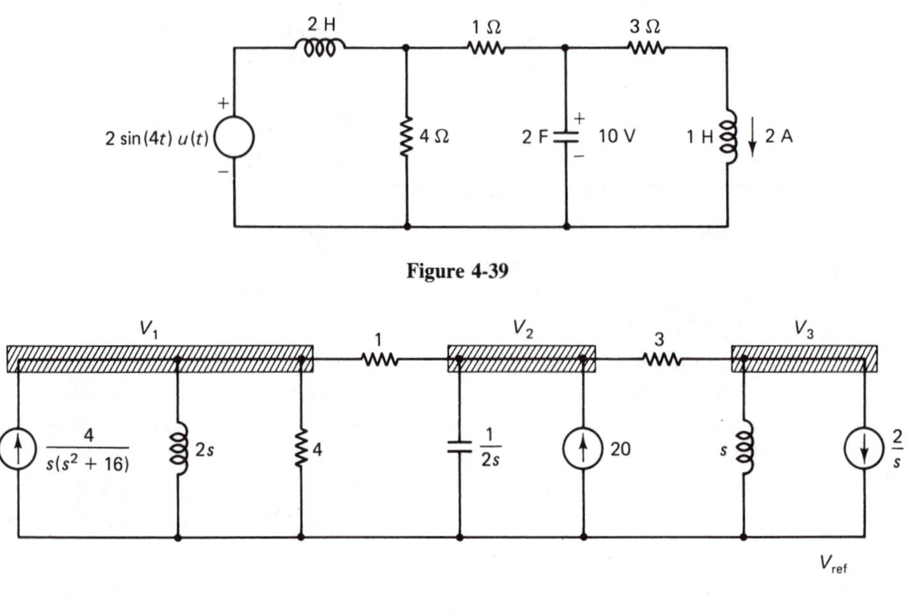

Figure 4-39

Figure 4-40

source has one terminal connected to the reference node, we may treat it as another voltage node. Since we know the value of this voltage, we do not write an equation at this known voltage node.

EXAMPLE 4-14

For the Laplace circuit shown in Fig. 4-41, write the node equations using the modified node voltage method.

SOLUTION This circuit is written as a four-node circuit. The equations generated from the circuit are

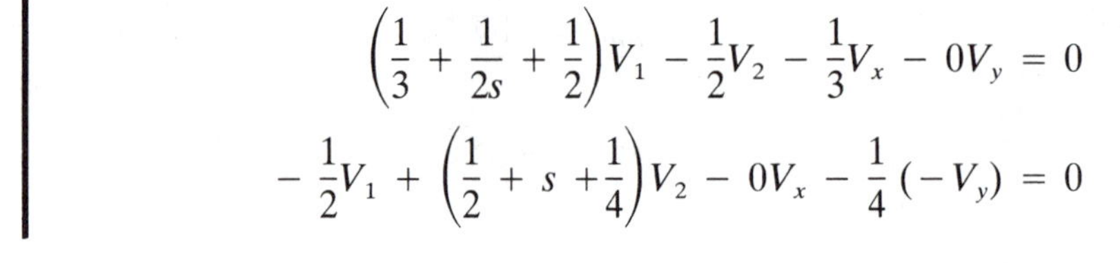

$$\left(\frac{1}{3} + \frac{1}{2s} + \frac{1}{2}\right)V_1 - \frac{1}{2}V_2 - \frac{1}{3}V_x - 0V_y = 0$$

$$-\frac{1}{2}V_1 + \left(\frac{1}{2} + s + \frac{1}{4}\right)V_2 - 0V_x - \frac{1}{4}(-V_y) = 0$$

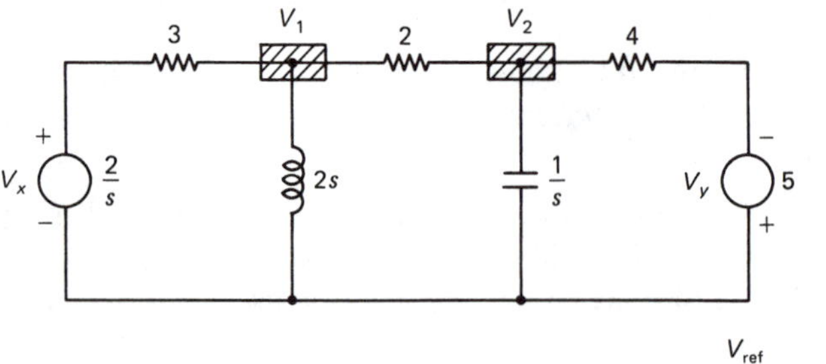

Figure 4-41

In the second equation V_y is entered as negative since the node voltages are always measured with respect to the reference.

The values of V_x and V_y are then substituted into the equations. Since these terms will now become constants, we move them to the right side of the equation.

$$\left(\frac{1}{3} + \frac{1}{2s} + \frac{1}{2}\right)V_1 - \frac{1}{2}V_2 = \frac{2}{3s}$$

$$-\frac{1}{2}V_1 + \left(\frac{1}{2} + s + \frac{1}{4}\right)V_2 = \frac{-5}{4}$$

These equations are the same equations as we would find if we did source conversion and then wrote the equations. With practice we can go directly to these equations by doing the algebra mentally.

4-6 SOLVING DEPENDENT SOURCE CIRCUITS

Whenever we work with an active circuit, we typically have a dependent source. A dependent source has a value dependent on a voltage or current from another location, usually in the same circuit.

A dependent source may be either a voltage source or a current source whose output value is equal to a multiplier, K, times a voltage or current from another location. There are four different types of dependent sources shown in Fig. 4-42: voltage-controlled voltage source (VCVS), current-controlled volt-

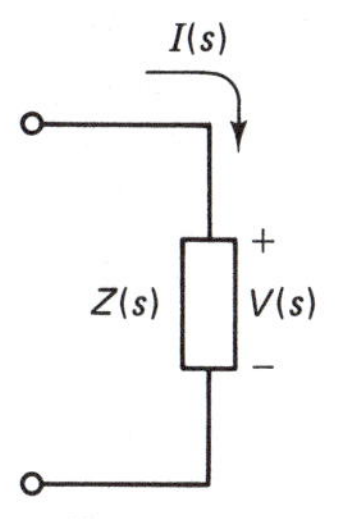

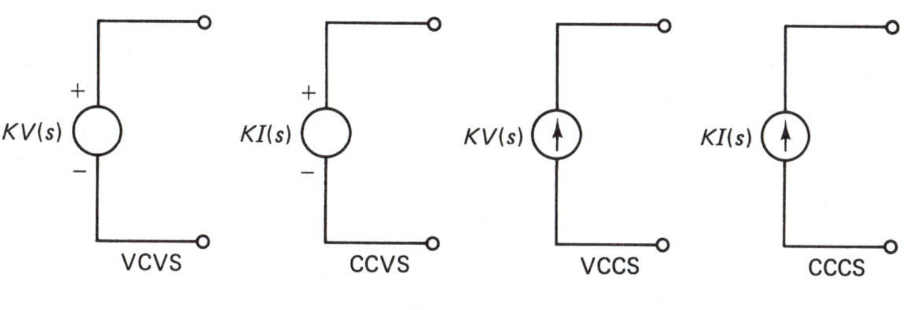

Figure 4-42

age source (CCVS), voltage-controlled current source (VCCS), and current-controlled current source (CCCS). By using Ohm's law and source conversion, we can change one type of source to any other type of source.

EXAMPLE 4-15

Convert the VCCS in Fig. 4-43 to a CCVS.

SOLUTION First we convert the current source to a voltage source using source conversion as shown in Fig. 4-44a. This makes it a VCVS.

We then convert from a voltage-controlled voltage source to a current-controlled voltage source. This is done by developing an equation for $V_3(s)$ in terms of a current. Unless otherwise stated, the current will be at the same component. In this case Ohm's law will work.

$$V_3(s) = \frac{1}{8s} I_3(s)$$

We then substitute this value of $V_3(s)$ into the expression for the VCVS of Fig. 4-44a.

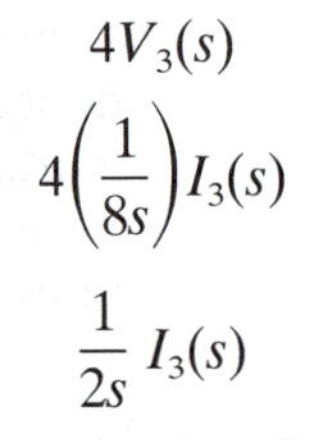

$$4V_3(s)$$

$$4\left(\frac{1}{8s}\right) I_3(s)$$

$$\frac{1}{2s} I_3(s)$$

Figure 4-44b shows the resulting circuit using a CCVS.

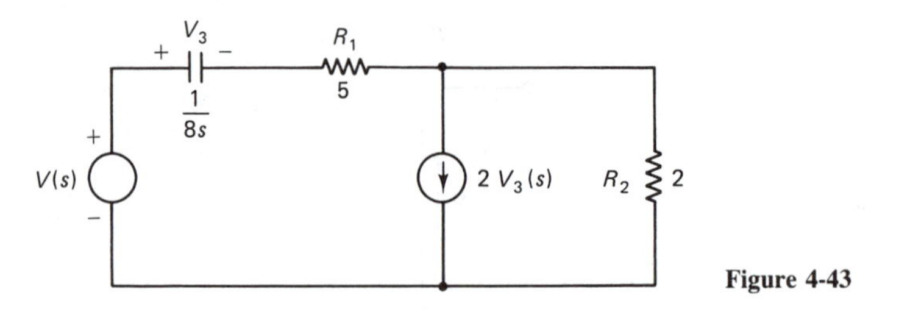

Figure 4-43

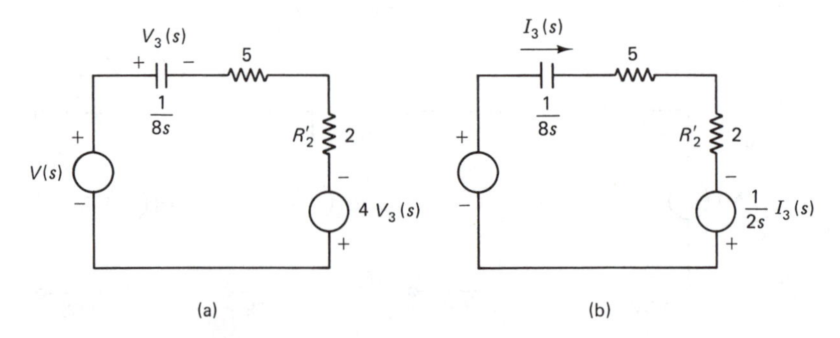

Figure 4-44

In this example we could have converted first to a CCCS and then used source conversion, which would result in the same final circuit. In any case, since source conversion was used, we must remember that the voltage across R_2 is now the voltage across R_2' plus the voltage across the CCVS.

EXAMPLE 4-16

Find $I(s)$ in terms of $V_{in}(s)$ for Fig. 4-45.

SOLUTION We must first find $V_2(s)$ using the voltage divider rule.

$$V_2(s) = \frac{V_{in}(10/s)}{2 + (10/s)} = \frac{5V_{in}}{s + 5}$$

Now we can find the value of the VCVS by substitution.

$$10V_2(s) = \frac{50V_{in}}{s + 5}$$

By using Ohm's law we find the current through the inductor.

$$I(s) = \frac{\frac{50V_{in}}{s + 5}}{5s} = \frac{10V_{in}}{s\,(s + 5)}$$

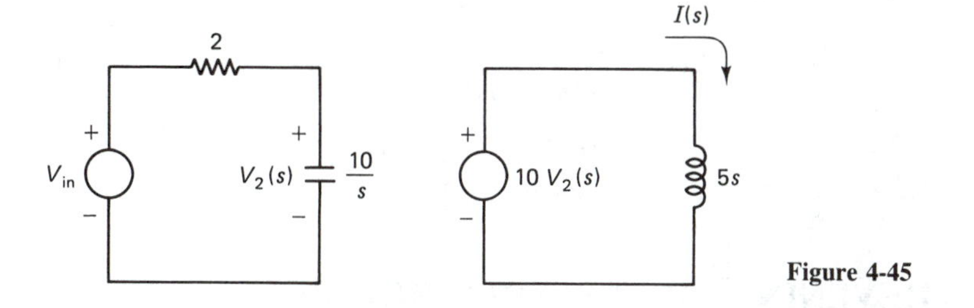

Figure 4-45

When we have multiple dependent sources, we use the multiple-source techniques as seen in previous sections. Since the dependent sources have a variable associated with them, we will have to move their values algebraically to the correct location for that variable.

EXAMPLE 4-17

Find the current through the inductor in Fig. 4-46 using the mesh current method.

SOLUTION Figure 4-47 shows the Laplace circuit set up for mesh analysis. We must get the dependent sources in terms of variables in the mesh equations. Notice that I_x is the same but the opposite of I_1.

$$4I_x = -4I_1$$

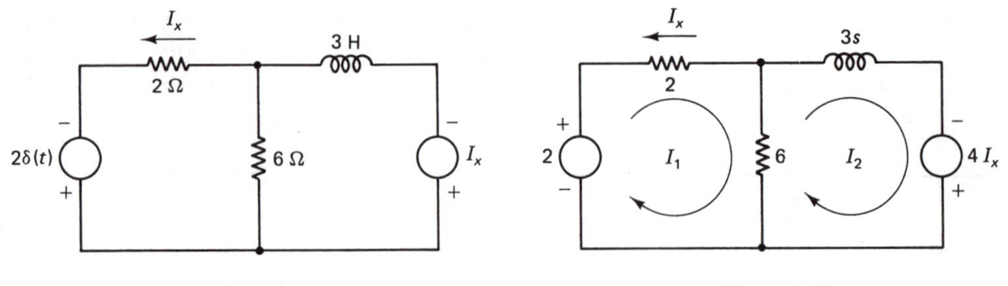

Figure 4-46

Figure 4-47

Writing the loop equations, we have

$$(2 + 6)I_1 + -6I_2 = 2$$

$$-(6)I_1 + (3s + 6)I_2 = -4I_1$$

In the second equation we need to move $-4I_1$ to the left side of the equation. Doing this and combining terms, the two equations become

$$8I_1 - 6I_2 = 2$$

$$-2I_1 + (3s + 6)I_2 = 0$$

At this point we work the problem as any other mesh problem. The Laplace current and resulting time-domain current are

$$I_2(s) = \frac{1}{6(s + 1.5)}$$

$$i(t) = 0.16667e^{-1.5t}u(t)$$

4-7 SOLVING THÉVENIN AND NORTON CIRCUITS

Thévenin and Norton circuits are frequently used to simplify analysis of networks. As with all Laplace circuits, Thévenin and Norton are done exactly as if they were dc circuits. Since the following procedure works in both the time domain and the Laplace domain, it may be used to calculate values on paper as well as physically applied to a circuit in the lab.

THÉVENIN OR NORTON EQUIVALENT IMPEDANCE

1. Remove the component or network for which you are finding the Thévenin or Norton equivalent circuit. The point of removal must have two and only two terminals.
2. Make all independent sources zero. A zero-volt voltage source is a short, and a zero-ampere current source is an open.
3. Drive the terminal where the component or network was removed with

a voltage source V_z, and determine the impedance that V_z sees. For a circuit in the lab we measure I_z and divide it into V_z to determine the impedance.

4. **(a)** To find the Thévenin voltage, find the open-circuit voltage with the component or network and V_z removed.
   (b) To find the Norton current, short the terminals and find the current in the short with the component or network and V_z removed.

The impedance we will find will be a function of s. The function may not be an easily recognizable combination of resistance, inductance, and/or capacitance. We therefore put the function next to a box and treat the box as a component. Do not get this confused with block diagrams, which are entirely different.

EXAMPLE 4-18

Using a Thévenin equivalent circuit, find $I_1(s)$ in Fig. 4-48 when $R_1 = 1\ \Omega$ and $4\ \Omega$.

SOLUTION Figure 4-49a shows the circuit with the source at zero, the component removed, and a source V_z driving the two terminals where the component was removed. The source sees a series network of a capacitor and resistor, which is the Thévenin equivalent impedance.

$$Z_{\text{TH}}(s) = \frac{1}{5s} + 4 = \frac{4(s + 0.05)}{s}$$

Using Fig. 4-49b, we can find the open-circuit voltage at the terminals a–b, which is the Thévenin equivalent voltage. The terminals are marked a–b so that we know which terminal is which when we draw the Thévenin equivalent circuit. We usually measure the voltage at a with respect to b. When we measure this way, the plus side of the source in the equivalent circuit will be closest to the a terminal.

Using Kirchhoff's voltage law, the voltage across the open terminals is

$$V_{\text{TH}} = V_C + V_{R2}$$

and using Ohm's law, the voltage across the capacitor and resistor is

$$V_c = \frac{2}{5s} \qquad V_{R2} = 2(4) = 8$$

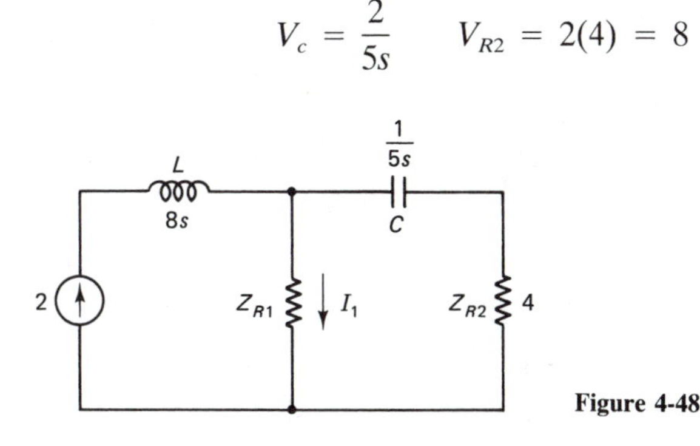

Figure 4-48

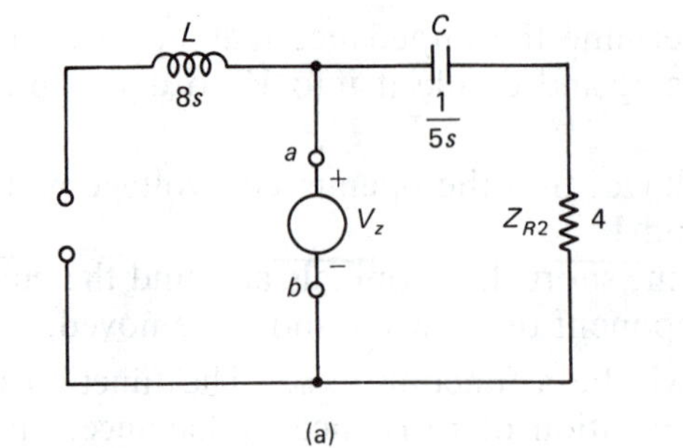

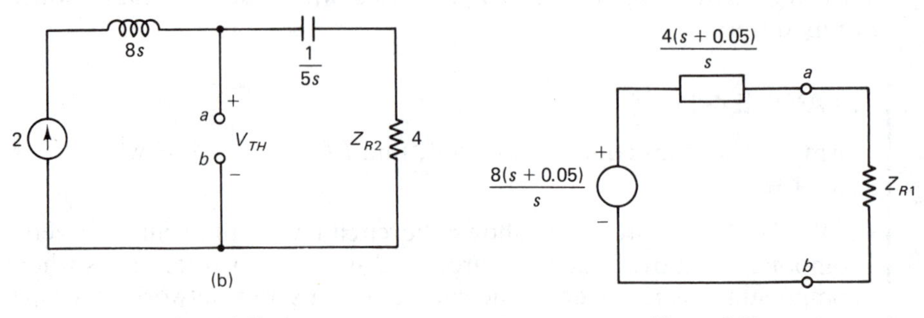

Figure 4-49

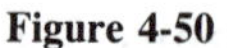

Figure 4-50

Substituting into the Kirchhoff's voltage law equation gives

$$V_{\mathrm{TH}} = \frac{2}{5s} + 8 = \frac{8(s + 0.05)}{s}$$

The equivalent circuit, and the component for which the Thévenin equivalent circuit was found, is shown in Fig. 4-50. Since it is not a pure resistance, capacitance, or inductance, a block is used to indicate a component, but it is still treated as any other Laplace impedance.

We can now easily find the current through the resistor for any value it may become. We could put in a value and solve, or we could use the variable Z_{R1} and solve for a general equation. Using the second choice and Ohm's law, we obtain

$$I(s) = \frac{\dfrac{8(s + 0.05)}{s}}{\dfrac{4(s + 0.05)}{s} + Z_{R1}}$$

$$= \frac{\dfrac{8}{4 + Z_{R1}}(s + 0.05)}{s + \dfrac{0.2}{4 + Z_{R1}}}$$

Substituting in $Z_{R1} = 1$ yields

$$I(s) = \frac{1.6(s + 0.05)}{s + 0.04}$$

and $i(t)$ is

$$i(t) = 1.6\,\delta(t) + 0.016e^{-0.04t}u(t)$$

Substituting in $Z_{R1} = 4$, we have

$$I(s) = \frac{s + 0.05}{s + 0.025}$$

and $i(t)$ is

$$i(t) = \delta(t) + 0.025e^{-0.025t}u(t)$$

When we have dependent sources in the circuit, we cannot remove them. With dependent sources involved, we must use one of two methods to solve the Thévenin or Norton equivalent impedance.

For both methods we still remove all independent sources, remove the network for which we want to find the Thévenin or Norton equivalent circuit, and drive that point with a source, V_z. The first method is to find an equation having only the variable V_z and I_z, and solve for V_z/I_z, which is the equivalent impedance. The second method is to find an equation for V_z and I_z in terms of the same variable, divide V_z by I_z, and cancel out the variable that V_z and I_z are in terms of.

EXAMPLE 4-19

Find the Norton equivalent circuit for the network C_2 and L in Fig. 4-51.

SOLUTION Figure 4-52 shows the independent source removed, the network removed, and the V_z source inserted. Normally, one method is used, but we will use both methods and solve for the equivalent impedance twice.

Method 1. There are a number of good approaches to this, but let's use Kirchhoff's voltage law around the outside loop.

$$V_z - 10I_x - 5I_z - 5I_x = 0$$

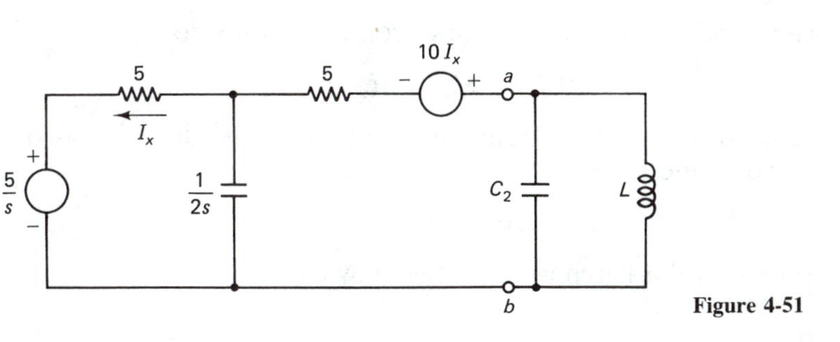

Figure 4-51

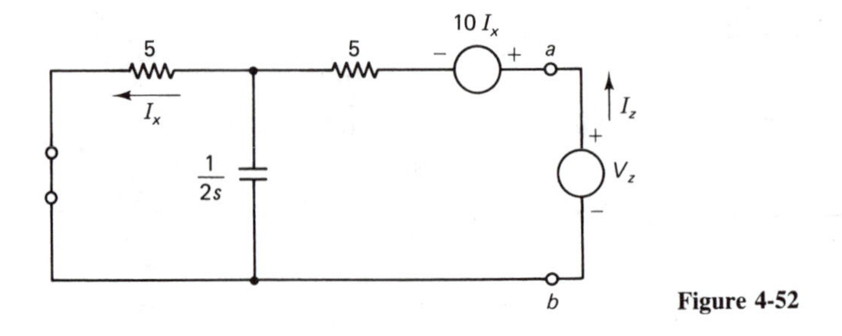

Figure 4-52

This equation has three variables in it, but we want only V_z and I_z. We can use current divider rule to find I_x in terms of I_z.

$$I_x = \frac{I_z\left(\frac{1}{2s}\right)}{5 + \frac{1}{2s}} = \frac{0.1I_z}{s + 0.1}$$

When this is substituted in to Kirchhoff's voltage law equation, we have

$$V_z - 5I_z - 15\left(\frac{0.1I_z}{s + 0.1}\right) = 0$$

Solving for V_z/I_z gives

$$\frac{V_z}{I_z} = \frac{5(s + 0.4)}{s + 0.1}$$

Method 2. In this method we need to find I_z in terms of a variable, and find V_z in terms of the same variable. This variable can be any variable. Let's use I_x so that we can see the similarities between this method and the previous method.

First we will use the current divider to find I_x.

$$I_x = \frac{I_z\left(\frac{1}{2s}\right)}{5 + \frac{1}{2}s} = \frac{0.1I_z}{s + 0.1}$$

Next we use Kirchhoff's voltage law around the outside loop.

$$V_z - 10I_x - 5I_z - 5I_x = 0$$

This time we need to find I_z in terms of I_x. This can be done by solving the current divider rule for I_z:

$$I_z = 10I_x(s + 0.1)$$

and substituting into the Kirchhoff's voltage law equation.

$$V_z - 10I_x - 5[10I_x(s + 0.1)] - 5I_x = 0$$

$$V_z = I_x(50s + 20)$$

Finally, we divide V_z by I_z and cancel the I_x.

$$\frac{V_z}{I_z} = \frac{I_x(50s + 20)}{10I_x(s + 0.1)} = \frac{5(s + 0.4)}{s + 0.1}$$

which gives us the same result. You may find one or the other method easier depending on the particular circuit, but when done properly, you will get the same result with either method.

The next step is to find the Norton short-circuit current. We remove the source we added and short the terminals as shown in Fig. 4-53. Since this is a multiple-source circuit, we will use mesh.

Using Fig. 4-53, we can find the short-circuit current as indicated at the terminals a–b, which is the Norton equivalent current. The terminals are marked a–b so that we know which terminal is which in the equivalent circuit. We usually measure the current from a to b. When we measure this way the current source in the equivalent circuit points toward the a terminal. In this case I_N will be equal to I_2.

The mesh equations from the circuit are

$$\left(5 + \frac{1}{2s}\right)I_1 - \frac{1}{2s}I_2 = \frac{5}{s}$$

$$-\frac{1}{2s}I_1 + \left(5 + \frac{1}{2s}\right)I_2 = 10I_x \quad \text{or} \quad = 10(-I_1)$$

Notice that $-I_1$ is substituted for I_x.

Moving the I_1 term to the left and removing fractions, we get

$$(10s + 1)I_1 - I_2 = 10$$

$$(20s - 1)I_1 + (10s + 1)I_2 = 0$$

We only need to solve for I_2, which is

$$I_2 = I_N = \frac{-2(s - 0.05)}{s(s + 0.4)}$$

Figure 4-54 shows the Norton equivalent circuit.

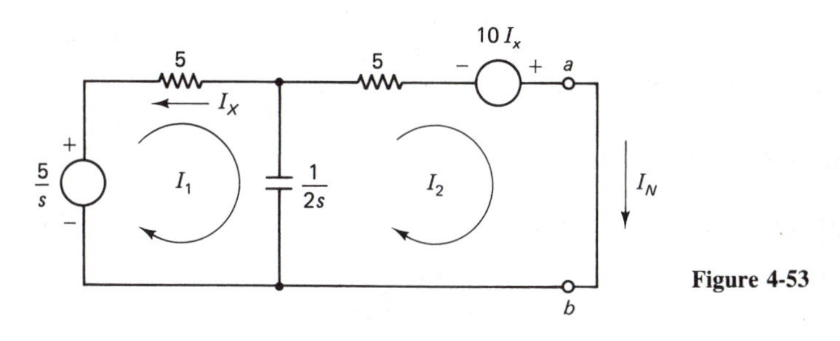

Figure 4-53

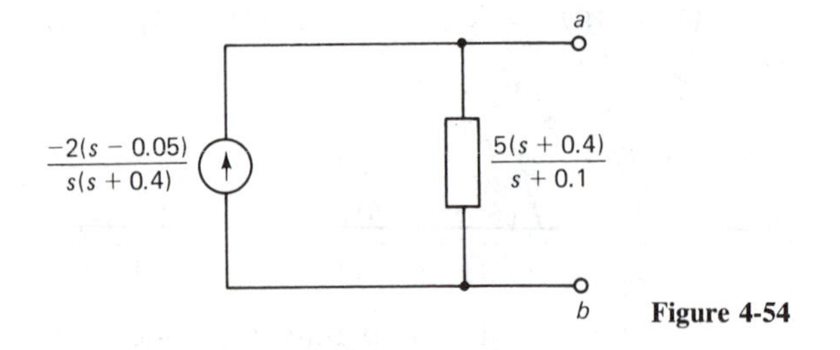

Figure 4-54

4-8 CIRCUIT ORDER

The circuit order is the order of the time-domain differential equation required to solve the circuit. Therefore, a first-order circuit generates a first-order differential equation, a second-order circuit generates a second-order differential equation, and so on. When using Laplace transforms it is unnecessary to write the differential equation since different orders of circuit are easily identified in the Laplace domain or from the circuit.

The order of the circuit is determined by the number of nonredundant reactive components (inductors and capacitors) in the circuit. Since the number of nonredundant reactive components also determines the poles, the number of circuit poles is equal to the circuit order. The order of the differential equation, the circuit order, and the poles are, therefore, due to the nature of the circuit, which is how and where the R, L, and C's are used.

To help distinguish the circuit poles from the source poles either an impulse source or step source with a magnitude of 1 is used. When an impulse source is used, there will be no poles due to the source in the Laplace equation since the Laplace transform of the impulse is a constant. When the impulse is used, the response is usually called the impulse response.

In the lab it is easier to generate a step function. The response is therefore called the step response. The step function will put a pole value of zero in the Laplace function, making it easy to identify.

4-8-1 Redundancy

The most common type of redundancy occurs in a circuit when two or more of one type of component (R, L, or C) can be combined into a single component of the same type. For example, two series capacitors can be combined into a single equivalent capacitor. To determine the circuit's order you must first replace all redundant components with a single equivalent component. This is only necessary for the reactive components since the number of nonredundant reactive components is equal to the order of the circuit.

EXAMPLE 4-20

Find the order of the circuit in Fig. 4-55.

SOLUTION C_1 and C_2 are in series since the same current flows through them, and they can therefore be combined into a single value. L_1 and L_2

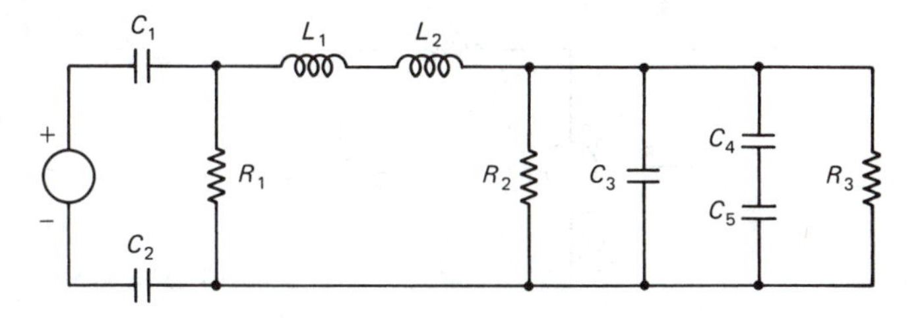

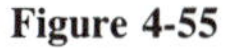

Figure 4-55

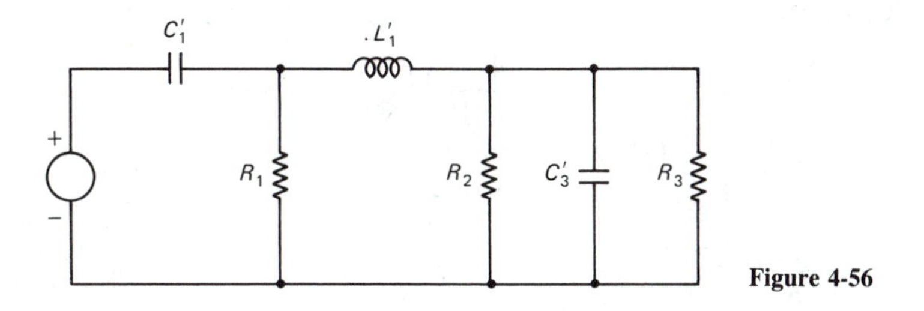

Figure 4-56

are in series and can be combined. C_4 and C_5 are in series, and when combined the resultant capacitor can be combined with C_3, which is in parallel. Figure 4-56 shows the circuit with the redundancy removed. From this circuit we count the number of reactive components, which is three. This is, therefore, a third-order circuit.

4-8-2 First-Order Circuits

A first-order circuit can contain only one nonredundant reactive component. The circuit can contribute only one real pole to the response. This means that the inverse will contain a $Ke^{-\alpha t}$ term due to the circuit. This will always be transient unless the circuit contains only pure reactance.

EXAMPLE 4-21

For the first-order circuit shown in Fig. 4-57, find the Laplace voltage across the capacitor and the current through it. Identify the term that is due to the nature of the circuit.

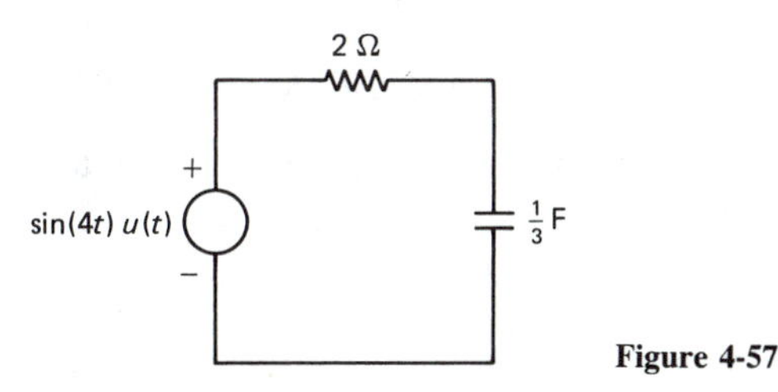

Figure 4-57

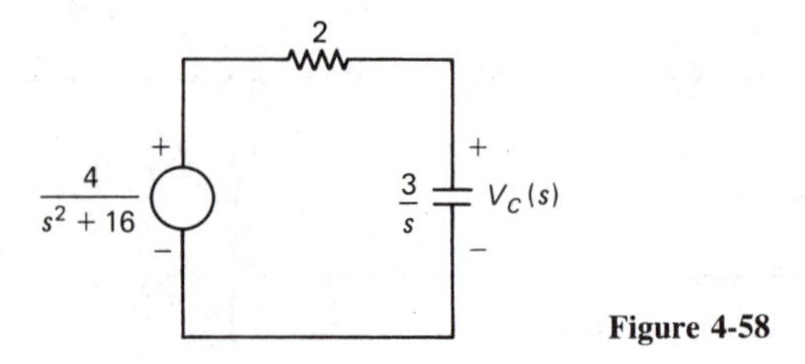

Figure 4-58

SOLUTION The Laplace circuit is shown in Fig. 4-58. Using the voltage divider rule, we obtain

$$V_C(s) = \frac{\left(\dfrac{4}{s^2 + 16}\right)\left(\dfrac{3}{s}\right)}{2 + \dfrac{3}{s}}$$

$$= \frac{6}{(s^2 + 16)(s + 1.5)}$$

The pole due to the factor $(s^2 + 16)$ is from the source, and the pole due to the factor $(s + 1.5)$ is from the nature of the circuit.

4-8-3 Second-Order Circuits

A second-order circuit will have two nonredundant reactive components. Consequently, the circuit response will have two poles due to the circuit. With two poles there are three possible cases for the inverse. Each of the three cases may have different responses when combined with the effects of the other poles due to the sources. Figure 4-59a shows the impulse response where the poles are due only to the circuit. Figure 4-59b shows the step response where the circuit poles are combined with a pole of $s = 0$ (a step function). Each of the three cases has a name associated with it.

1. *Underdamped.* The two poles are complex conjugates, and the inverse of the complex conjugates has the form

$$Ke^{-\alpha t} \sin (\omega t + \theta)\, u(t)$$

2. *Critically damped.* The two poles are real values and equal. The inverse has the form

$$(K_1 t + K_2)\, e^{-\alpha t} u(t)$$

3. *Overdamped.* The two poles are real values and unequal. The inverse has the form

$$K_1 e^{-\alpha_1 t} + K_2 e^{-\alpha_2 t}$$

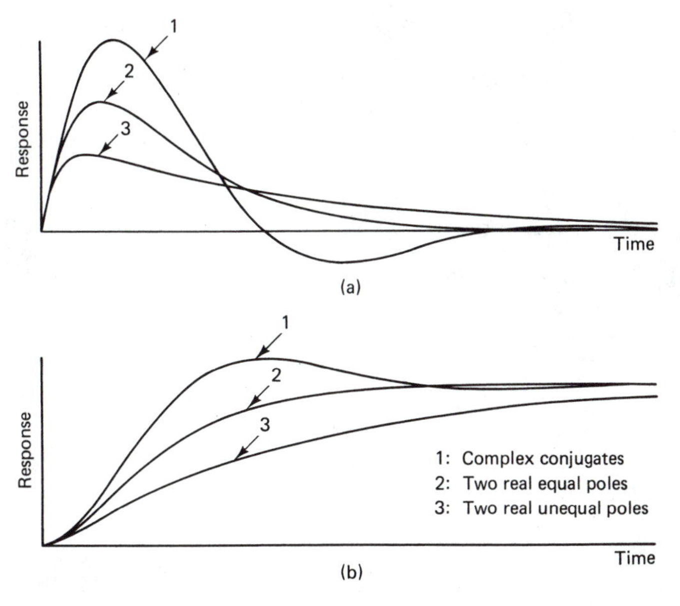

Figure 4-59

EXAMPLE 4-22

For the equation shown, is the quadratic in the denominator underdamped, critically damped, or overdamped?

$$V(s) = \frac{s + 4}{s(s^2 + 6s + 9)}$$

SOLUTION The factors of the quadratic are $s = -3, -3$. Therefore, we may write the equation

$$V(s) = \frac{s + 4}{s(s + 3)^2}$$

Since the quadratic has two equal roots, the system is critically damped.

When designing second-order circuits we may only need to determine if it is underdamped, critically damped, or overdamped. To do this, we use another form of the quadratic:

$$s^2 + 2\zeta\omega_n s + \omega_n^2 \qquad (4\text{-}5)$$

where ω_n = the natural resonant frequency, in rad/sec (ω is the Greek lowercase letter omega)

ζ = the damping ratio (ζ is the Greek lowercase letter zeta)

If we solve for the roots of the function in Eq. (4-5), we have

$$s = -\frac{2\zeta\omega_n}{2} \pm \sqrt{\left(\frac{2\zeta\omega_n}{2}\right)^2 - \omega_n^2} \tag{4-6a}$$

$$s = -\zeta\omega_n \pm \omega_n \sqrt{\zeta^2 - 1} \tag{4-6b}$$

Notice that ζ determines which of the three possible cases we have.

1. $\zeta < 1$: *underdamped.* Since $\zeta^2 - 1$ will be negative, we may rewrite the roots as

$$s = -\zeta\omega_n \pm \omega_n \sqrt{(-1)(1 - \zeta^2)}$$

$$s = -\zeta\omega_n \pm j\omega_n \sqrt{1 - \zeta^2} \tag{4-7}$$

2. $\zeta = 1$: *critically damped.* Since $\zeta^2 - 1 = 0$, we may rewrite the function as

$$s = -\omega_n, -\omega_n \tag{4-8}$$

3. $\zeta > 1$: *overdamped.* Since $\zeta^2 - 1$ will be positive, we will have two unequal real numbers.

$$s = -\zeta\omega_n + \omega_n \sqrt{\zeta^2 - 1}, \qquad -\zeta_n - \omega_n \sqrt{\zeta^2 - 1} \tag{4-9}$$

We notice that ω_n appears in Eqs. (4-8) and (4-9) where the inverse does not have a sinusoidal term. This frequency does not apply directly to this inverse, but in Eq. (4-8) it applies to the gain and phase response curves in Chapter 5. In the overdamped case, ω_n does not have any significance here or in Chapter 5, but is used simply to avoid introduction of another variable.

Previously when we factored a complex quadratic in the denominator, the poles were called $s = -\alpha \pm j\omega$. The ω became the frequency of the sinusoidal function in the inverse Laplace transform. This is actually called the damped resonant frequency, and we should add a d subscript to distinguish it from the natural resonant frequency ω_n.

Only in the underdamped case can we show a relationship of ω_d, the damped resonant frequency, and α, the damping constant, to Eq. (4-7). Since the poles are at

$$s = -\alpha \pm j\omega_d$$

and since Eq. (4-7) is

$$s = -\zeta\omega_n \pm j\omega_n \sqrt{1 - \zeta^2}$$

we can find the relation by setting the real parts of each equation equal to each other and setting the imaginary parts equal to each other.

$$\alpha = \zeta\omega_n \tag{4-10a}$$

$$\omega_d = \omega_n \sqrt{1 - \zeta^2} \tag{4-10b}$$

EXAMPLE 4-23

For the second-order pole, find ζ, ω_n, α, and ω_d.

$$\frac{N(s)}{s^2 + 4s + 13}$$

SOLUTION We set the third term, 13, equal to the third term in Eq. (4-5) to find ω_n:

$$\omega_n^2 = 13 \Rightarrow \omega_n = \sqrt{13} = 3.6056$$

and then set the second terms equal to solve for ζ.

$$2\zeta\omega_n s = 4s$$

$$\zeta = \frac{4}{2\omega_n} = \frac{4}{2(3.6056)} = 0.55470$$

We can find α and ω_d by factoring the quadratic or by using Eq. (4-10).

$$\alpha = \zeta\omega_n = (0.55470)(3.6056) = 2$$

$$\omega_d = \omega_n\sqrt{1 - \zeta^2} = (3.6056)\sqrt{1 - 0.55470^2} = 3$$

This means we could write the quadratic in the form

$$(s + 2)^2 + 9$$

EXAMPLE 4-24

For Fig. 4-60, write the equation for the Laplace voltage across the capacitor and find ζ, ω_n, α, and ω_d.

SOLUTION Since this is a current source driving a parallel circuit, we can find the Laplace impedance of R, L, and C and then use Ohm's law to find the voltage across the network, which is also the voltage across the capacitor.

$$Z(s) = \frac{1}{\frac{1}{4} + \frac{1}{2s} + 2s} = \frac{0.5s}{s^2 + 0.125s + 0.25}$$

$$V(s) = Z(s)I(s) = \frac{0.5s}{s^2 + 0.125s + 0.25}\frac{4}{s}$$

$$= \frac{2}{s^2 + 0.125s + 0.25}$$

Using Eq. (4-5), we find that

$$\omega_n = \sqrt{0.25} = 0.5$$

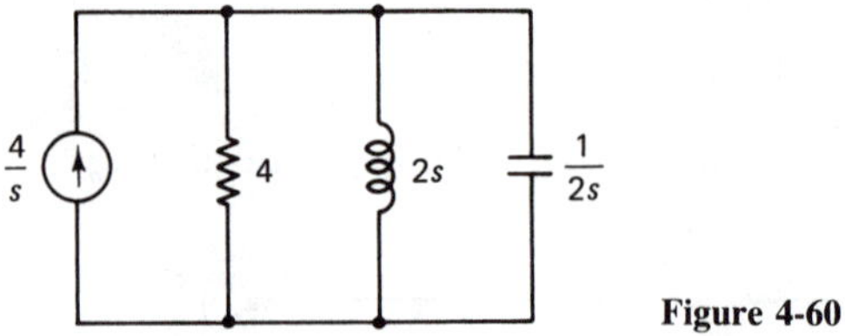

Figure 4-60

and

$$2\zeta\omega_n = 0.125 \Rightarrow \zeta = \frac{0.125}{(2)(0.5)} = 0.125$$

To find α and ω_d we can use Eq. (4-10) or we can just factor. This time let's factor.

$$s = -0.0625 \pm j0.49608$$

Therefore,

$$\alpha = 0.0625$$

$$\omega_d = 0.49608$$

EXAMPLE 4-25

For the circuit shown in Fig. 4-61, find the value of R in order for the circuits current to be over-, under-, and critically damped.

SOLUTION This can be solved by finding how ζ varies with changes in R. Let's first find the equation for the current.

$$I(s) = \frac{V(s)}{Z(s)} = \frac{V(s)}{Z_R + sL + \dfrac{1}{sC}}$$

$$= \frac{\left(\dfrac{s}{L}\right) V(s)}{s^2 + \dfrac{Z_R}{L} s + \dfrac{1}{LC}}$$

Using Eq. (4-5), we obtain

$$\omega_n = \frac{1}{\sqrt{LC}} = \frac{1}{\sqrt{3\left(\dfrac{1}{6}\right)}} = 1.4142$$

$$2\zeta\omega_n = \frac{Z_R}{L} \Rightarrow \zeta = \frac{Z_R}{2\omega_n L} = \frac{Z_R}{2(1.4142)(3)}$$

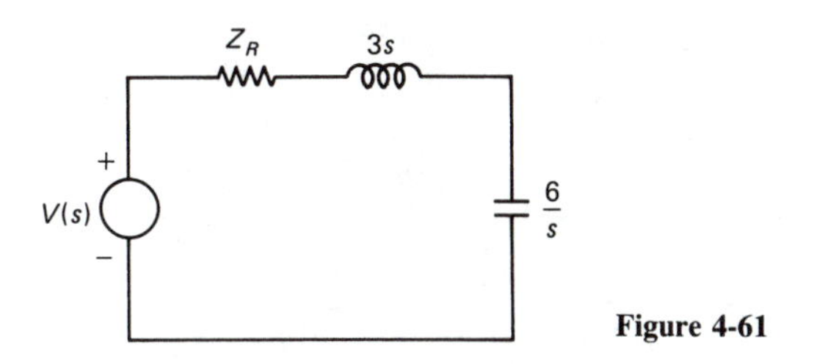

Figure 4-61

Therefore,

$$Z_R = \zeta(8.4853)$$

From this result we can determine R by substituting different values of ζ. Therefore,

$$\zeta = 1 \Rightarrow R = 8.4853 \Rightarrow \text{critically damped}$$

$$\zeta > 1 \Rightarrow R > 8.4853 \Rightarrow \text{overdamped}$$

$$\zeta < 1 \Rightarrow R < 8.4853 \Rightarrow \text{underdamped}$$

PROBLEMS

Section 4-1

1. Redraw the circuit shown in Fig. 4-62 without switches and show the initial conditions for the reactive component.

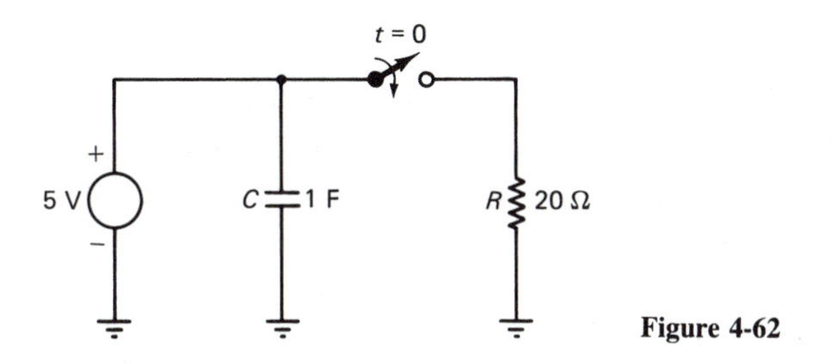

Figure 4-62

2. Redraw the circuit shown in Fig. 4-63 without switches and show the initial conditions for the reactive components.

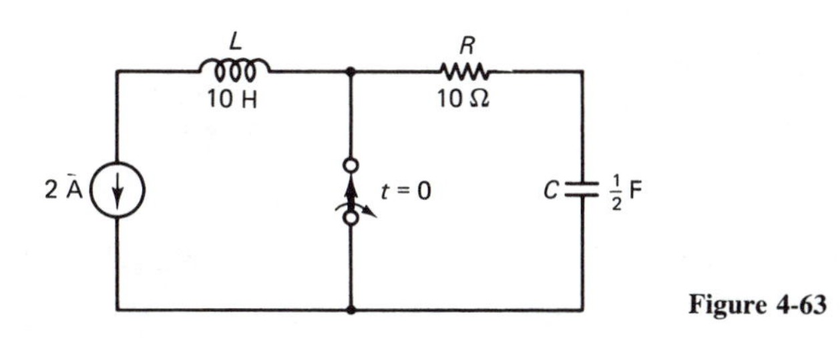

Figure 4-63

3. Redraw the circuit shown in Fig. 4-64 without switches and show the initial conditions for the reactive components.

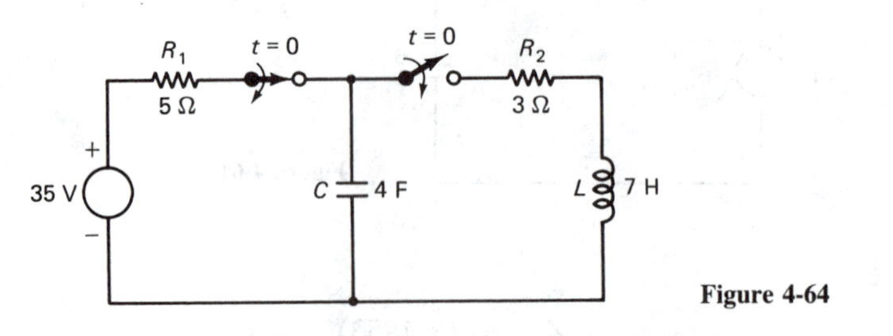

Figure 4-64

4. Redraw the circuit shown in Fig. 4-65 without switches and show the initial conditions for the reactive components.

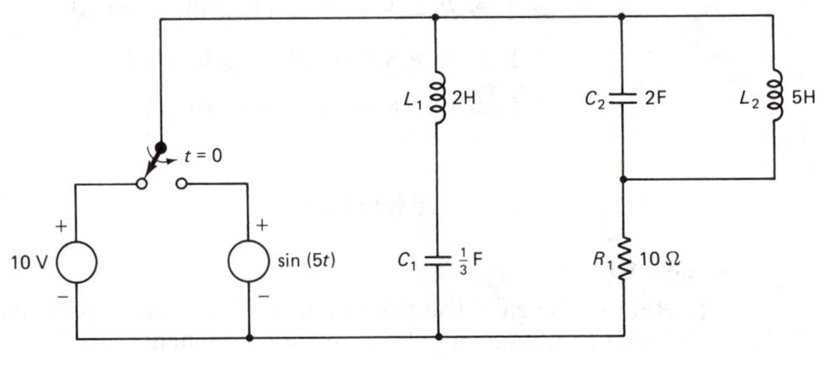

Figure 4-65

Section 4-2

5. Draw the Laplace Norton equivalent circuit (natural equivalent circuit) for Fig. 4-66a, and then using source conversion, show the Thévenin equivalent form.

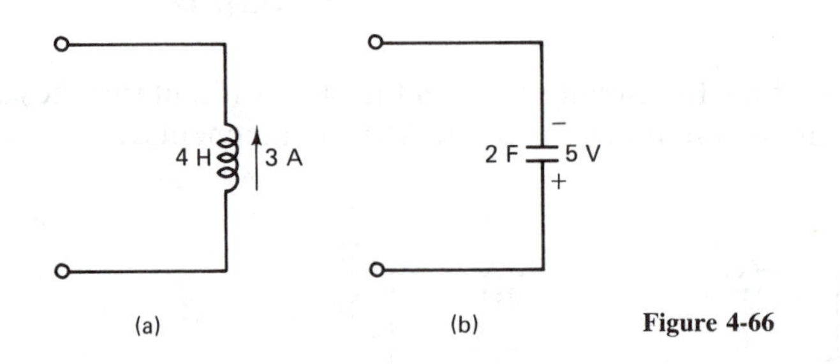

Figure 4-66

6. Draw the Laplace Thévenin equivalent circuit (natural equivalent circuit) for Fig. 4-66b, and then using source conversion, show the Norton equivalent form.

Section 4-3

7. Convert Fig. 4-62 to the Laplace equivalent circuit with the reactive component in its natural equivalent form.
8. Convert Fig. 4-63 to the Laplace equivalent circuit with the reactive components in their natural equivalent form.
9. Convert Fig. 4-64 to the Laplace equivalent circuit with the reactive components in their natural equivalent form.
10. Convert Fig. 4-65 to the Laplace equivalent circuit with the reactive components in their natural equivalent form.
11. Draw the Laplace equivalent circuit for Fig. 4-67 using:
    (a) All voltage sources.
    (b) All current sources. (Don't forget the applied voltage source.)

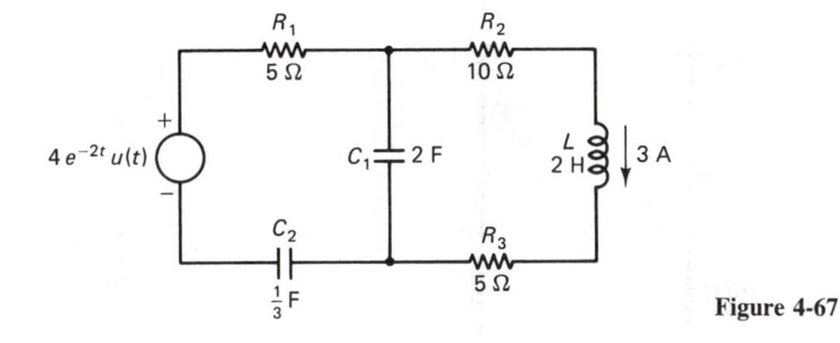

Figure 4-67

Section 4-4

12. For Fig. 4-68:
    (a) Find the Laplace voltage indicated.
    (b) Find the time-domain voltage.
    (c) Identify the transient and steady-state terms.

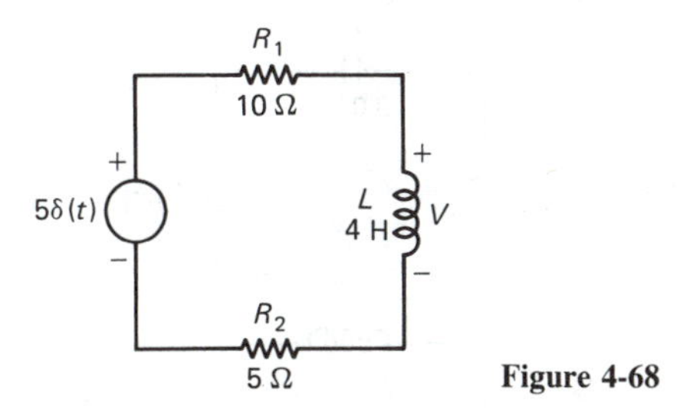

Figure 4-68

13. For Fig. 4-69:
    (a) Find the Laplace current indicated.
    (b) Find the time-domain current.
    (c) Identify the transient and steady-state terms.

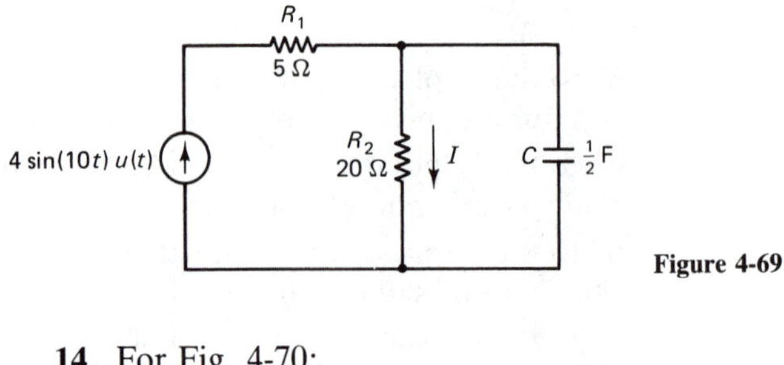

Figure 4-69

14. For Fig. 4-70:
(a) Find the Laplace voltage V_{a-b}.
(b) Find the time-domain voltage.

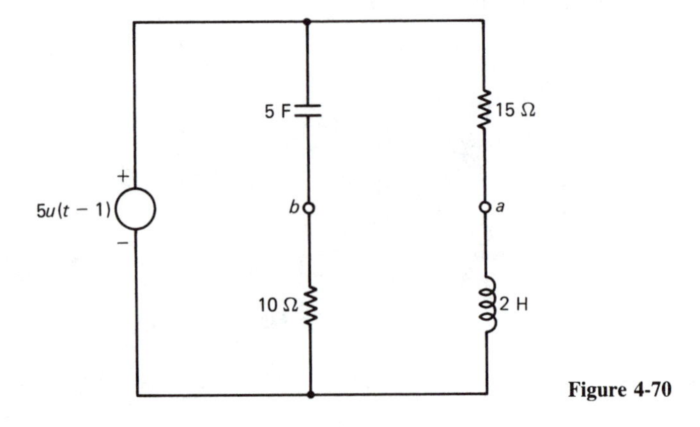

Figure 4-70

15. For Fig. 4-71:
(a) Find the Laplace current indicated.
(b) Find the time-domain current.
(c) Identify the transient and steady-state terms.

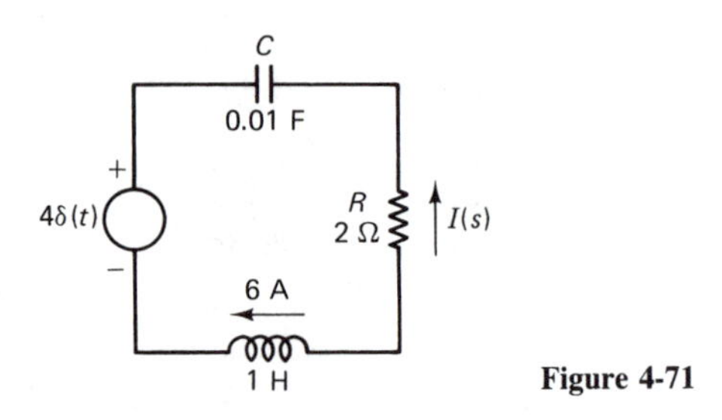

Figure 4-71

Section 4-5

16. For Fig. 4-72, using superposition, find the time-domain current through the inductor in the direction indicated.

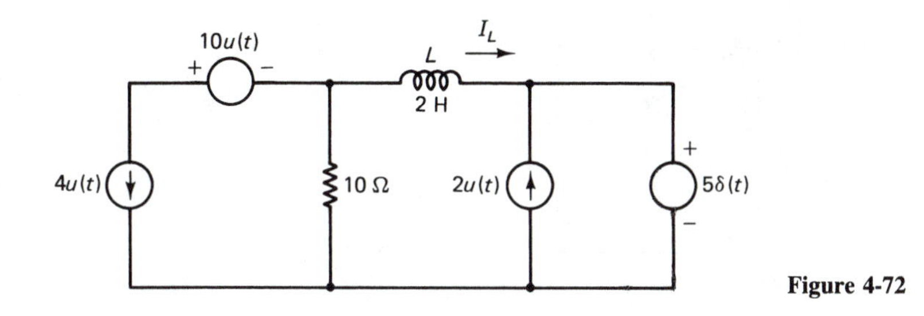

Figure 4-72

17. Using superposition, find the Laplace voltage as indicated in Fig. 4-73.

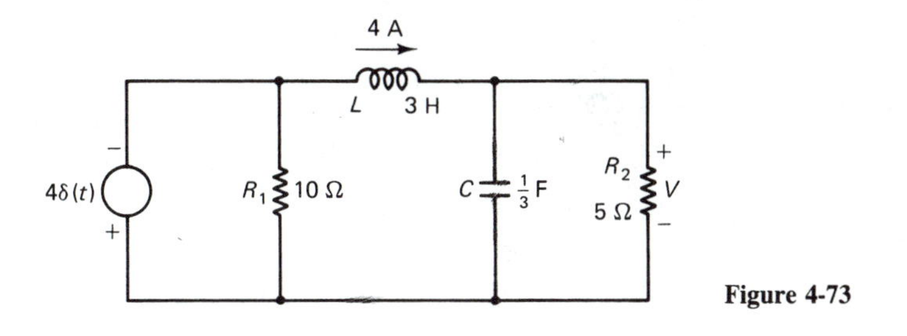

Figure 4-73

18. For Fig. 4-74, write the mesh current equation in a form suitable for solving the currents by determinants.

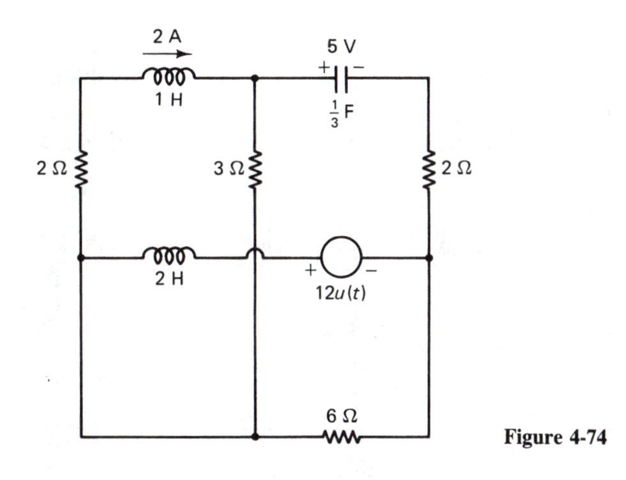

Figure 4-74

19. For Fig. 4-75:

(a) Use the mesh current method to find the Laplace current I_a and I_b. (Don't forget to use all voltage sources.)

(b) Find the inverse Laplace transform of these currents.

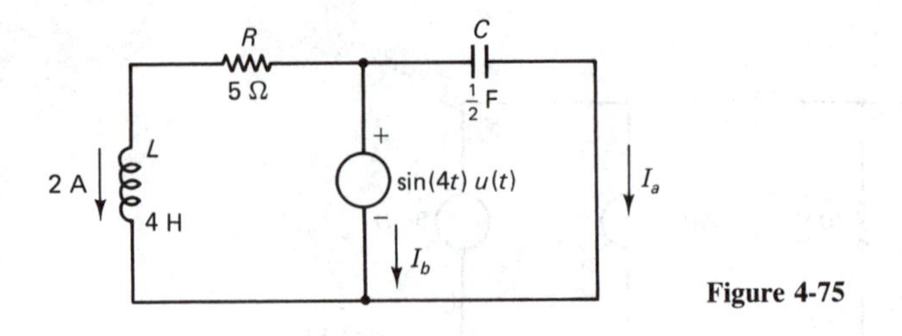

Figure 4-75

20. For Fig. 4-76:

(a) Use the mesh current method to find the Laplace current I_a and I_b.

(b) Find the inverse Laplace transform for I_a and I_b.

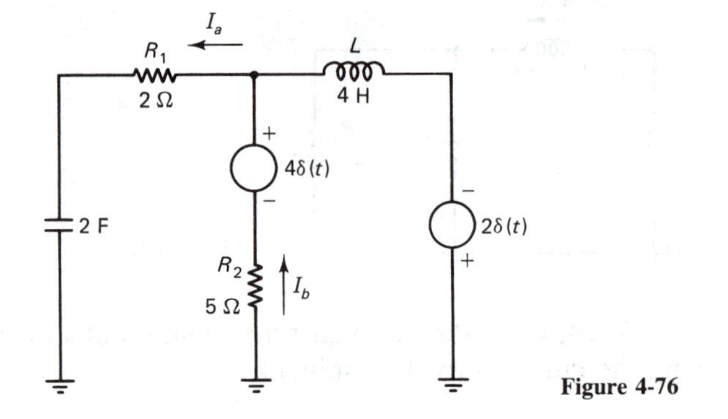

Figure 4-76

21. For Fig. 4-77:

(a) Write the node voltage equations. (Don't forget to use all current sources.)

(b) Solve the node voltage equations and find V_{a-b}.

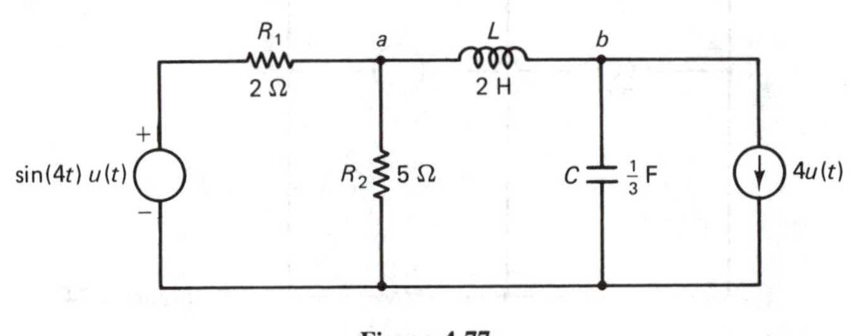

Figure 4-77

22. For Fig. 4-78:

(a) Write the node voltage equations.

(b) Using the node equations solve for the Laplace voltage across R_3.

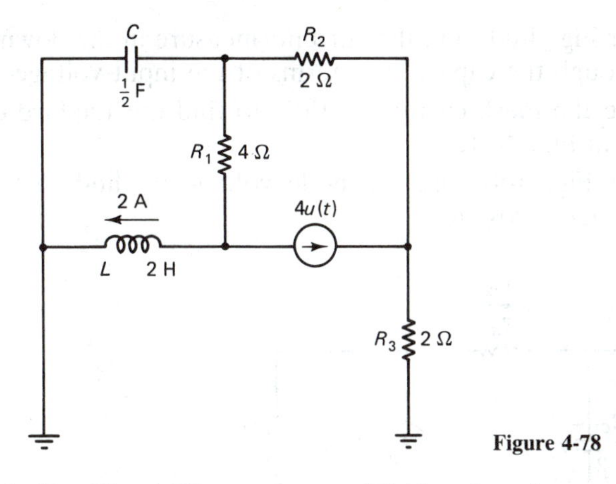

Figure 4-78

23. For Fig. 4-77, use the modified node voltage equations as described in Section 4-5-3 to:
    (a) Write the node voltage equations.
    (b) Solve the node voltage equations to find the current measured from a to b.
24. For Fig. 4-79, set up the modified node voltage equations as shown in Section 4-5-3.

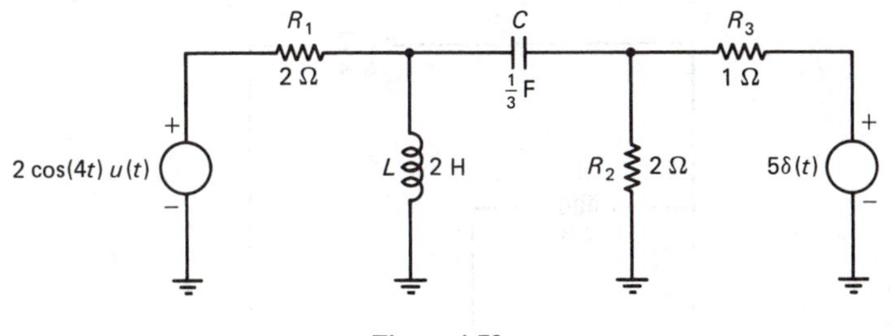

Figure 4-79

Section 4-6

25. For Fig. 4-80, convert the VCVS to a:
    (a) CCVS
    (b) VCCS
    (c) CCCS

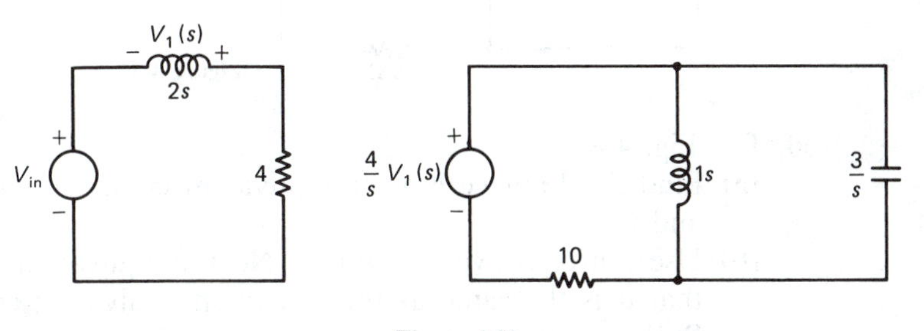

Figure 4-80

26. For Fig. 4-80, find the current (measure in the downward direction) through the capacitor in terms of the input voltage.

27. Use the mesh current method to find the Laplace current through R_2 in Fig. 4-81.

28. For Fig. 4-81, use the node voltage method to find the Laplace voltage across R_2.

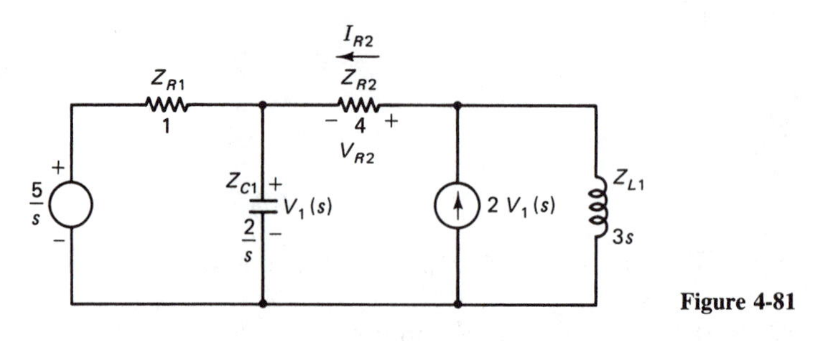

Figure 4-81

Section 4-7

29. For Fig. 4-82, find the Laplace Thévenin equivalent circuit for the network R_1 and L_1.

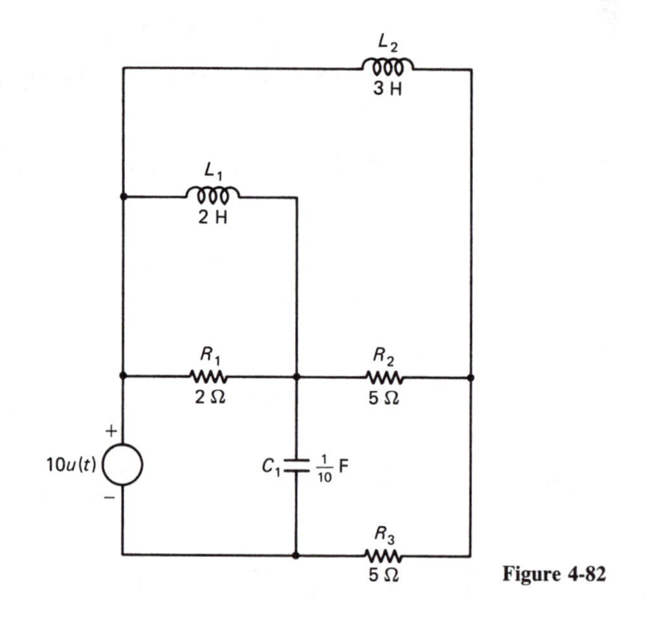

Figure 4-82

30. For Fig. 4-82:

(a) Find the Laplace Norton equivalent circuit for the network R_1 and L_1.

(b) Use source conversion on the Norton equivalent circuit to show that it is the same as the Thévenin equivalent circuit found in Problem 29.

31. For Fig. 4-83:

(a) Find the Laplace Thévenin equivalent circuit for R_3.

(b) Find the Laplace voltage across R_3 when it is changed to 2 Ω, and then when it is 10 Ω.

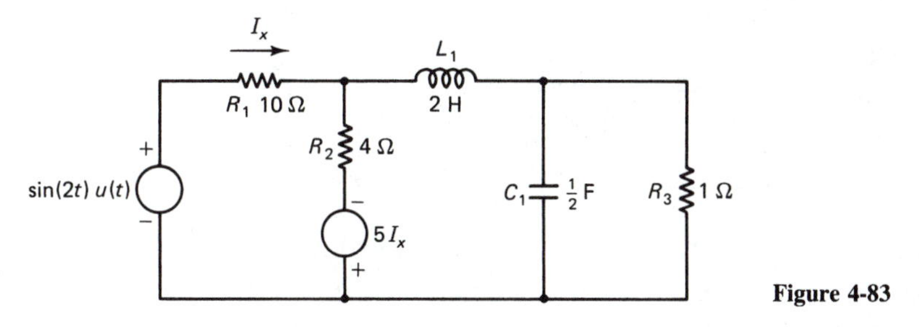

Figure 4-83

32. For Fig. 4-83:

(a) Find the Norton equivalent circuit for L_1.

(b) Find the current through L_1 when it is 2 H, and then when it is 5 H.

Section 4-8

33. Find the order of the circuits in Fig. 4-84.

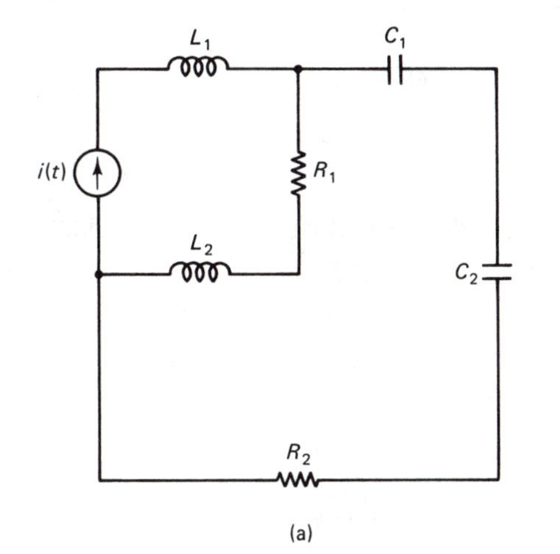

(a)

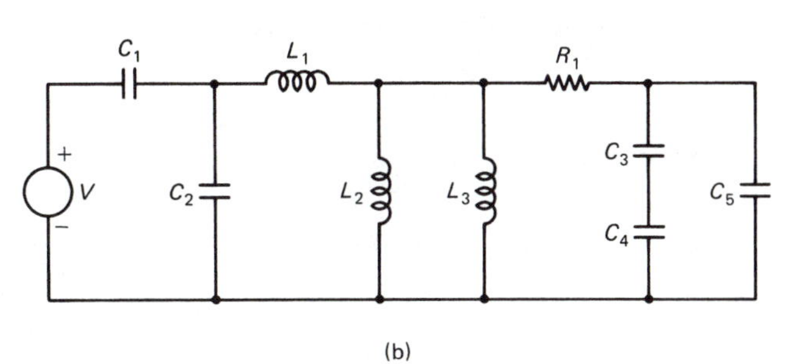

(b)

Figure 4-84

34. Find the step response for the voltage across R_2 in Fig. 4-85, and find the value of the pole due to the circuit.

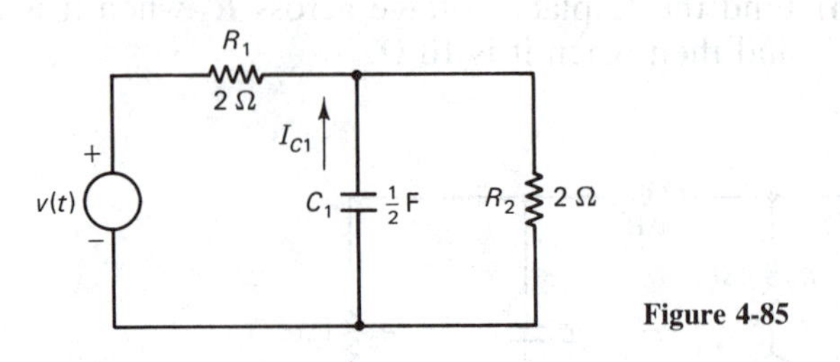

Figure 4-85

35. Find the impulse response for the current through C_1 in Fig. 4-85.

36. For the following second-order equations, find ζ, ω_n, α, and ω_d.

(a) $s^2 + 10s + 25$

(b) $4s^2 + 6s + 36$

(c) $s^2 + 29s + 100$

(d) $s^2 + 9$

37. Find the Laplace current in Fig. 4-86. Is the circuit under-, over-, or critically damped?

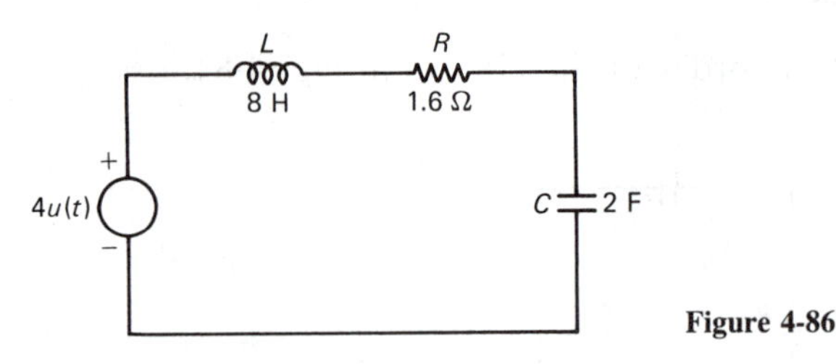

Figure 4-86

38. Find the values of C in Fig. 4-87 to cause the voltage, $v(t)$, to be under-, over-, and critically damped.

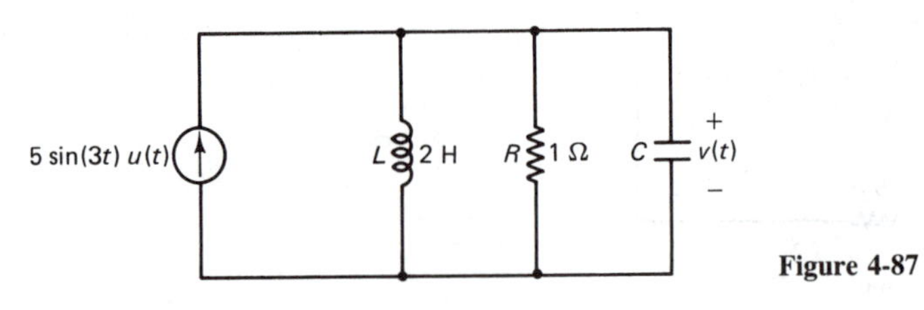

Figure 4-87

39. Is the current through R in Fig. 4-88 under-, over-, or critically damped?

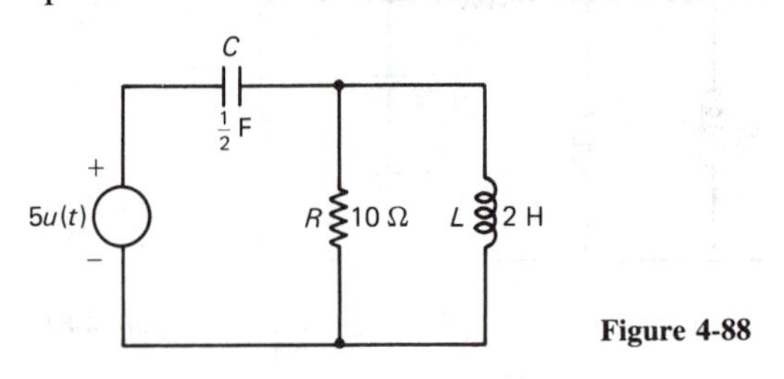

Figure 4-88

CHAPTER 5

Sinusoidal Steady State

5-0 INTRODUCTION

Sinusoidal steady-state analysis starts with the transfer function. For an electronic circuit, the transfer function is the ratio of the output voltage or current to the input voltage or current. There are a number of ways to use the transfer function in analyzing a circuit. We can use it for stability analysis or network analysis, but the most obvious way to use the transfer function is to find the output for a given input. If we find the ratio of the output magnitude and phase to the input magnitude and phase, we can graph the gain (magnitude function) and the phase (phase function) for a range of frequencies. These graphs can be used to determine which frequencies the circuit amplifies and which frequencies it attenuates.

In this chapter we concentrate on the magnitude function and phase function of the sinusoidal steady-state response. To do this analysis, the system must be linear. For a linear system the components R, L, and C must be fixed values or vary in time only.

5-1 TRANSFER FUNCTION

In general, the transfer function is the ratio of the output to the input. The transfer function is not restricted to voltage and current. In control systems it may represent the ratio of the output rpm of a motor to the input voltage, but the transfer function is always the ratio of the output function and the input function.

The ratio of output voltage to input voltage is the voltage gain, but this is a special case of the transfer function which has no units associated with it. In electronics the input might be current and the output a voltage that causes the transfer function to have units. In this case it is not a gain. So we should avoid calling the ratio a gain, and instead refer to it as a transfer function.

In Laplace, the transfer function is usually denoted by $G(s)$, the output is denoted by $C(s)$, and the input is denoted by $R(s)$. Stated in an equation,

$$G(s) = \frac{C(s)}{R(s)} \tag{5-1}$$

Once we have the transfer function we can calculate the output for any given input by multiplying the transfer function by the input.

$$C(s) = R(s)G(s) \tag{5-2}$$

We must remember that multiplication in Laplace does not correspond to multiplication in the time domain.

In a case where we have an unknown circuit we can apply an input and measure the output. From the output we can determine the transfer function. Two inputs are typically used to do this: the step input and the impulse input. The output is called the step response and the impulse response, respectively.

Both the step and impulse response have advantages and disadvantages. The step response is easiest to generate in the lab; however, we must ignore the zero-value pole generated by the input to determine the transfer function. Therefore, the output from the step is the transfer function with an added pole. For the impulse response, the output is the complete transfer function; however, we can only approximate the impulse function in the lab.

Since an impulse does not add any poles or zeros, the impulse response is the inverse of the transfer function. To show this, we start with Eq. (5-2):

$$C(s) = R(s)G(s)$$

and substitute the impulse input

$$R(s) = \mathcal{L}[\delta\ (t)] = 1$$

into it:

$$C(s) = 1G(s)$$

and find the inverse

$$c(t) = g(t) \tag{5-3}$$

The impulse response is therefore the inverse Laplace of the transfer function.

There are two mathematical methods used to find the transfer function. These two methods are similar to the methods used in Section 4-7 when we were finding the Thévenin impedance for circuits with dependent sources. The only difference is that instead of using V_z/I_z, we will use "output/input," where the "output" or "input" may be voltage or current.

The first method is to find an equation having only the output and input variable, and solve for output/input, which is the transfer function. The second method is to find an equation for the output and input in terms of the same variable, divide the output by the input, and cancel out the variable that the output and input are in terms of.

EXAMPLE 5-1

For the circuit in Fig. 5-1, find the transfer function where $V_{in}(s)$ is the input and $V(s)$ is the output. Find the unit step response.

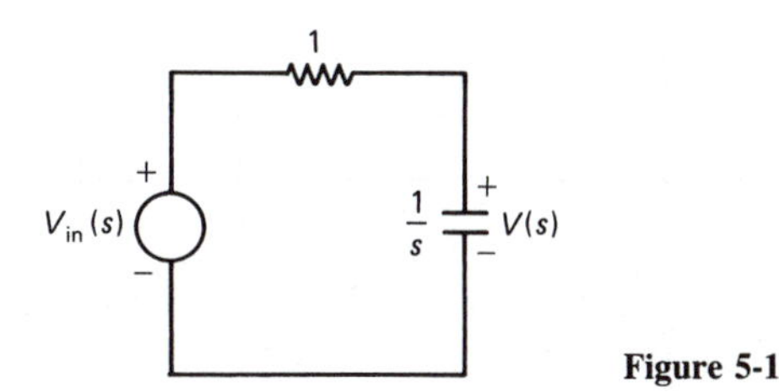

Figure 5-1

SOLUTION Let's use the first method. In this circuit we can use the voltage divider rule, which will give us one equation involving only the two variables.

$$V(s) = \frac{V_{in}(s)\,\frac{1}{s}}{1 + \frac{1}{s}}$$

Simplifying and solving for output over input gives

$$G(s) = \frac{V(s)}{V_{in}(s)} = \frac{1}{s + 1}$$

To find the step response, we must first find the Laplace transform of the input, which is

$$R(s) = \mathscr{L}[1u(t)] = \frac{1}{s}$$

Then using Eq. (5-2), we have

$$C(s) = R(s)G(s)$$

$$= V(s) = \frac{1}{s}\frac{1}{s + 1} = \frac{1}{s(s + 1)}$$

Last, we find the inverse Laplace.

$$c(t) = v(t) = 1u(t) - e^{-t}u(t)$$

EXAMPLE 5-2

For Fig. 5-2 find the transfer function using I as the output and V_{in} as the input. Find the impulse response.

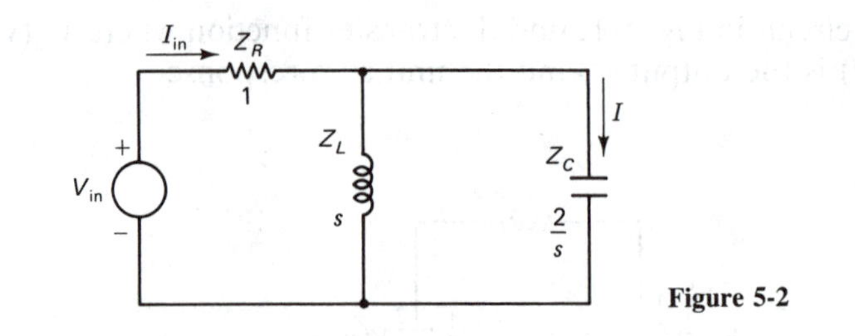

Figure 5-2

SOLUTION Let's use the second method to solve this circuit. We can find the input and output in terms of any variable. Let's use I_{in}. First we will find V_{in} in terms of I_{in} using Ohm's law.

$$V_{in}(s) = I_{in}(s)\left(1 + \frac{s\frac{2}{s}}{s + \frac{2}{s}}\right)$$

Simplifying this yields

$$V_{in}(s) = I_{in}\left(\frac{s^2 + 2s + 2}{s^2 + 2}\right)$$

Next we need to find I in terms of I_{in}. We will use the current divider rule.

$$I = \frac{I_{in}s}{s + \frac{2}{s}}$$

Simplifying this, we have

$$I = \frac{I_{in}s^2}{s^2 + 2}$$

Now we divide the output by the input to find the transfer function.

$$\frac{I(s)}{V_{in}(s)} = \frac{\frac{I_{in}s^2}{s^2 + 2}}{I_{in}\left(\frac{s^2 + 2s + 2}{s^2 + 2}\right)}$$

Canceling terms, we have

$$G(s) = \frac{I(s)}{V_{in}(s)} = \frac{s^2}{s^2 + 2s + 2}$$

Now that we have the transfer function, we can find the inverse of it to get the impulse response.

$$g(t) = \mathcal{L}^{-1}[G(s)] = \mathcal{L}^{-1}\left[\frac{s^2}{s^2 + 2s + 2}\right]$$

The inverse Laplace transform is therefore

$$g(t) = \delta(t) + 2e^{-t} \sin (t - 90°)$$

5-2 POLE–ZERO PLOT AND STABILITY

Once the transfer function is found, we can plot the poles and zeros on a Cartesian plane. A lot of material has been written for analyzing systems using pole–zero plots. The pole–zero plot is also used extensively in active-filter texts. We will look at an introduction to this approach.

5-2-1 Pole–Zero Plot

The poles and zeros of the transfer function are plotted on the s-plane. The s-plane is a Cartesian plane with the x-axis representing real values (σ-axis) and the y-axis representing imaginary values ($j\omega$-axis). We plot the values of s that make the transfer function zero (zeros) or infinity (poles). Poles are indicated by an $\times$ and zeros are indicated with a $\circ$.

EXAMPLE 5-3

Draw the pole–zero plot for the transfer function shown.

$$G(s) = \frac{(s + 2)[(s + 1)^2 + 9]}{s^2[(s + 4)^2 + 4]}$$

SOLUTION First we find the roots of the numerator, which are the zeros.

$$s = -2, -1 + j3, -1 - j3$$

These are the $\circ$'s plotted on the graph in Fig. 5-3.

Next we find the roots of the denominator, which are the poles.

$$s = 0, 0, -4 + j2, -4 - j2$$

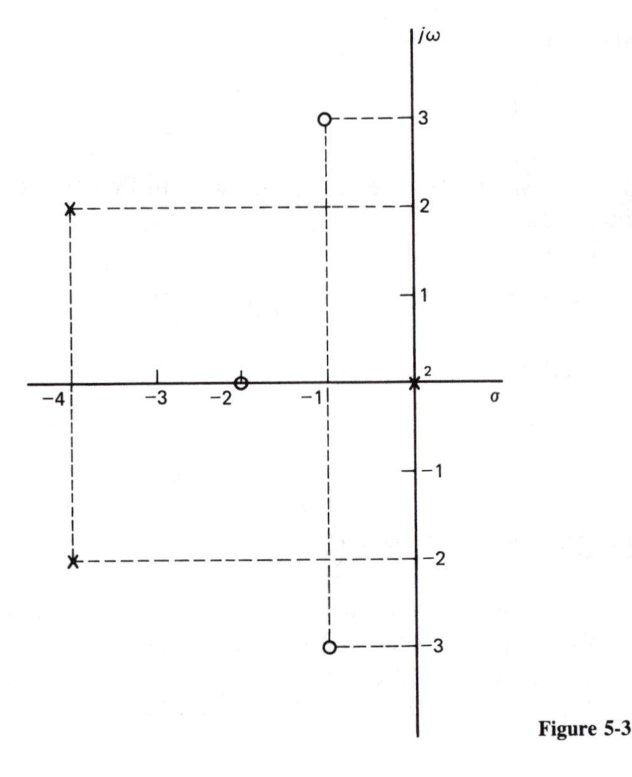

Figure 5-3

These are the ×'s plotted on the graph. Notice that there is a multiple pole at the origin. The number of poles are indicated by a superscript.

5-2-2 Stability Using Pole–Zero Plot

For an amplifier to work properly, the output must respond to the input. When the output does not, the system is unstable. We can determine the stability from the pole–zero plot. There are three possible cases for stability: stable, unstable, or marginally stable. These can be determined by the output when the system's input is an impulse function.

1. The system is stable if the impulse response goes to zero in a finite amount of time.
2. The system is unstable if the impuse response increases without bound.
3. The system is marginally stable if the impulse response goes to a finite value or moves between finite limits.

Since the poles of a Laplace function determine the form of the time response, the stability depends on the poles and not the zeros. Figure 5-4 shows the three regions: to the left of the $j\omega$-axis, on the $j\omega$-axis, and to the right of the $j\omega$-axis.

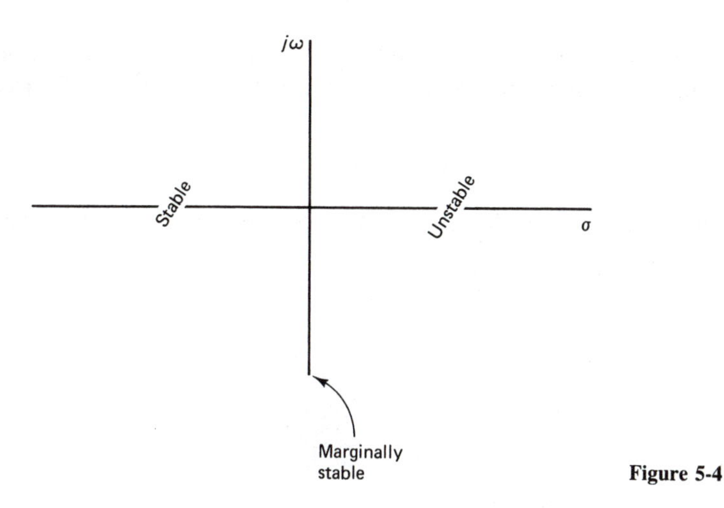

Figure 5-4

The rightmost pole of the transfer function determines the stability of the circuit or system. This is similar to a chain being as strong as its weakest link. When all poles of the transfer function are to the left of the $j\omega$-axis, the circuit is stable. When the rightmost pole is single order and on the $j\omega$-axis, the circuit is marginally stable. A multiple-order pole on the $j\omega$-axis makes the circuit unstable. When there are any poles to the right of the $j\omega$-axis, the circuit is unstable.

When we observe the form of the inverse Laplace for each type of pole, it becomes obvious why the s-plane is divided into these three regions. The key to understanding why these regions exist is in the exponential multiplier's power. We will see that for real and complex poles, the power will go through three changes: e^{-at}, e^{0t}, and e^{+at}.

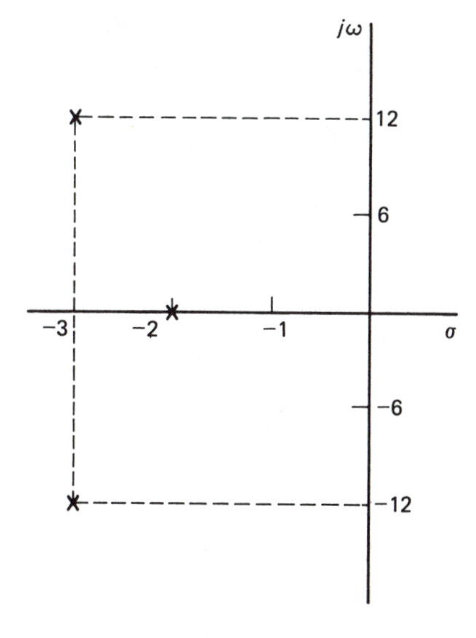

Figure 5-5

When real or complex poles are on the left of the $j\omega$-axis, the time response always contains an e^{-at} multiplier. For example, if the transfer function is

$$G(s) = \frac{N(s)}{(s + 2)[(s + 3)^2 + 144]}$$

then it has all poles to the left of the $j\omega$-axis, as shown in Fig. 5-5. The inverse due to the $(s + 2)$ factor is

$$f_1(t) = Ke^{-2t}u(t)$$

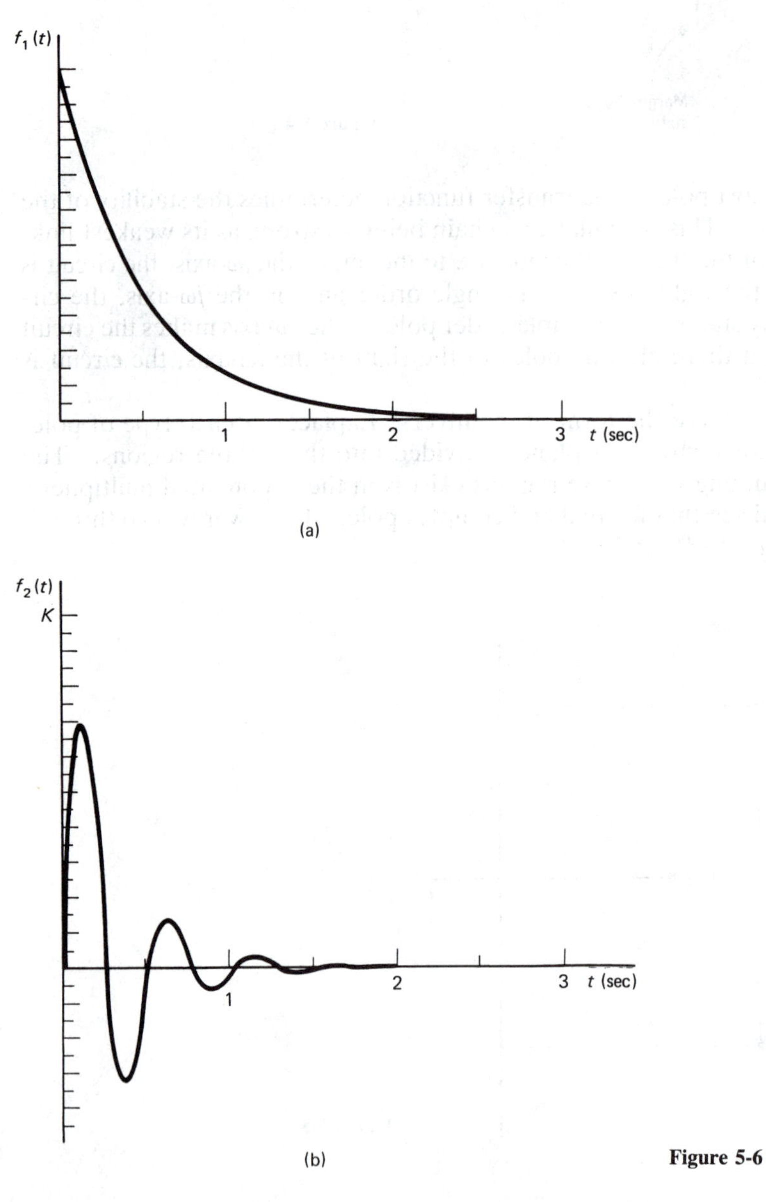

Figure 5-6

This part of the inverse Laplace is transient and will go away in steady state, as shown in Fig. 5-6a. The inverse due to the factor $[(s + 3)^2 + 144]$ is

$$f_2(t) = Ke^{-3t} \sin (12t + \theta)\, u(t)$$

This is also transient and will go to zero in steady state, as shown in Fig. 5-6b.

Single-order poles on the $j\omega$-axis are called marginally stable. The inverse of poles on the origin may be considered to contain an exponential factor raised to the $-0t$ power. Since this term is equal to 1, we usually do not write it in the inverse equation. Figure 5-7 shows the poles for the transfer function

$$G(s) = \frac{N(s)}{s(s^2 + 144)}$$

The pole due to s gives us a response of

$$f_1(t) = Ku(t)$$

which is shown in Fig. 5-8a, and the pole due to $(s^2 + 144)$ gives us

$$f_2(t) = K \sin (12t + \theta) u(t)$$

which is shown in Fig. 5-8b. This circuit may or may not be stable. If the circuit is supposed to be an oscillator and we have only the complex poles, it will be stable, but it will not be useful as an amplifier. When there is a pole at the origin and if the input is also a step function, the inverse will be

$$f_1(t) = (K_1 t + K_2) u(t)$$

which is unstable.

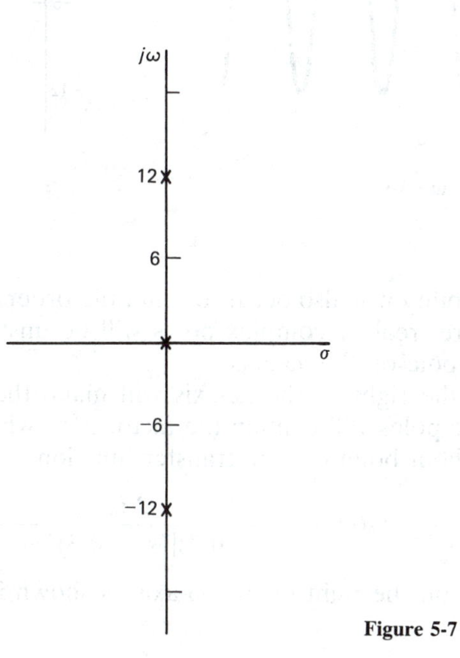

Figure 5-7

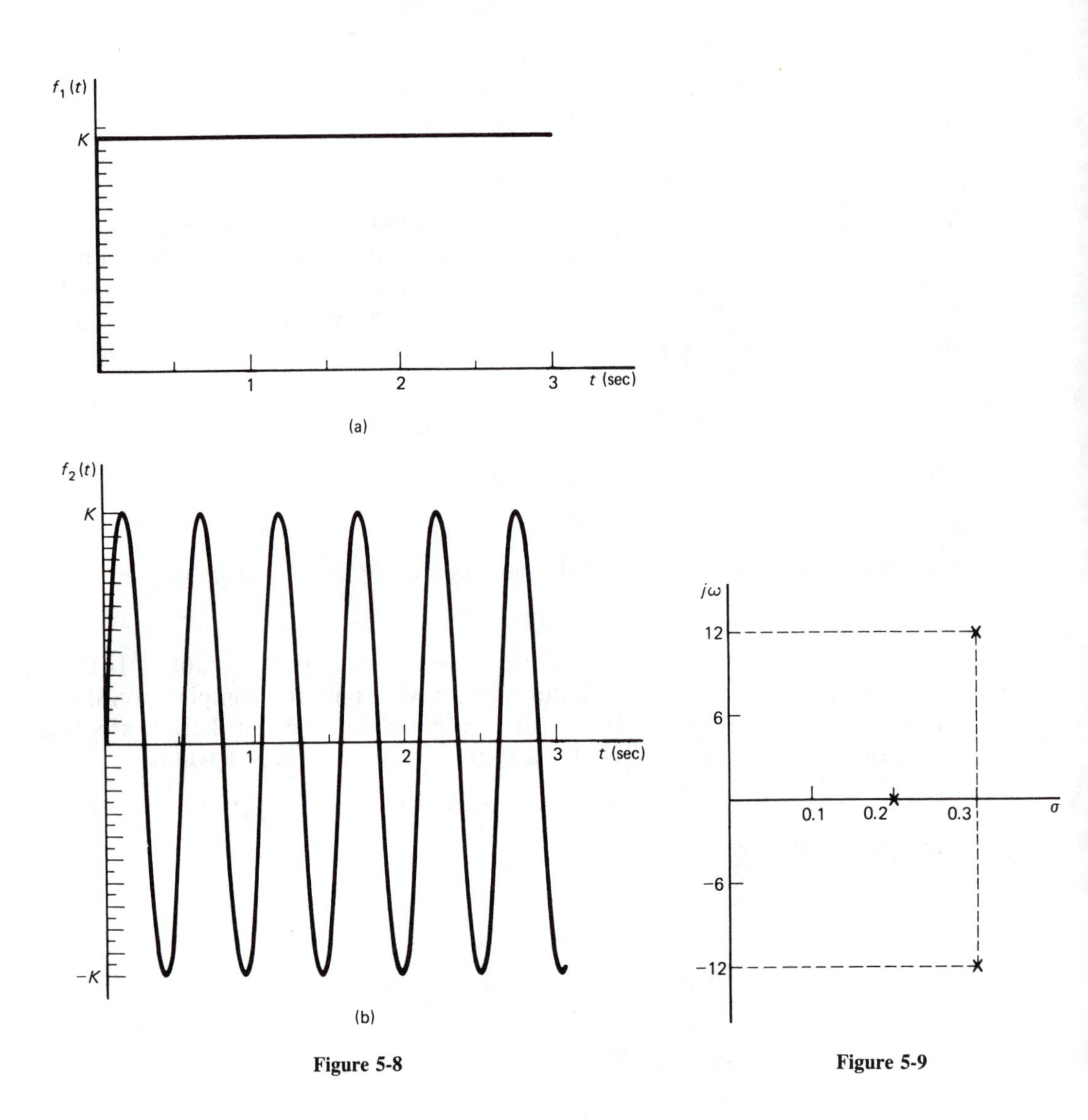

Figure 5-8

Figure 5-9

This instability will also occur for multiple-order complex poles on the $j\omega$-axis. Therefore, real or complex poles will be unstable if they appear as a multiple-order pole on the $j\omega$-axis.

Poles on the right of the $j\omega$-axis will make the system unstable. The inverse of these poles will contain the factor e^{+at}, which will cause the output to increase without bound. The transfer function

$$G(s) = \frac{N(s)}{(s - 0.2)[(s - 0.3)^2 + 144]}$$

has both poles on the right of the $j\omega$-axis as shown in Fig. 5-9. The inverse

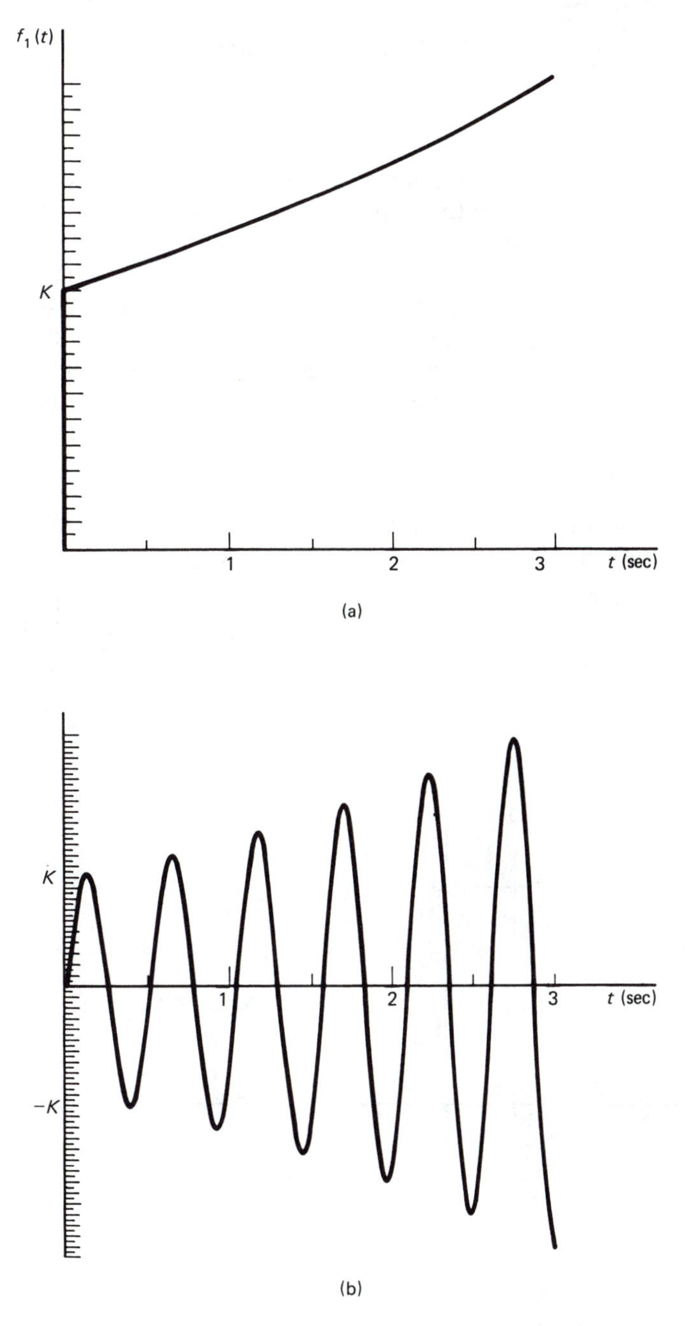

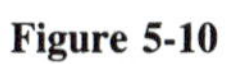

Figure 5-10

of the poles due to $(s - 0.2)$ and $[(s - 0.3)^2 + 144]$ are

$$f_1(t) = Ke^{+0.2t}u(t)$$

$$f_2(t) = Ke^{+0.3t}\sin(12t + \theta)u(t)$$

respectively, and are shown in Fig. 5-10.

Another point of interest is the speed of decay for the inverse function and its relationship to the $j\omega$-axis. If we start to the far left of the $j\omega$-axis, the function's exponential multiplier will have a very large negative power and will decay very rapidly. The closer we move to the $j\omega$-axis, the slower the function will decay since the exponential multiplier becomes smaller. When we reach the $j\omega$-axis, the function will not decay at all and will become a steady-state value. When we move to the right, the function will increase without bound. The farther to the right of the $j\omega$-axis we go, the faster this increase will become.

EXAMPLE 5.4

Draw the pole–zero plot and determine the stability of the circuit given the following transfer function.

$$G(s) = \frac{(s - 3)[(s + 1)^2 + 16]}{s(s + 1)^2\,[(s + 2)^2 + 9]}$$

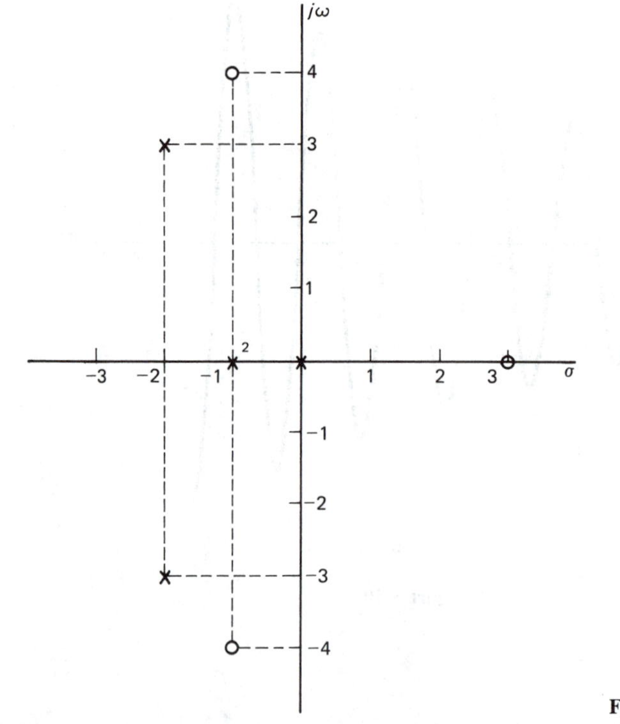

Figure 5-11

SOLUTION The pole–zero plot is shown in Fig. 5-11. This system is marginally stable since the rightmost pole is on the $j\omega$-axis and is single order. There is a zero in the right half of the s-plane, but since the zeros do not determine the form of the inverse, they do not affect the stability of the system.

5-3 STEADY-STATE FREQUENCY RESPONSE AND THE BODE PLOT

An amplifier or filter will have many different types of input signals. These inputs may be complex and constantly changing. It would be impossible to analyze the output response for each different signal. Fortunately, most signals can be approximated by a combination of sinusoidal components. If we know the output frequency response for a sinusoidal input, we can break any input waveform into sinusoidal components and determine its response.

Normally, we draw two graphs—the magnitude function and the phase function—on semilog graph paper. (A review of semilog graph paper is given in Appendix D.) There are two popular methods used to draw the graphs. One is the exact plot, which usually requires a computer. The other way is the Bode plot, which is an approximation. We will look at both of these methods.

5-3-1 Converting the Transfer Function to the Time Domain

In the time domain the magnitude function and phase function are found by the ratio of the ac steady-state output to input. The transfer function is also the ratio of the output to the input as stated in Eq. (5-1), but this is in Laplace and not in the time domain. To convert this Laplace ratio to the time-domain ratio, we simply replace s with $j\omega$.

To understand why this works, we need to look at what we did to the circuit component values when going to Laplace. For the inductor we had

$$L \Rightarrow sL$$

If we replace s by $j\omega$, we have

$$sL \Rightarrow j\omega L$$

The j is a phase shift of 90° and ω is equal to $2\pi f$. We also know that the inductive reactance is $X_L = 2\pi fL$. Therefore, we have

$$L \Rightarrow sL \Rightarrow j\omega L = X_L \angle 90^\circ \tag{5-4}$$

With a similar procedure the capacitor is

$$C \Rightarrow \frac{1}{sC} \Rightarrow \frac{1}{j\omega C} = X_C \angle -90^\circ \tag{5-5}$$

The resistance is the same in each step.

$$R \Rightarrow R \Rightarrow R = R \tag{5-6}$$

When we let s become $j\omega$ we have, in a sense, been doing steady-state ac circuit analysis, but this is true if, and only if, we let $s = j\omega$ in the final step. When we use Laplace this way, we have a function in terms of the frequency. By substituting in different frequencies, we can graph the magnitude and phase as frequency changes without doing the circuit analysis at each frequency.

The procedure to find the time-domain magnitude function and phase function is as follows:

1. Find the Laplace transfer function.
2. Replace s by $j\omega$ to find the magnitude function, $M(j\omega)$, and the phase function, $P(j\omega)$.

$$M(j\omega) \angle P(j\omega) = G(s)\,|_{s=j\omega} \tag{5-7}$$

3. Plug in different values of ω. This will result in a numerical magnitude and phase value.
4. On semilog graph paper plot the dB value of the magnitude, $M_{dB}(j\omega)$, and the phase, $P(j\omega)$.

From our experience with the transfer function we know that we will have either real or complex conjugate roots in the denominator and numerator along with a constant multiplier. Since this is true, we can break up the frequency analysis of the transfer function into:

1. A gain constant
2. Real poles
3. Real zeros
4. Complex poles
5. Complex zeros

To make the analysis easier, we need to rewrite the transfer function into a different form. We want a single numerator constant, K, and all factors having the constant part equal to 1. Don't be disturbed about not understanding why this is easier because this point may not make sense until we go through Section 5-3-7.

Therefore, real poles and zeros of the form

$$(s + \omega_c)^r \tag{5-8a}$$

become

$$\omega_c^r\left(\frac{s}{\omega_c} + 1\right)^r \tag{5-8b}$$

We use ω_c because this value tells us the critical frequency (also called the break frequency) of this factor. The ω_c^r multiplier is part of the gain constant K.

The complex conjugate poles and zeros in the form of

$$(s^2 + 2\zeta\omega_c s + \omega_c^2)^r \tag{5-9a}$$

become

$$\omega_c^{2r}\left(\frac{s^2}{\omega_c^2} + \frac{s\zeta}{\omega_c}s + 1\right)^r \tag{5-9b}$$

Here again ω_c tells us the critical frequency. This value is still ω_n, but we will use ω_c since in this context we are looking at magnitude functions and phase functions. The ω_c^{2r} multiplier will become part of the gain constant K.

EXAMPLE 5-5

Write the following transfer function in a form suitable for frequency analysis.

$$G(s) = \frac{87(s + 2)[(s + 1)^2 + 4]}{(s^2 + 6s + 9)(s^2 + 4s + 29)}$$

SOLUTION We must first factor everything into the form shown in Eqs. (5-8a) and (5-9a) so that we can determine which roots are real and which roots are complex.

$$G(s) = \frac{87(s + 2)(s^2 + 2s + 5)}{(s + 3)^2(s^2 + 4s + 29)}$$

The next step is to make each factor's constant part equal to 1 by putting each factor in the form of Eq. (5-8b) or (5-9b).

$$G(s) = \frac{87(2)\left(\frac{s}{2} + 1\right)(5)\left(\frac{s^2}{5} + \frac{2}{5}s + 1\right)}{(3)^2\left(\frac{s}{3} + 1\right)^2(29)\left(\frac{s^2}{29} + \frac{4}{29}s + 1\right)}$$

And finally, we combine all the constant multipliers in the numerator to form the gain constant, K.

$$G(s) = \frac{\frac{10}{3}\left(\frac{s}{2} + 1\right)\left(\frac{s^2}{5} + \frac{2}{5}s + 1\right)}{\left(\frac{s}{3} + 1\right)^2\left(\frac{s^2}{29} + \frac{4}{29}s + 1\right)}$$

5-3-2 Gain Constant

When we put the transfer function in a form suitable for drawing the gain function and phase function, there will be a constant in the numerator called the gain constant which will be denoted by the variable K.

$$M_{\text{dB}}(j\omega)\,\angle P(j\omega) = G(s)\Big|_{s=j\omega} = \frac{KN(s)}{D(s)}\Big|_{s=j\omega}$$

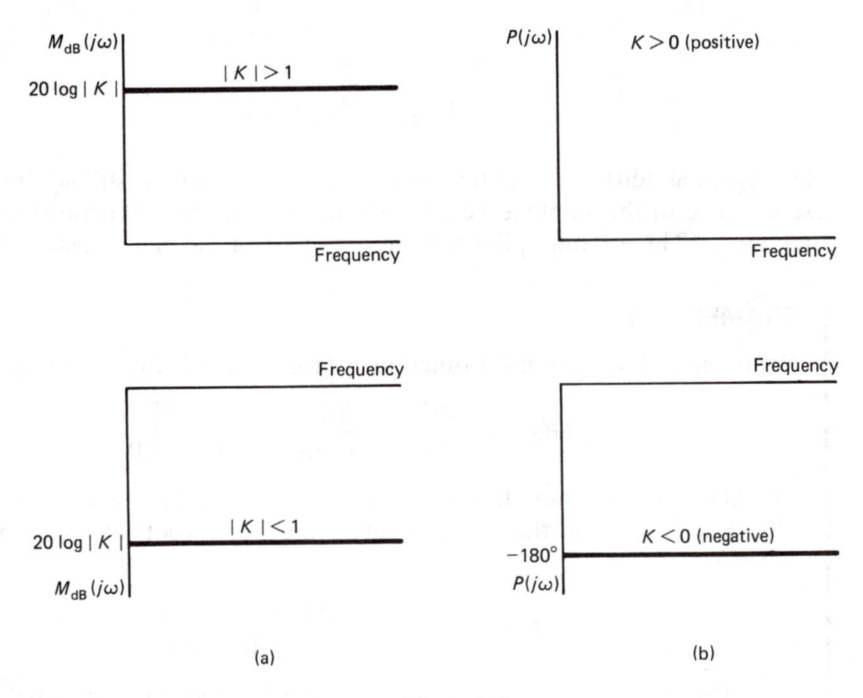

Figure 5-12

The gain constant can be expressed as a magnitude and phase.

$$M(j\omega)\ \angle P(j\omega) = |K|\ \angle 0^\circ \quad \text{or} \quad \angle -180^\circ \tag{5-10}$$

The phase angle will be 0° when K is positive, and the phase angle will be −180° when K is negative. We typically choose −180° instead of using +180°.

The magnitude function is almost exclusively a plot on semilog graph paper with the vertical axis in dB. (Appendix D reviews semilog graph paper.) The gain constant, unaffected by changes in frequency, will have a constant dB value given by

$$M_{dB}(j\omega) = 20 \log |K| \tag{5-11}$$

The magnitude in dB caused by the gain constant will be positive for $|K| > 1$, and the magnitude in dB will be negative for $|K| < 1$. Figure 5-12a shows both cases for $|K|$.

For the phase function, we look at the sign of K instead of the magnitude of K. If K is positive, the phase angle is 0°, and if K is negative, the phase angle is −180°. Figure 5-12b shows the two possible cases for the phase angle.

5-3-3 Real Poles

We must separate real poles into nonzero and zero poles. The analysis of these two value types is different. First we will look at nonzero poles.

The nonzero pole in the correct form for frequency analysis using Eq. (5-8b) with $r = 1$ is

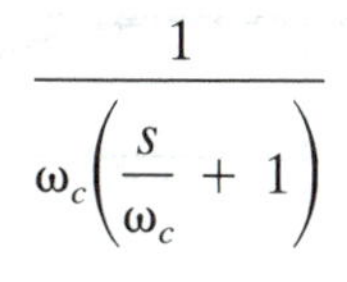

$$\frac{1}{\omega_c\left(\dfrac{s}{\omega_c} + 1\right)}$$

The multiplier ω_c is part of the constant, K. Therefore, we will work with

$$\frac{1}{\dfrac{s}{\omega_c} + 1}$$

We can determine the magnitude and phase by letting $s = j\omega$:

$$M(j\omega)\ \angle P(j\omega) = \left(j\frac{\omega}{\omega_c} + 1\right)^{-1} \tag{5-12}$$

Keep in mind that ω_c is a fixed constant value and ω is the variable we will substitute into with different frequency values. Using rectangular-to-polar conversion, we obtain

$$M(j\omega) = \left[\sqrt{\left(\frac{\omega}{\omega_c}\right)^2 + 1^2}\right]^{-1} \tag{5-13a}$$

$$M_{\text{dB}}(j\omega) = -20 \log \sqrt{\left(\frac{\omega}{\omega_c}\right)^2 + 1^2} \tag{5-13b}$$

$$P(j\omega) = \frac{1}{\tan^{-1}\left(\dfrac{\omega/\omega_c}{1}\right)}$$

$$P(j\omega) = -\tan^{-1}\left(\frac{\omega}{\omega_c}\right) \tag{5-14}$$

Substituting values of ω into Eqs. (5-13b) and (5-14) that range from below ω_c to above ω_c, we can plot the graphs shown in Fig. 5-13. In Fig. 5-13a we get approximately 0 dB below ω_c, since in Eq. (5-13b) ω/ω_c becomes much less than 1 and we are essentially finding the logarithm of 1. When ω is above ω_c, in Eq. (5-13b), then ω/ω_c becomes much greater than 1 and we essentially have $-20 \log(\omega/\omega_c)$. When $\omega = \omega_c$, we calculate $-20 \log(2^{1/2})$, which is equal to -3.0103 dB.

Roll-off is the rate or slope of the magnitude as frequency changes. For the magnitude function, when we are above ω_c, we find that the magnitude approaches $-20 \log(\omega/\omega_c)$. The more we increase the frequency, the closer we get to this asymptote. Using this asymptote, the magnitude changes by 20

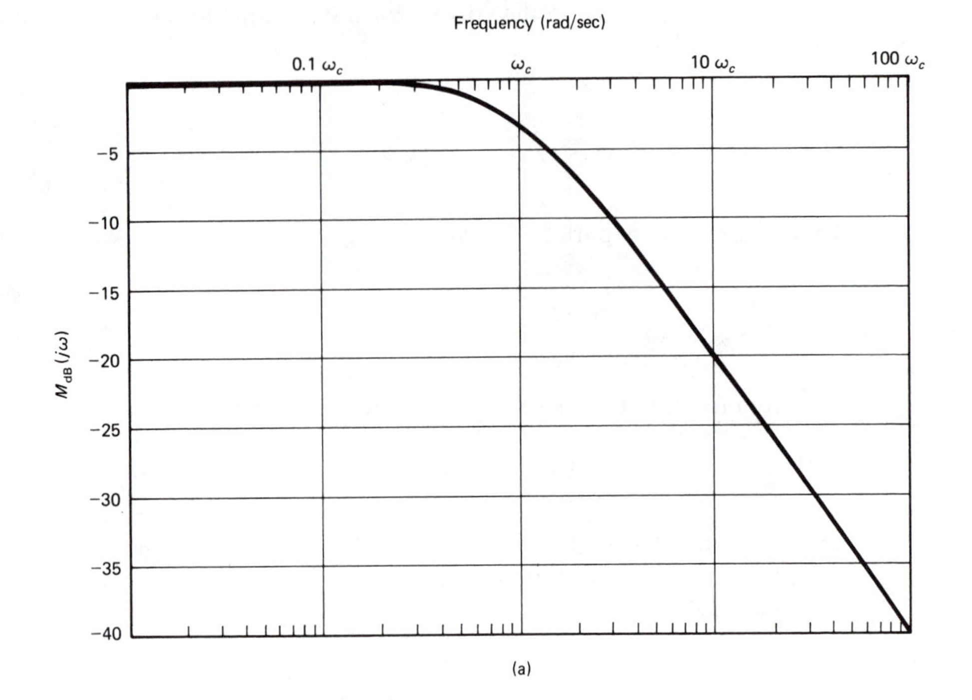

(a)

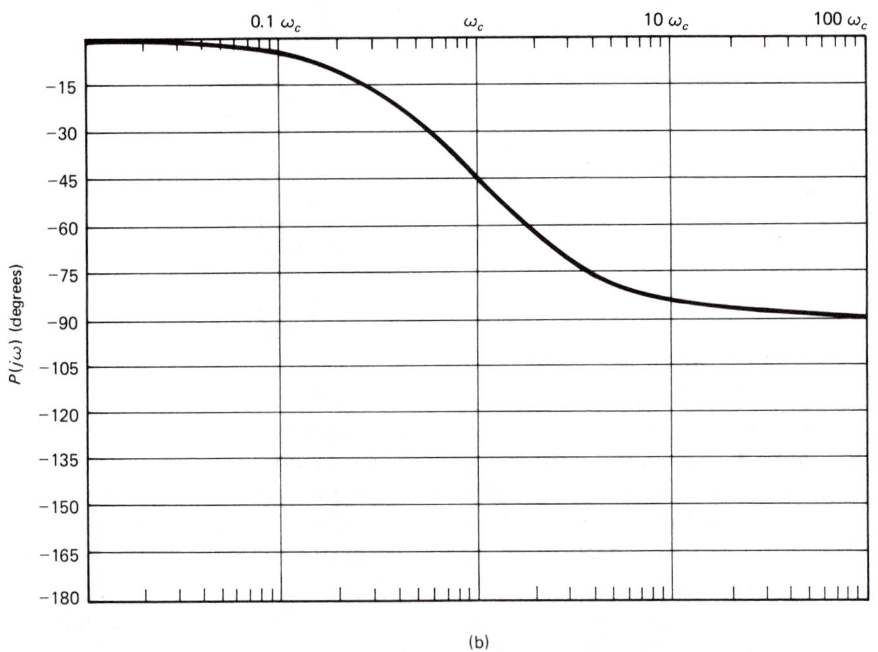

(b)

Figure 5-13

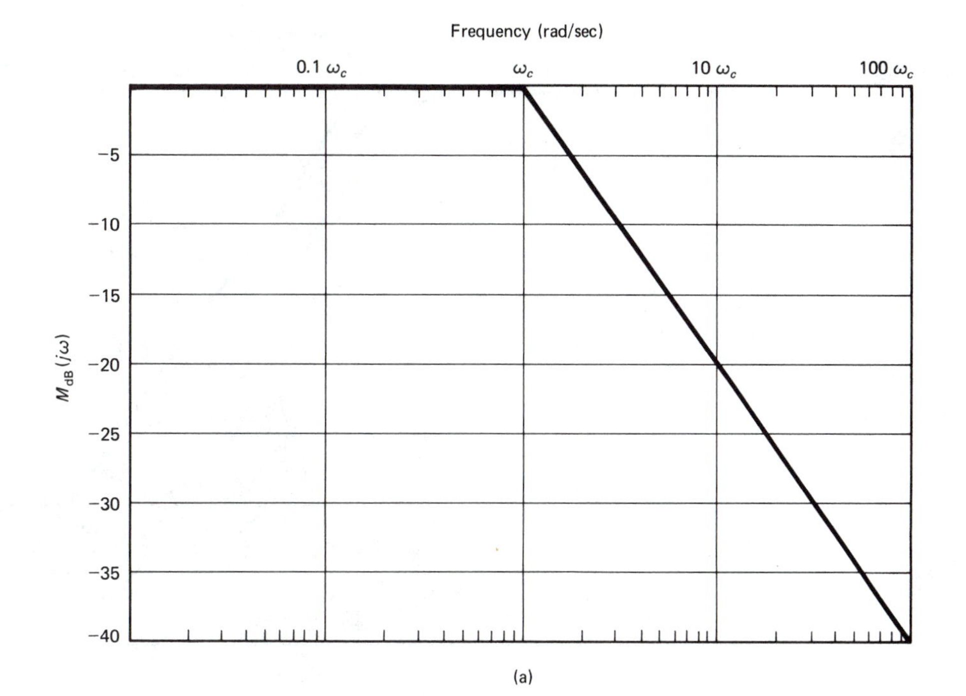

(a)

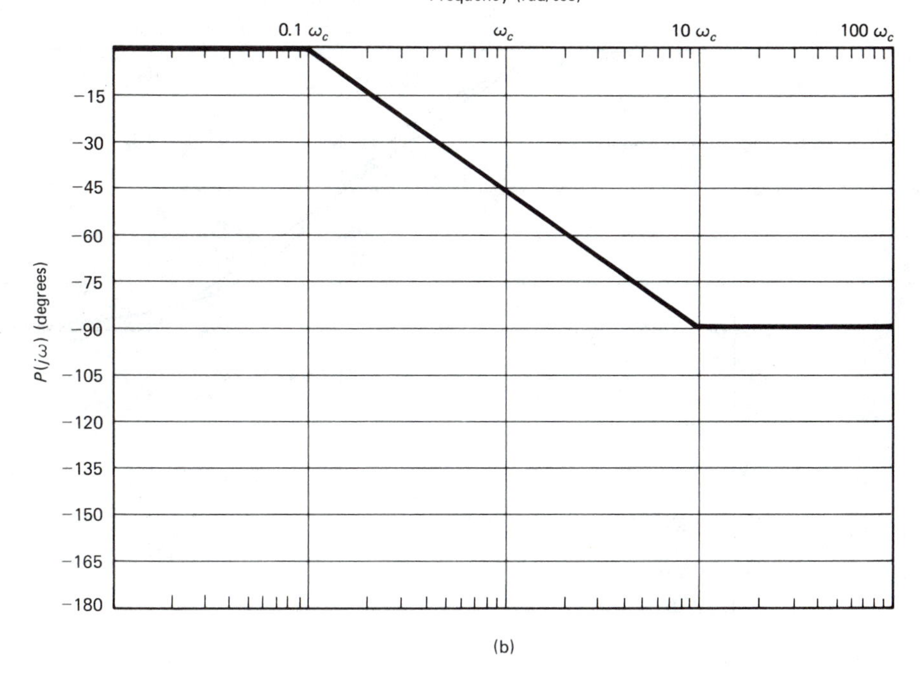

(b)

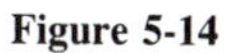

Figure 5-14

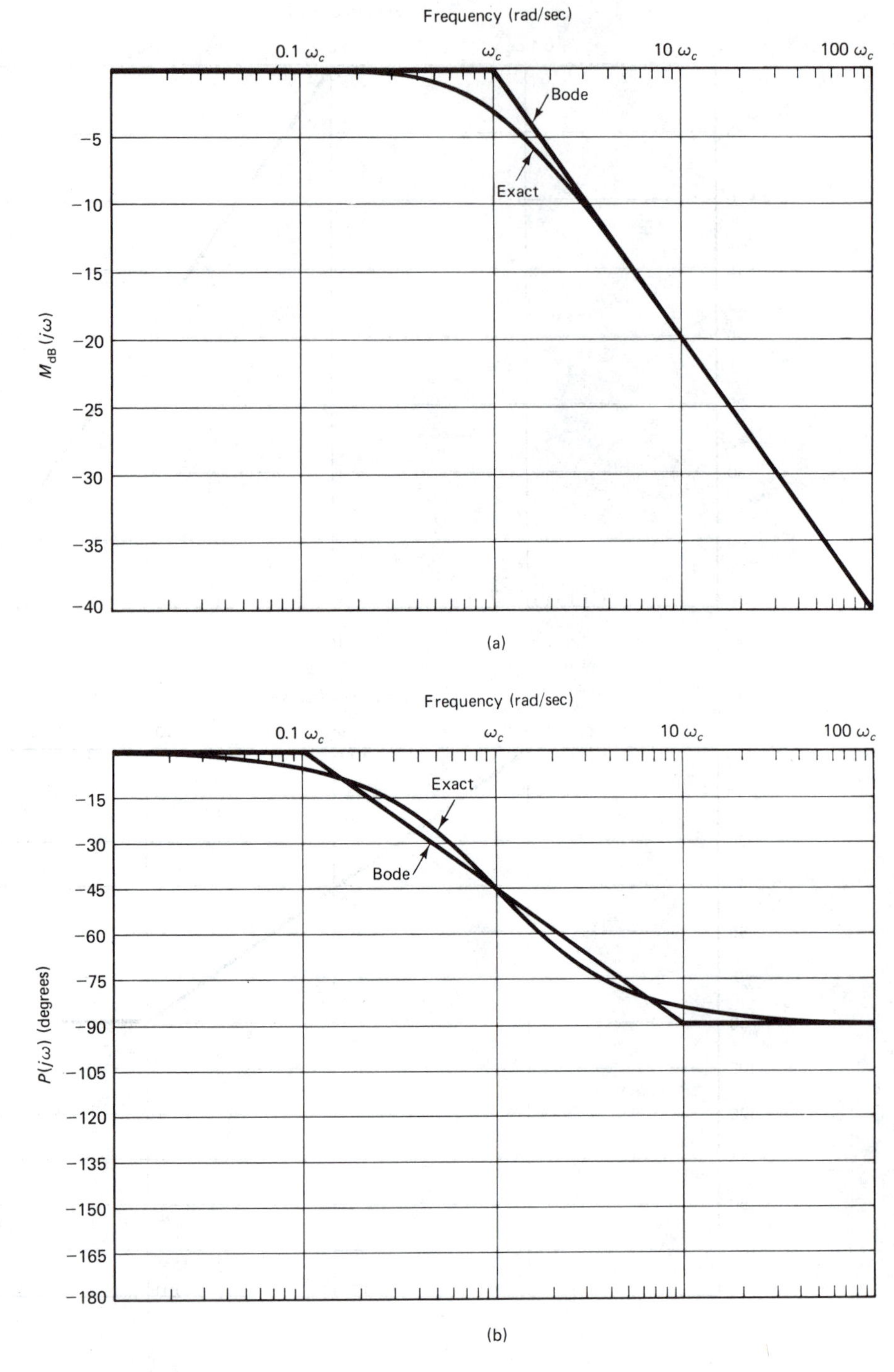
Frequency (rad/sec)
0.1 ω_c
ω_c
10 ω_c
100 ω_c
Bode
Exact
−5
−10
−15
−20
−25
−30
−35
−40
$M_{dB}(j\omega)$
(a)
Frequency (rad/sec)
0.1 ω_c
ω_c
10 ω_c
100 ω_c
Exact
Bode
−15
−30
−45
−60
−75
−90
−105
−120
−135
−150
−165
−180
$P(j\omega)$ (degrees)
(b)

Figure 5-15

dB for each change in the frequency by a factor of 10 (which is a decade). Therefore, we have a roll-off rate of -20 dB/decade.

Similarly, we can find significant points on the phase function. In Fig. 5-13b, using Eq. (5-14), we see that when $\omega = \omega_c$, we calculate $-\tan^{-1}(1)$. This value is exactly $-45°$. Once ω is less than $0.1\omega_c$, the phase shift is approximately 0° since we are calculating the arctangent of approximately 0. When ω is greater than $10\omega_c$, we are finding the arctangent of a large number, which is approximately $-90°$.

These observations lead us to the Bode plot, which is an approximation using straight lines. Figure 5-14 shows the Bode plot for the single pole:

$$\frac{1}{\dfrac{s}{\omega_c} + 1}$$

For the Bode magnitude function in Fig. 5-14a, the gain is 0 dB up to and including the break frequency ω_c. At ω_c we must remember that the exact value is -3.0103 dB, even though we draw it at 0 dB. Above ω_c we draw a line with a slope of -20 dB/decade. So we draw the asymptote instead of the exact values.

For the Bode phase function we assume that there is no phase shift for frequencies less than a decade below the break frequency, and we assume $-90°$ for frequencies greater than a decade above the break frequency. Between a decade below and a decade above the break frequency, we draw a line at a slope of $-45°$ per decade. The only place where this approximation is exactly equal to the exact plot is at the break frequency, where both types of plots are $-45°$.

Figure 5-15 shows both the Bode approximation and the exact plot of the single pole. The frequency plot in Fig. 5-15a shows that the maximum error occurs at ω_c, which is only 3.0103 dB.

EXAMPLE 5-6

Draw the Bode plot with the frequency scale in hertz. Find the exact value of magnitude and phase at 10 Hz.

$$\frac{1}{\dfrac{s}{20} + 1}$$

SOLUTION The break frequency is 20 rad/sec. Using

$$\omega = 2\pi f$$

we can solve for the break frequency in hertz.

$$f = \frac{\omega}{2\pi} = \frac{20}{2\pi} = 3.1831 \text{ Hz}$$

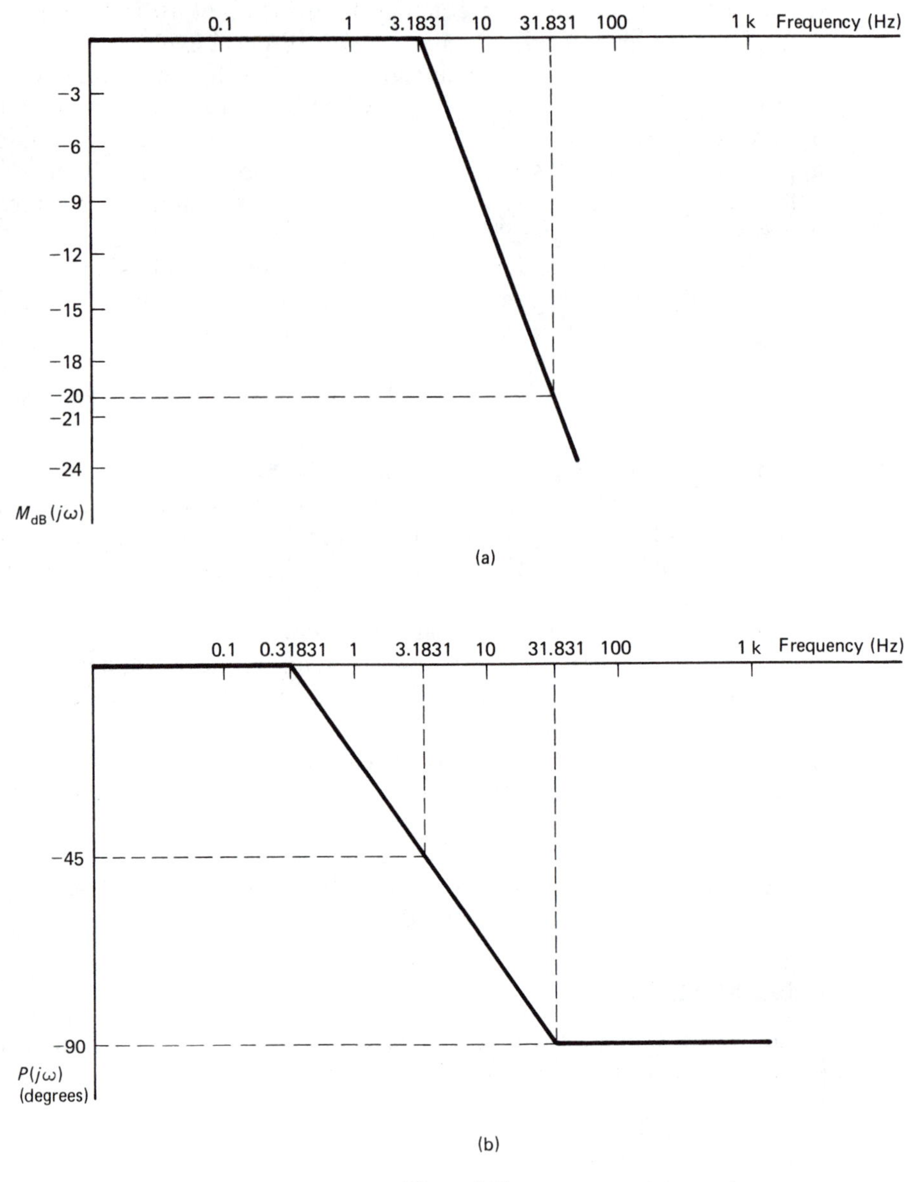

Figure 5-16

We must include at least one decade below 3.1831 Hz (which is 0.31831 Hz) and one decade above (which is 31.831 Hz) in order to show all the phase shift. Figure 5-16 shows the magnitude function and phase function.

To find the exact values at a frequency we substitute $s = j\omega$ into the transfer function. We were given a frequency in hertz, so we must first convert it to rad/sec.

$$\omega = 2\pi f = 2\pi(10) = 62.832 \text{ rad/sec}$$

Substituting into the function, we have

$$\frac{1}{j\dfrac{62.832}{20} + 1}$$

Using polar–rectangular conversions, we solve for the magnitude and phase.

$$M(j\omega)\ \underline{/P(j\omega)} = 0.30331\ \underline{/-72.343}$$

Finding the magnitude in dB, we get

$$M_{\text{dB}}(j\omega) = 20\log(0.30331)$$

$$= -10.362 \text{ dB}$$

A pole value of zero is dealt with differently since it has a break frequency at 0 rad/sec or at dc. In this case the exact plot and the Bode plot are identical. Using the procedure for nonzero poles, we have

$$M(j\omega)\ \underline{/P(j\omega)} = \left.\frac{1}{s}\right|_{s=j\omega}$$

$$M(j\omega)\ \underline{/P(j\omega)} = (j\omega)^{-1}$$

Splitting this into a gain function and a phase function, we have

$$M(j\omega) = \omega^{-1} \tag{5-15a}$$

$$M_{\text{dB}}(j\omega) = -20\log\omega \tag{5-15b}$$

$$P(j\omega) = \frac{1}{\tan^{-1}\left(\dfrac{\omega}{0}\right)} = -\tan^{-1}\infty = -90^\circ \tag{5-16}$$

The magnitude function is shown in Fig. 5-17a. Notice that it has a value of 0 dB at 1 rad/sec, and the slope is -20 dB/decade.

Figure 5-17b shows the phase function. Since for any value of ω we are finding the arctangent of ∞, we will always calculate -90°.

5-3-4 Real Zeros

Since the real zeros are very similar to real poles, we will not spend as much time on them. We will look at nonzero real zeros first, and then look at zero-value zeros.

The nonzero zero in proper form is

$$\frac{s}{\omega_c} + 1 \tag{5-17}$$

Substituting in $j\omega$ for s gives

$$M(j\omega)\ \underline{/P(j\omega)} = \frac{j\omega}{\omega_c} + 1$$

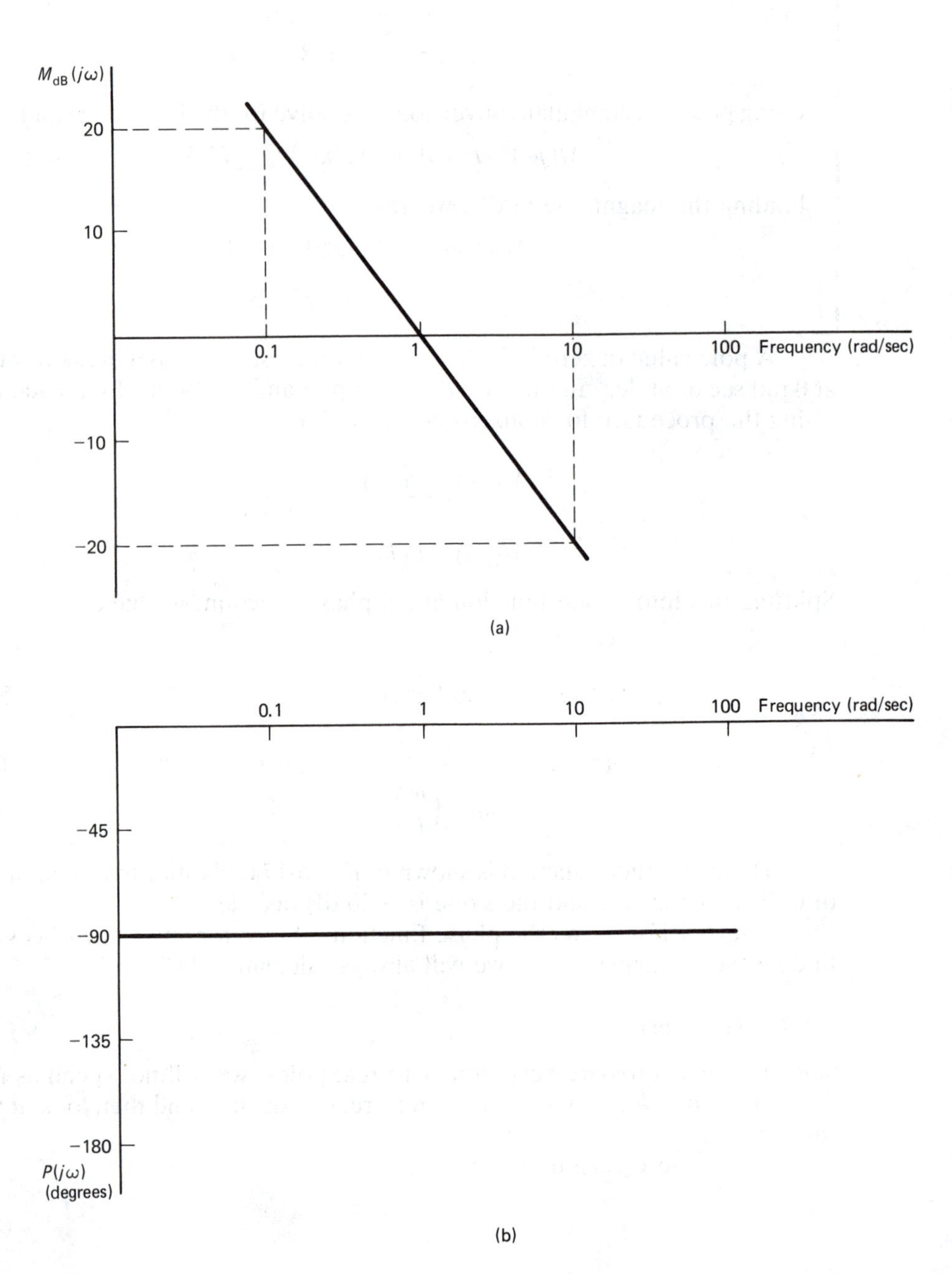

$M_{dB}(j\omega)$
20
10
−10
−20
0.1
1
10
100
Frequency (rad/sec)
(a)
0.1
1
10
100
Frequency (rad/sec)
−45
−90
−135
−180
$P(j\omega)$
(degrees)
(b)

Figure 5-17

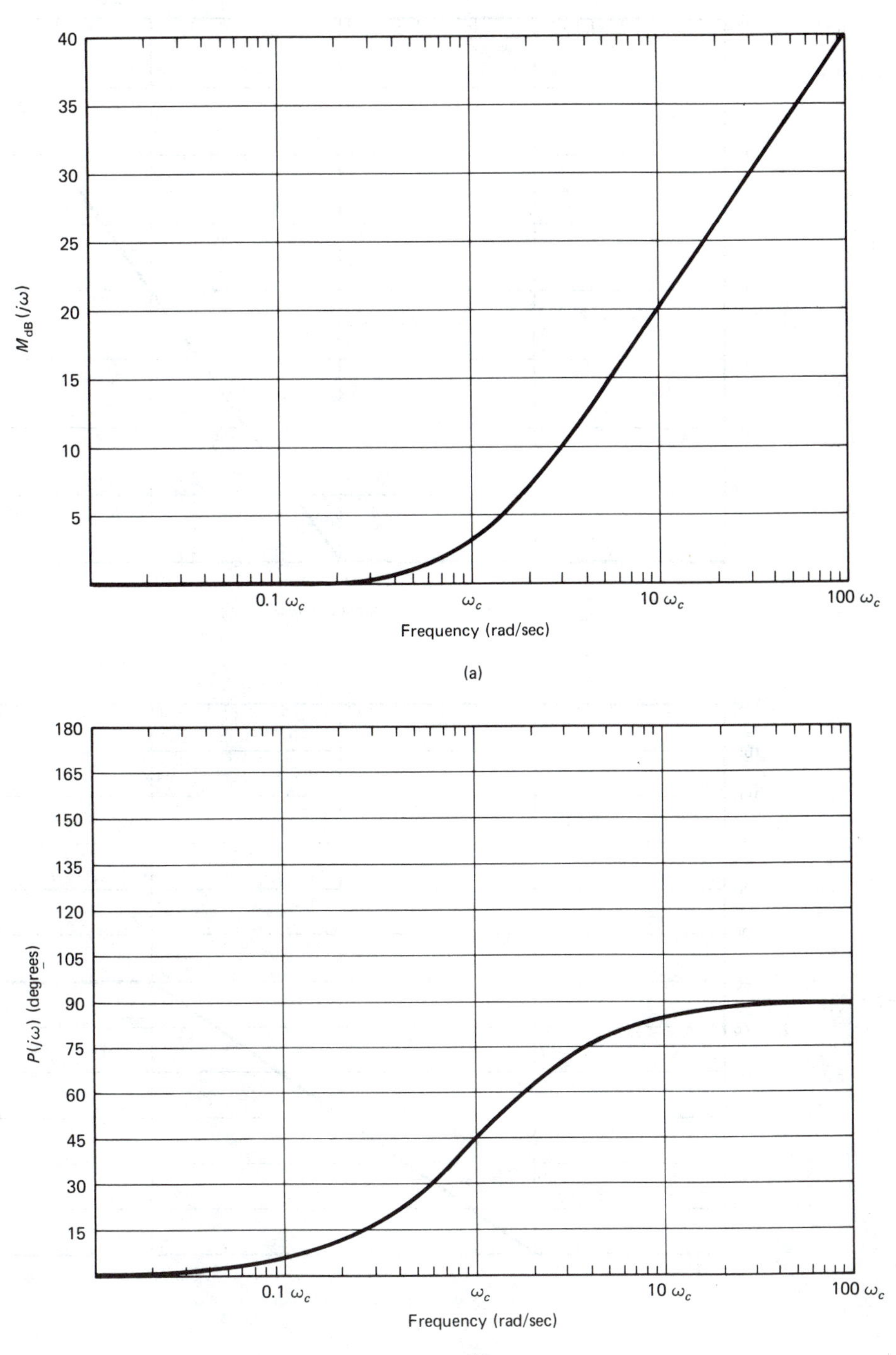
40
35
30
25
20
15
10
5
$M_{dB}(j\omega)$
0.1 ω_c
ω_c
10 ω_c
100 ω_c
Frequency (rad/sec)
(a)
180
165
150
135
120
105
90
75
60
45
30
15
$P(j\omega)$ (degrees)
0.1 ω_c
ω_c
10 ω_c
100 ω_c
Frequency (rad/sec)

(b)

Figure 5-18

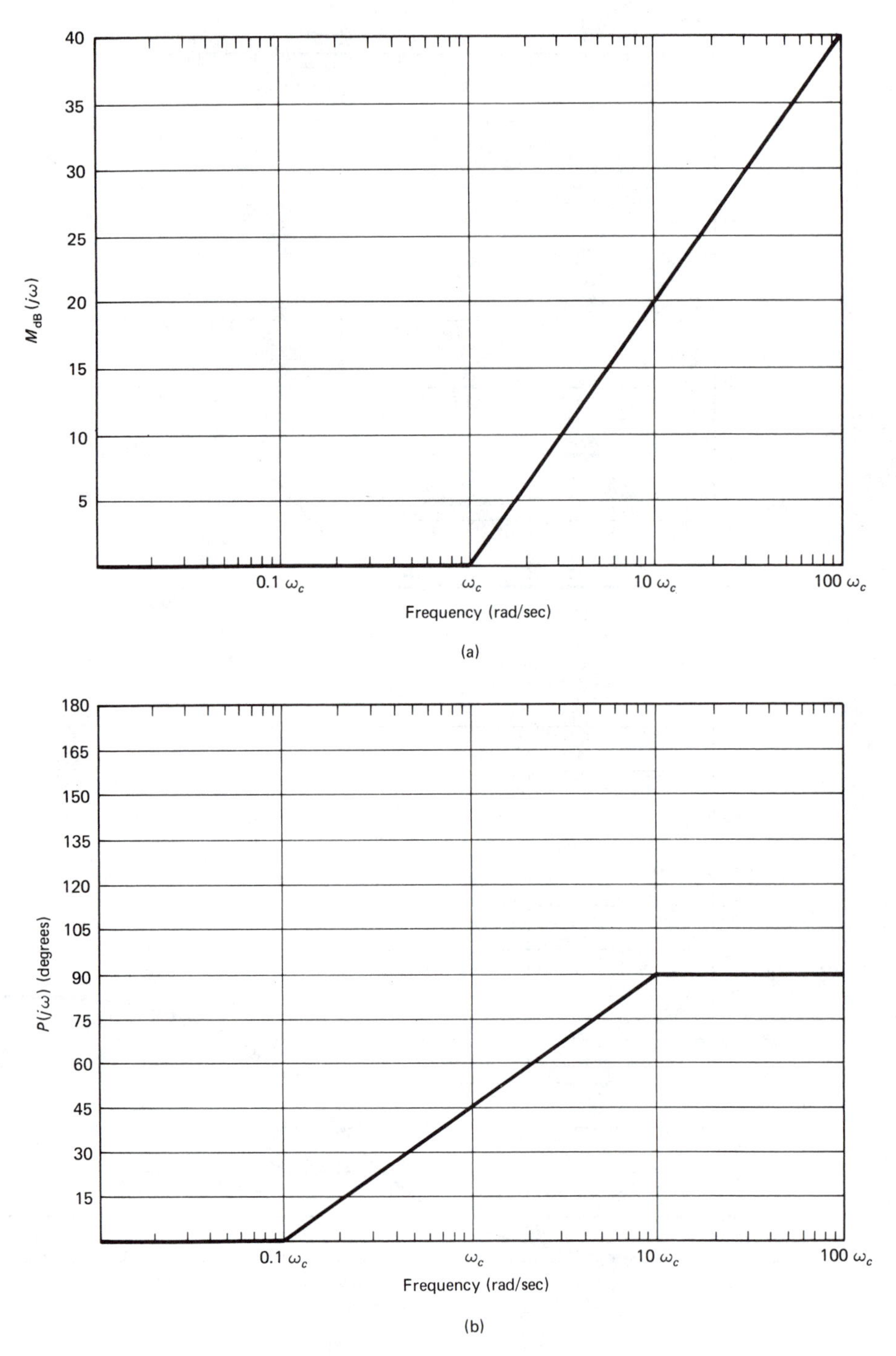
40
35
30
25
20
15
10
5
$M_{dB}(j\omega)$
0.1 ω_c
ω_c
10 ω_c
100 ω_c
Frequency (rad/sec)
(a)
180
165
150
135
120
105
90
75
60
45
30
15
$P(j\omega)$ (degrees)
0.1 ω_c
ω_c
10 ω_c
100 ω_c
Frequency (rad/sec)
(b)

Figure 5-19

and separating the magnitude and phase using rectangular-to-polar conversion yields

$$M(j\omega) = \sqrt{\left(\frac{\omega}{\omega_c}\right)^2 + 1^2} \tag{5-18a}$$

$$M_{dB} = 20 \log \sqrt{\left(\frac{\omega}{\omega_c}\right)^2 + 1^2} \tag{5-18b}$$

$$P(j\omega) = \tan^{-1}\left(\frac{\omega}{\omega_c}\right) \tag{5-19}$$

Notice that these are the same equations as real nonzero poles except that the minus sign is missing. The procedure for plotting the exact plot and the Bode plot is the same as the real nonzero poles. Figure 5-18 shows the exact plot, and Fig. 5-19 shows the Bode plot. Everything is exactly the same except that the slopes for the magnitude function and phase function are positive.

A real zero-value zero will be exactly like a zero-value pole (break frequency at 0 rad/sec) except that it will be positive. The exact plot and Bode plot for a zero-value zero is the same graph. Following the same approach as for the zero-value pole, the equations are

$$M(j\omega) \angle P(j\omega) = s \big|_{s=j\omega}$$

$$M(j\omega) \angle P(j\omega) = j\omega$$

$$M(j\omega) = \omega \tag{5-20a}$$

$$M_{dB}(j\omega) = 20 \log \omega \tag{5-20b}$$

$$P(j\omega) = \tan^{-1}\left(\frac{\omega}{0}\right) = \tan^{-1}(\infty) = 90^\circ \tag{5-21}$$

Notice that these are the same equations as the real zero-value pole except that these equations are positive. The exact and Bode plot are the same graphs and are shown in Fig. 5-20.

5-3-5 Complex Poles

A complex pole, with $r = 1$, in the transfer function will appear as

$$\frac{1}{s^2 + 2\zeta\omega_c s + \omega_c^2}$$

This function must have $\zeta < 1$ since $\zeta \geq 1$ indicates that the quadratic equation can be factored into real values. From here on we will assume that ζ is less

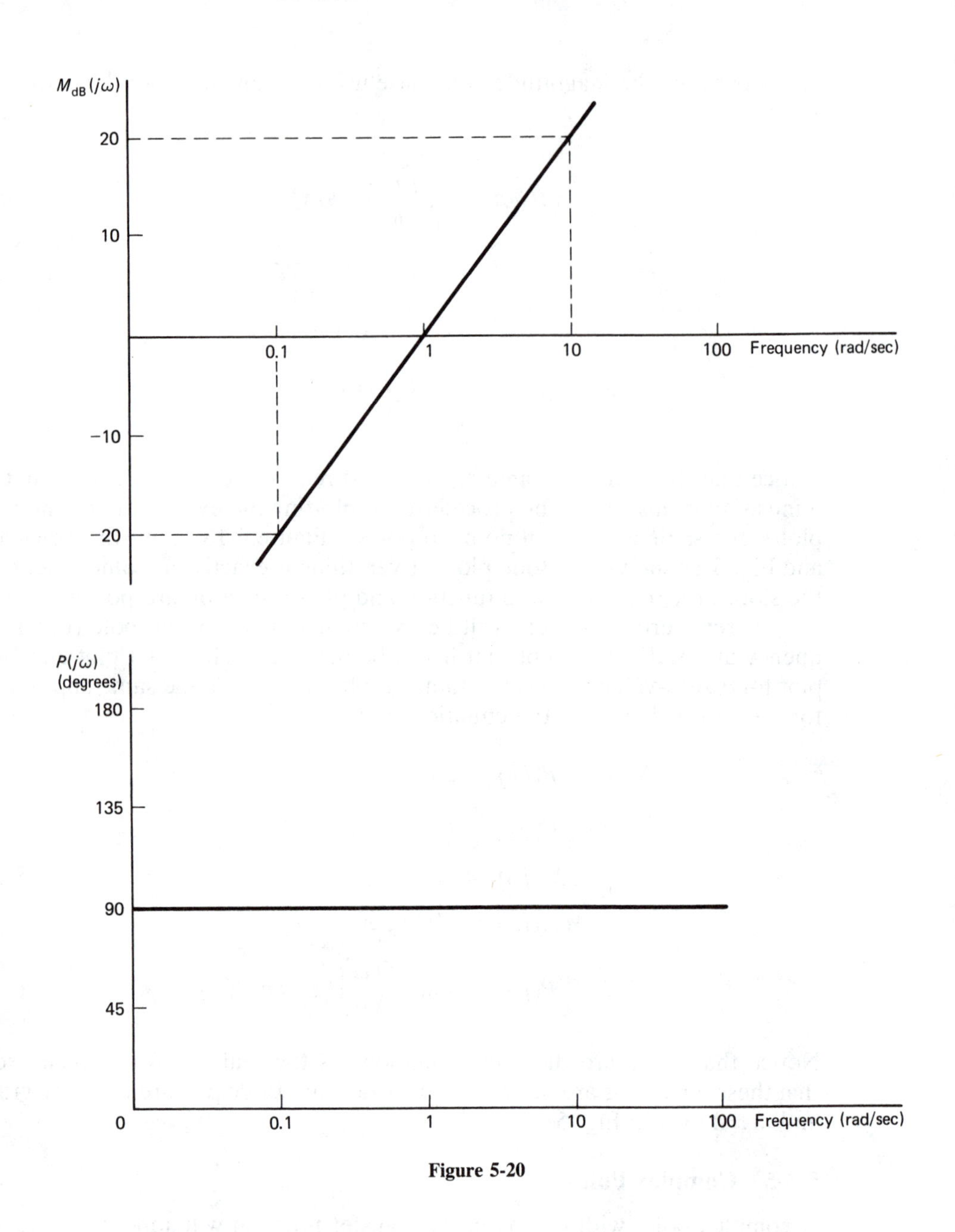

Figure 5-20

than 1. Arranging the function in the correct form for frequency analysis gives

$$\frac{1}{\omega_c^2\left(\dfrac{s^2}{\omega_c^2} + \dfrac{2\zeta}{\omega_c}s + 1\right)}$$

The ω_c^2 factor will become part of the gain constant. Therefore, we will work with

$$\frac{1}{\dfrac{s^2}{\omega_c^2} + \dfrac{2\zeta}{\omega_c} s + 1}$$

Substituting in $j\omega$ for s, we can find the magnitude function and the phase function.

$$M(j\omega) \angle P(j\omega) = \frac{1}{\dfrac{(j\omega)^2}{\omega_c^2} + \dfrac{2\zeta}{\omega_c} j\omega + 1} \tag{5-22}$$

$$M(j\omega) = \left\{ \sqrt{\left(\frac{2\zeta\omega}{\omega_c}\right)^2 + \left[1 - \left(\frac{\omega}{\omega_c}\right)^2\right]^2} \right\}^{-1} \tag{5-23a}$$

$$M_{\text{dB}}(j\omega) = -20 \log \sqrt{\left(\frac{2\zeta\omega}{\omega_c}\right)^2 + \left[1 - \left(\frac{\omega}{\omega_c}\right)^2\right]^2} \tag{5-23b}$$

$$P(j\omega) = -\tan^{-1}\left[\frac{\dfrac{2\zeta\omega}{\omega_c}}{1 - \left(\dfrac{\omega}{\omega_c}\right)^2}\right] \tag{5-24}$$

The magnitude and phase are graphed in Fig. 5-21. When ω is larger than ω_c, Eq. (5-23b) approaches $20 \log (\omega/\omega_c)^2$ and will asymptotically approach -40 dB/decade. When ω is less than ω_c, we will have approximately 0 dB, as shown in Fig. 5-21a. The phase function shown in Fig. 5-21b is similar to real poles except that it goes through 180° phase change.

Figure 5-21 shows a graph where $\zeta = 0.70711$ (exact value is $\sqrt{2}/2$). We can see from Eqs. (5-23) and (5-24) that the function is different for different values of ζ. Figure 5-22 shows the magnitude function and phase function for different values of ζ.

By observation, the magnitude function has a peak for $\zeta > 0.70711$, and the frequency where this peak occurs is slightly less than ω_c. Even though the peak is lower than ω_c, the asymptote will start at ω_c.

We can find at what frequency, ω_{pk}, this peak occurs by setting the derivative of Eq. (5-23a) equal to zero. (Actually, we take the derivative of the function under the radical since when this is at maximum, the square root will also be at maximum.) The result is

$$\omega_{\text{pk}} = \omega_c \sqrt{1 - 2\zeta^2} \tag{5-25}$$

This will have real values only for ζ less than 0.70711, since for values greater than or equal to 0.70711 there is no peak.

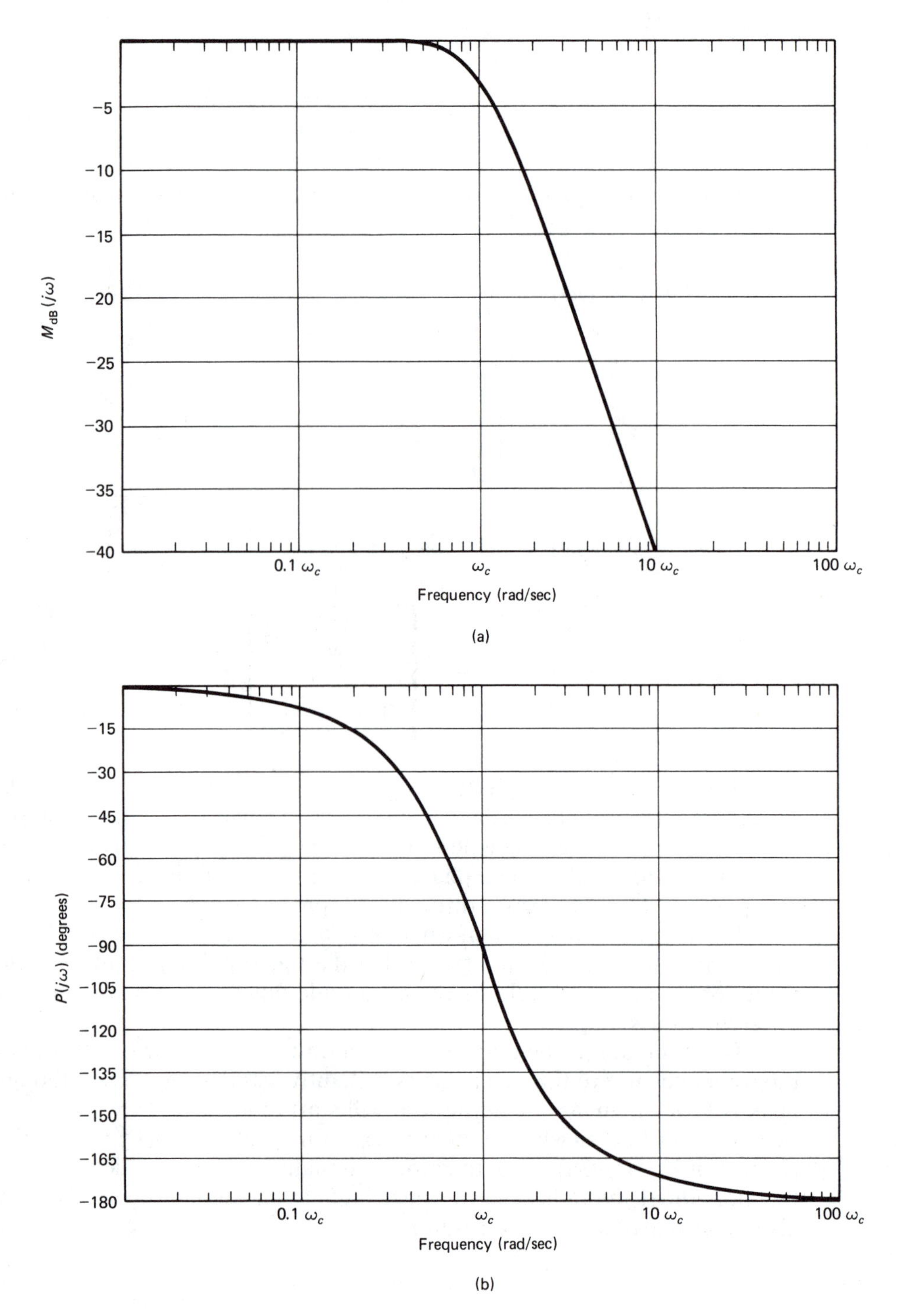

−5
−10
−15
−20
−25
−30
−35
−40
$M_{dB}(j\omega)$
0.1 ω_c
ω_c
10 ω_c
100 ω_c
Frequency (rad/sec)
(a)
−15
−30
−45
−60
−75
−90
−105
−120
−135
−150
−165
−180
$P(j\omega)$ (degrees)
0.1 ω_c
ω_c
10 ω_c
100 ω_c
Frequency (rad/sec)
(b)

Figure 5-21

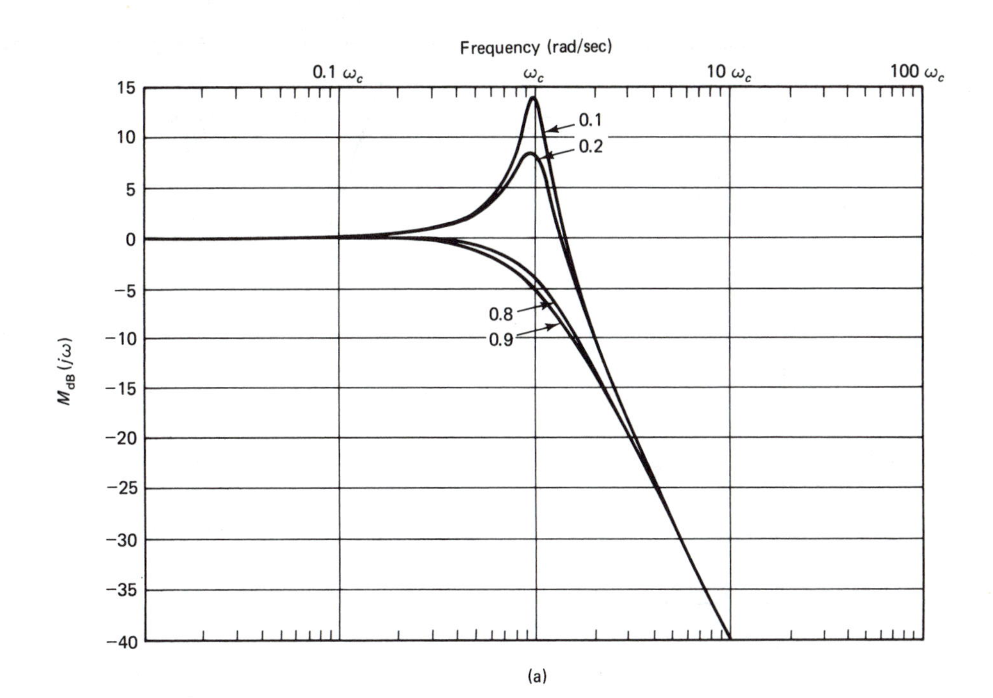

(a)

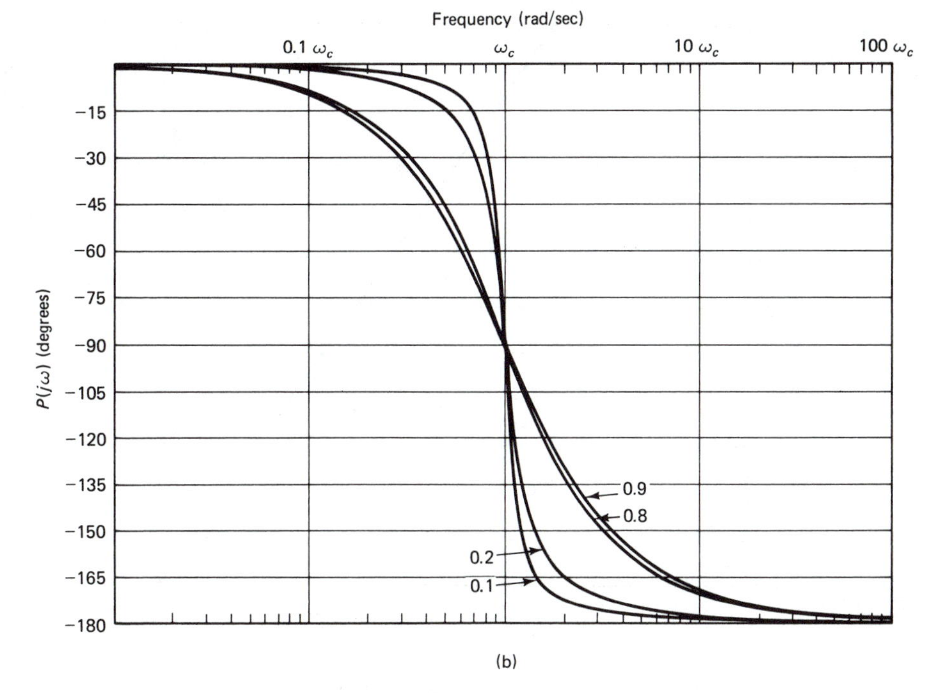

(b)

Figure 5-22

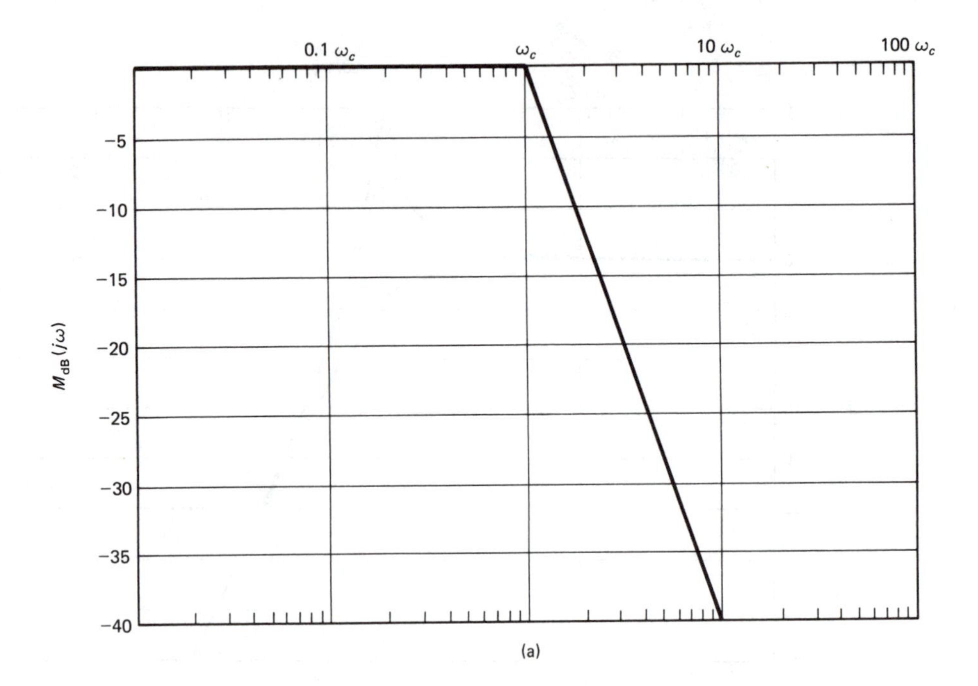

(a)

Frequency (rad/sec)

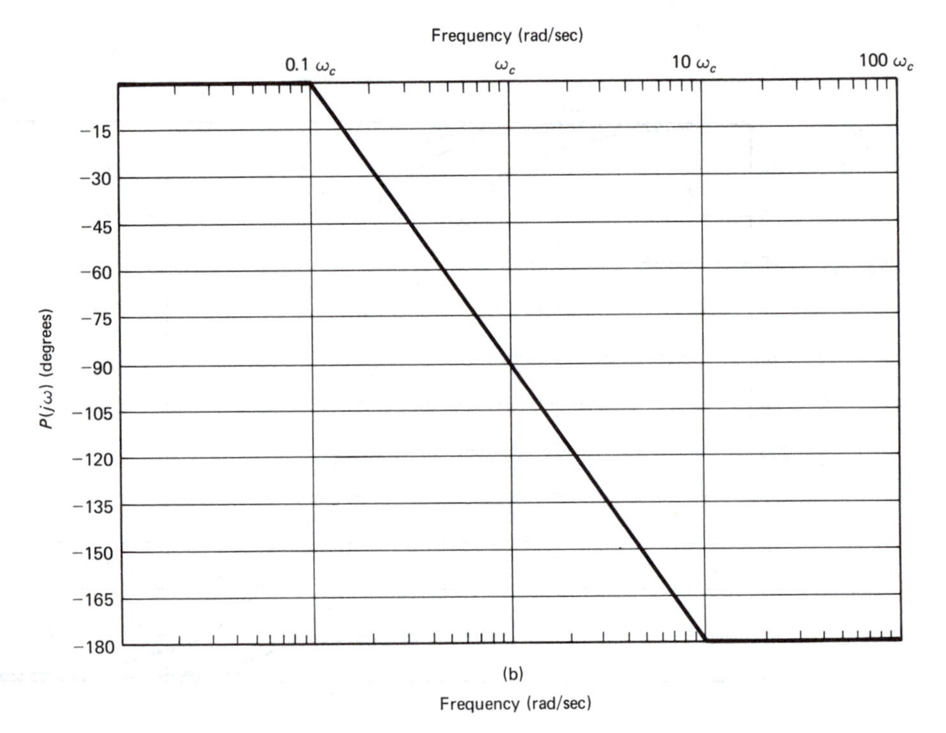

(b)

Frequency (rad/sec)

Figure 5-23

By substituting the value of ω_{pk} from Eq. (5-25) into Eq. (5-23b), the peak value of the magnitude function at this frequency is

$$\Delta M_{dB} = -20 \log 2\zeta\sqrt{1 - \zeta^2} \tag{5-26}$$

This equation does not really show the peak value, but it shows the difference between the magnitude value below ω_c and the peak.

Even though the complex poles can give an infinite peak, the Bode plot is based on $\zeta = 0.70711$, as shown in Fig. 5-21. We straighten the curve so that it is 0 dB up to and including ω_c and then follows the asymptote. Figure 5-23 shows the Bode plot. If we need to know what this peak value is, we use Eq. (5-26). If we need to know at what frequency the peak occurs, we use Eq. (5-25). However, when this accuracy is required, we need to use a computer and plot the exact functions.

5-3-6 Complex Zeros

We will find the same parallels between complex poles and complex zeros that we did between real poles and real zeros. A complex zero in the proper form is

$$\frac{s^2}{\omega_c^2} + \frac{2\zeta}{\omega_c} s + 1$$

This is the reciprocal of a complex pole, and the procedure to find the magnitude function and phase function is exactly the same.

$$M(j\omega) \angle P(j\omega) = \frac{(j\omega)^2}{\omega_c^2} + \frac{2\zeta}{\omega_c} j\omega + 1$$

$$M(j\omega) = \sqrt{\left(\frac{2\zeta\omega}{\omega_c}\right)^2 + \left[1 - \left(\frac{\omega}{\omega_c}\right)^2\right]^2} \tag{5-27a}$$

$$M_{dB}(j\omega) = 20 \log \sqrt{\left(\frac{2\zeta\omega}{\omega_c}\right)^2 + \left[1 - \left(\frac{\omega}{\omega_c}\right)^2\right]^2} \tag{5-27b}$$

$$P(j\omega) = \tan^{-1}\left[\frac{\dfrac{2\zeta\omega}{\omega_c}}{1 - \left(\dfrac{\omega}{\omega_c}\right)^2}\right] \tag{5-28}$$

Notice that Eqs. (5-27b) and (5-28) are the negative of Eqs. (5-23b) and (5-24).

Figure 5-24 shows the exact plot using different values of ζ. Notice that this is upside down of Fig. 5-22. Since these are so similar, it stands to reason that the peak occurs at the same frequency as the poles.

$$\omega_{pk} = \omega_c\sqrt{1 - 2\zeta^2} \tag{5-29}$$

As with the complex poles, this has real values only for $\zeta < 0.70711$.

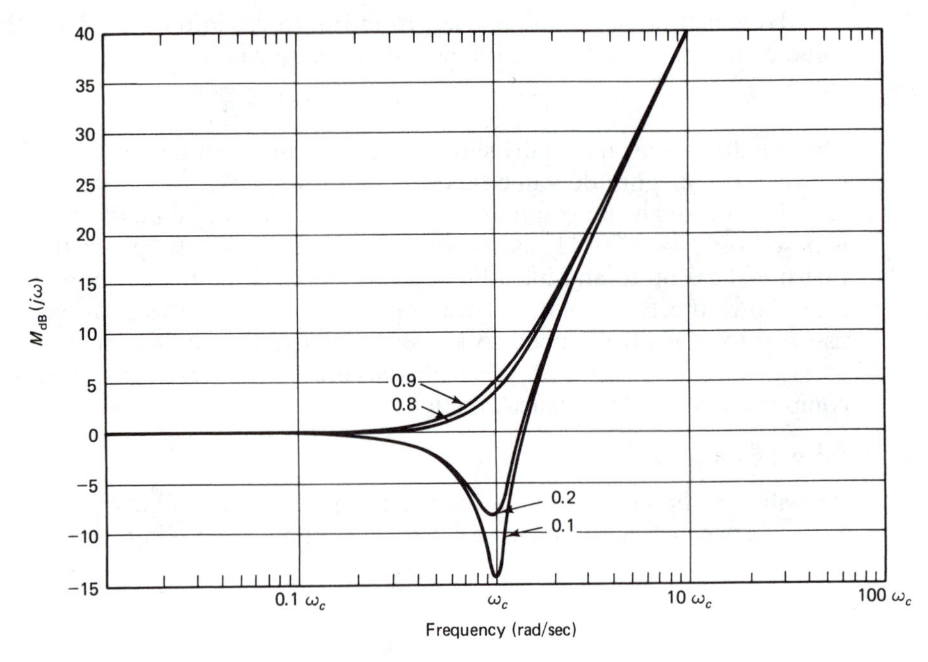

(a)

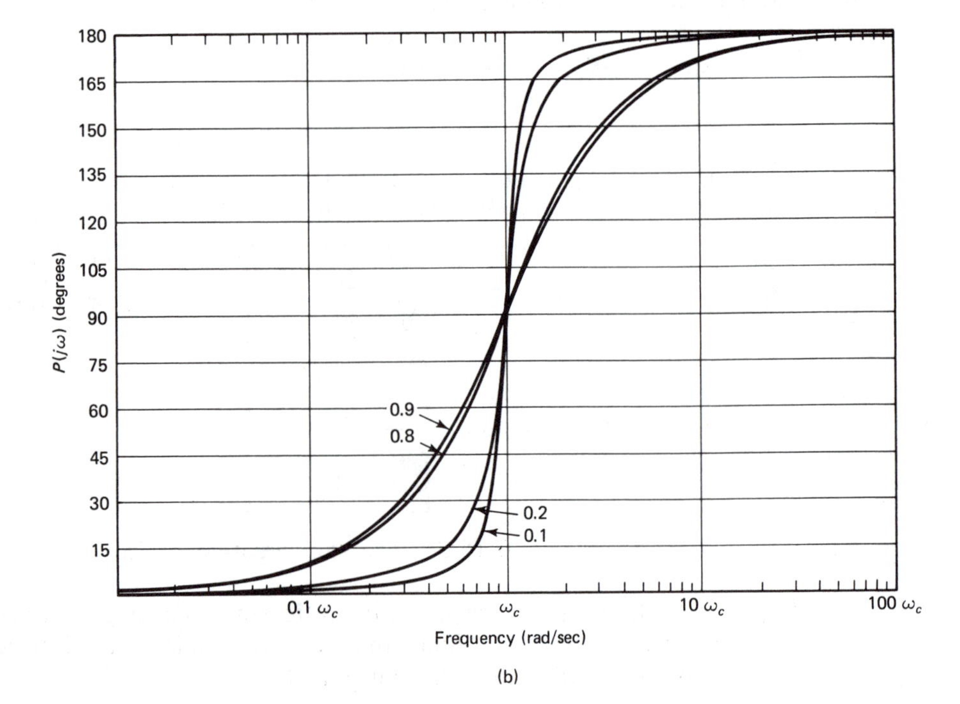

(b)

Figure 5-24

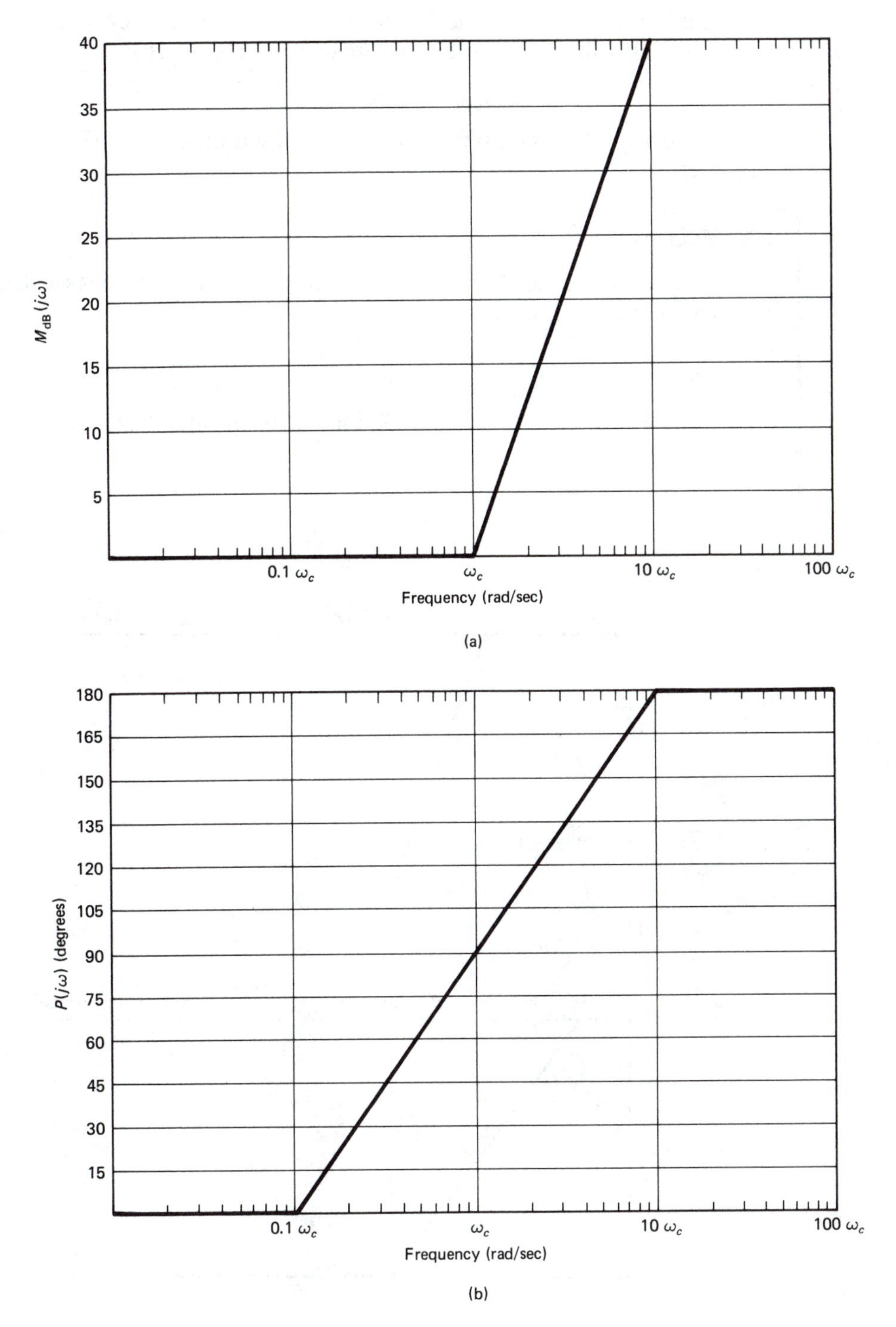

40
35
30
25
20
15
10
5
$M_{dB}(j\omega)$
0.1 ω_c
ω_c
10 ω_c
100 ω_c
Frequency (rad/sec)
(a)
180
165
150
135
120
105
90
75
60
45
30
15
$P(j\omega)$ (degrees)
0.1 ω_c
ω_c
10 ω_c
100 ω_c
Frequency (rad/sec)
(b)

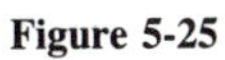
Figure 5-25

The difference between the function's value at a frequency much less than ω_c and the peak value is given by the opposite of Eq. (5-26).

$$\Delta M_{\text{dB}} = 20 \log 2\zeta\sqrt{1 - \zeta^2} \tag{5-30}$$

The Bode plot for complex zeros is also based on $\zeta = 0.70711$, as shown in Fig. 5-25.

EXAMPLE 5-7

Draw the Bode plot for the transfer function. What is the peak value, and at what frequency does it occur?

$$\frac{25}{s^2 + 4s + 25}$$

SOLUTION First we put the function in the proper form for frequency analysis

$$\frac{1}{\dfrac{s^2}{25} + \dfrac{4}{25}s + 1}$$

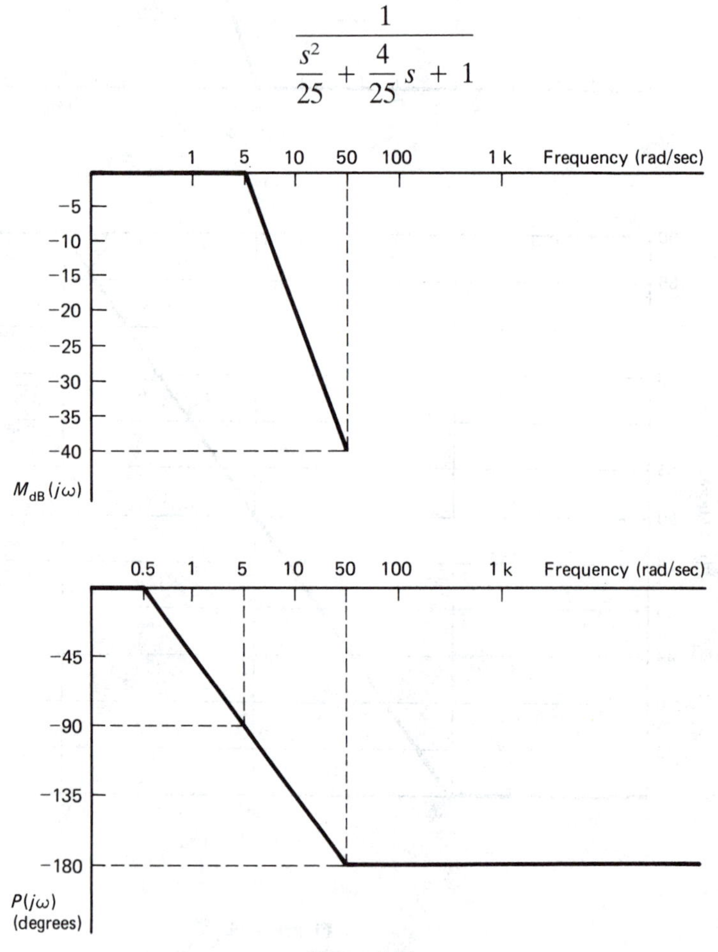

Figure 5-26

From this we see that ω_c^2 is 25 and ω_c is 5. The Bode plot is shown in Fig. 5-26.

Using the center coefficient, we see that

$$\frac{2\zeta}{\omega_c} = \frac{4}{25}$$

Therefore,

$$\zeta = \frac{4\omega_c}{2(25)} = \frac{4(5)}{2(25)} = 0.4$$

Using Eq. (5-26), we obtain

$$\begin{aligned}\Delta M_{\text{dB}} &= -20 \log 2\zeta \sqrt{1 - \zeta^2} \\ &= -20 \log 2(0.4) \sqrt{1 - (0.4)^2} \\ &= 2.6954 \text{ dB}\end{aligned}$$

In this case, this is the peak value, since the function is 0 dB at frequencies much less than ω_c.

Using Eq. (5-25), the peak occurs at

$$\begin{aligned}\omega_{\text{pk}} &= \omega_c \sqrt{1 - 2\zeta^2} \\ &= 5 \sqrt{1 - 2(0.4)^2} \\ &= 4.1231 \text{ rad/sec}\end{aligned}$$

5-3-7 Combining Poles and Zeros

When the transfer function is in the form described in Sec. 5-3-1, each pole and zero below its break frequency contributes 0 dB. We may therefore ignore the effect of a pole or zero until we reach its break frequency. This allows us to use a technique similar to the complex waveform analysis shown in Section 2-8, but instead of looking at a time shift, we look at the break frequency for when the pole or zero becomes "active."

The general steps to graphing the magnitude function and phase function are:

1. Put the transfer function in the proper form as described in Section 5-3-1.
2. Find the break frequencies of the poles and zeros.
3. Start at least one decade below the lowest break frequency.
4. Move up in frequency adding the effects of each break frequency until you are at least one decade above the highest break frequency. For multiple breaks occurring at the same frequency, the magnitude will actually be off by more than 3 dB, but we will ignore this since the Bode plot is an approximation. This will also occur with the phase angles, and we will ignore this error as well.

5. Find the dB value of the gain constant, K, and shift the graph by that amount.

EXAMPLE 5-8

Draw the magnitude function and the phase function for the transfer function shown.

$$\frac{170{,}000(s + 10)}{(s + 100)(s + 10000)}$$

SOLUTION First we put the transfer function in the proper form.

$$\frac{1.7\left(\dfrac{s}{10} + 1\right)}{\left(\dfrac{s}{100} + 1\right)\left(\dfrac{s}{10000} + 1\right)}$$

Next we list the break frequencies (the lowest break frequency first) and the gain constant in dB.

$$\omega_{C1Z} = 10 \text{ rad/sec}$$

$$\omega_{C2P} = 100 \text{ rad/sec}$$

$$\omega_{C3P} = 10000 \text{ rad/sec}$$

$$K_{\text{dB}} = 20 \log K$$

$$= 20 \log (1.7) = 4.6090 \text{ dB}$$

Notice the subscripting of the break frequencies (the critical frequencies). The subscript C1Z means that the first critical frequency is a zero, and the subscript C2P means that the second critical frequency is a pole.

Figure 5-27 shows the step-by-step procedure for drawing the magnitude function and the phase function. In Fig. 5-27a and b we draw the first and second break frequencies separately. Figure 5-27c shows the first two break frequencies added together. Figure 5-27d is the graph of the third break frequency by itself. Figure 5-27e is the third break frequency added to Fig. 5-27c. Finally, we shift the function by the gain constant K, which will not affect the phase function. The completed magnitude function and phase function are shown in Fig. 5-27f.

When we see how the slopes add and subtract similar to the ramp functions we did in Section 2-8, we can simplify the process used in Example 5-8. In the example we could have drawn only Fig. 5-27e first and then Fig. 5-27f.

One of the more difficult drawings is the phase function, where two break frequencies are close but not equal. When the break frequencies are exactly equal, they both start at the same point and end at the same point, with a constant slope. When close but unequal break frequencies occur, the slope

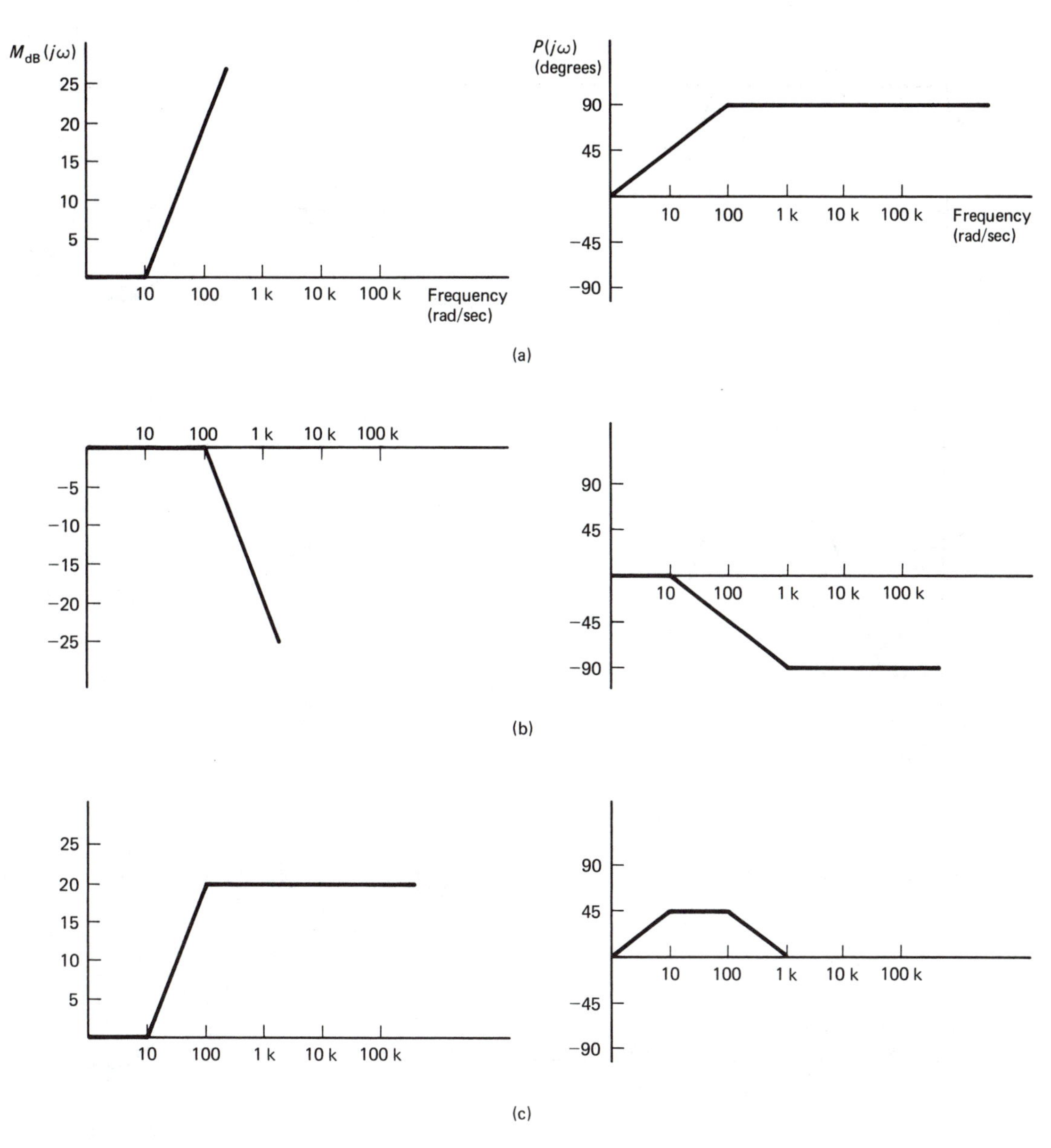

Figure 5-27

starts at the lowest break frequency's slope, changes to the combined slope, and then ends with the slope of the highest break frequency.

Figures 5-28a and b show two phase shifts occurring close to the same break frequency. The lower break frequency shown in Fig. 5-28a has a total phase angle change of P_1 degrees and the slope of the change is S_1 degrees/decade. The higher break frequency is shown in Fig. 5-28b, and it has a total phase angle change of P_2 degrees with a slope of S_2 degrees/decade.

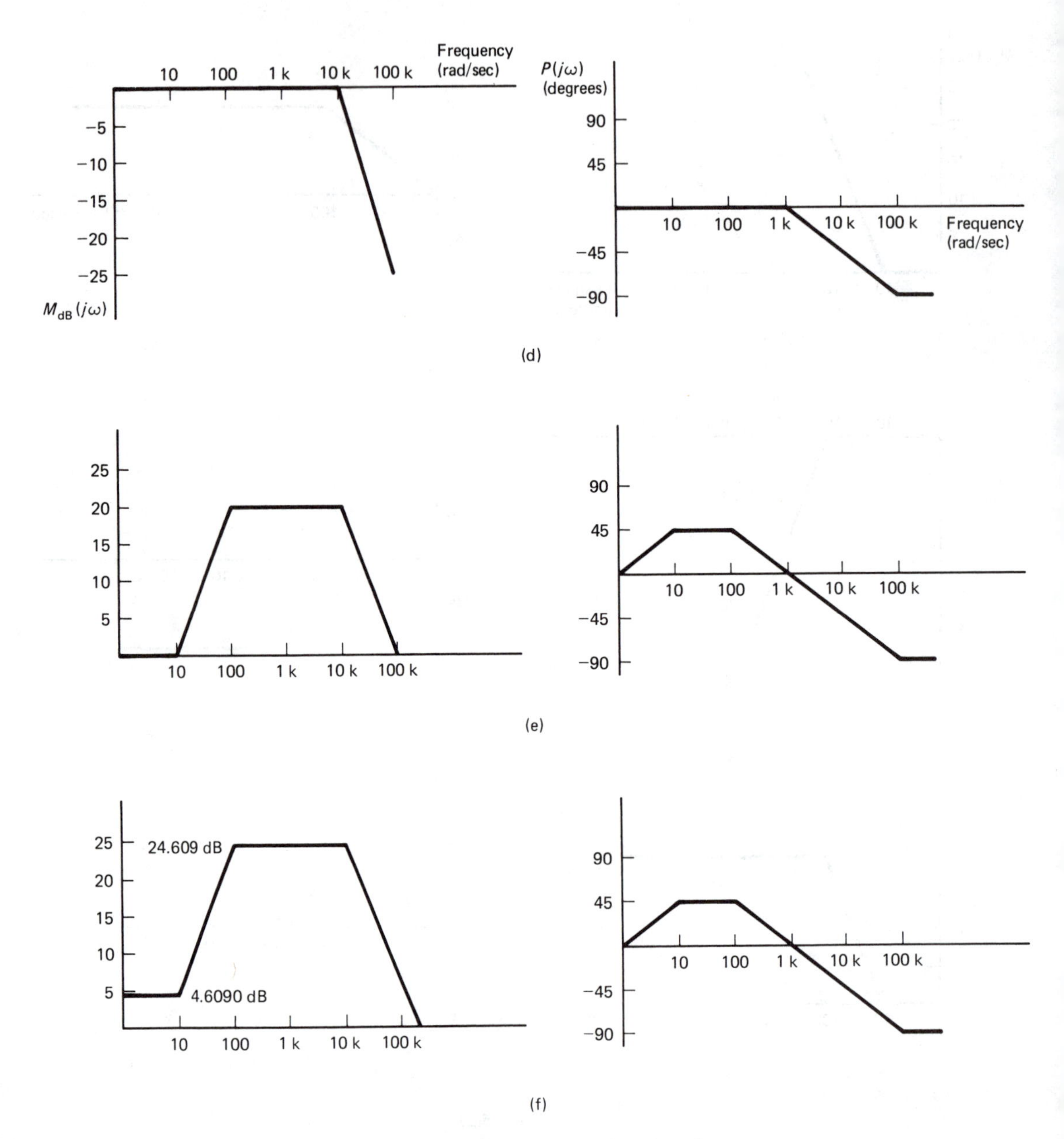

Figure 5-27 (Continued)

Figure 5-28c shows in solid line the part of the lowest break frequency's phase shift that is unaffected, and the dashed line shows how it would have continued if the second phase shift was not present. Since we know that the final phase shift will be $P_1 + P_2$, we draw the second phase shift from the endpoint we know, backward in frequency, to the beginning of this second

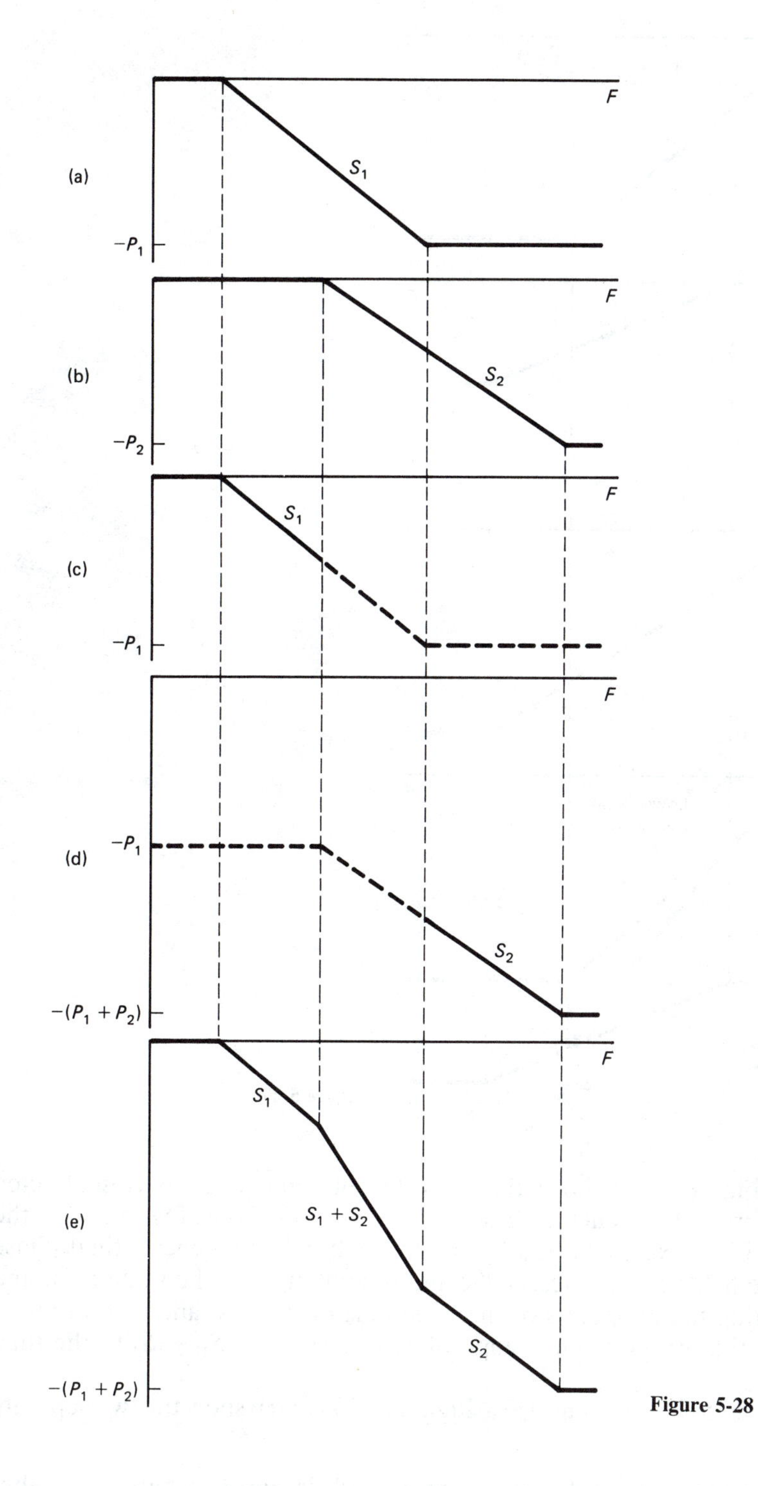

Figure 5-28

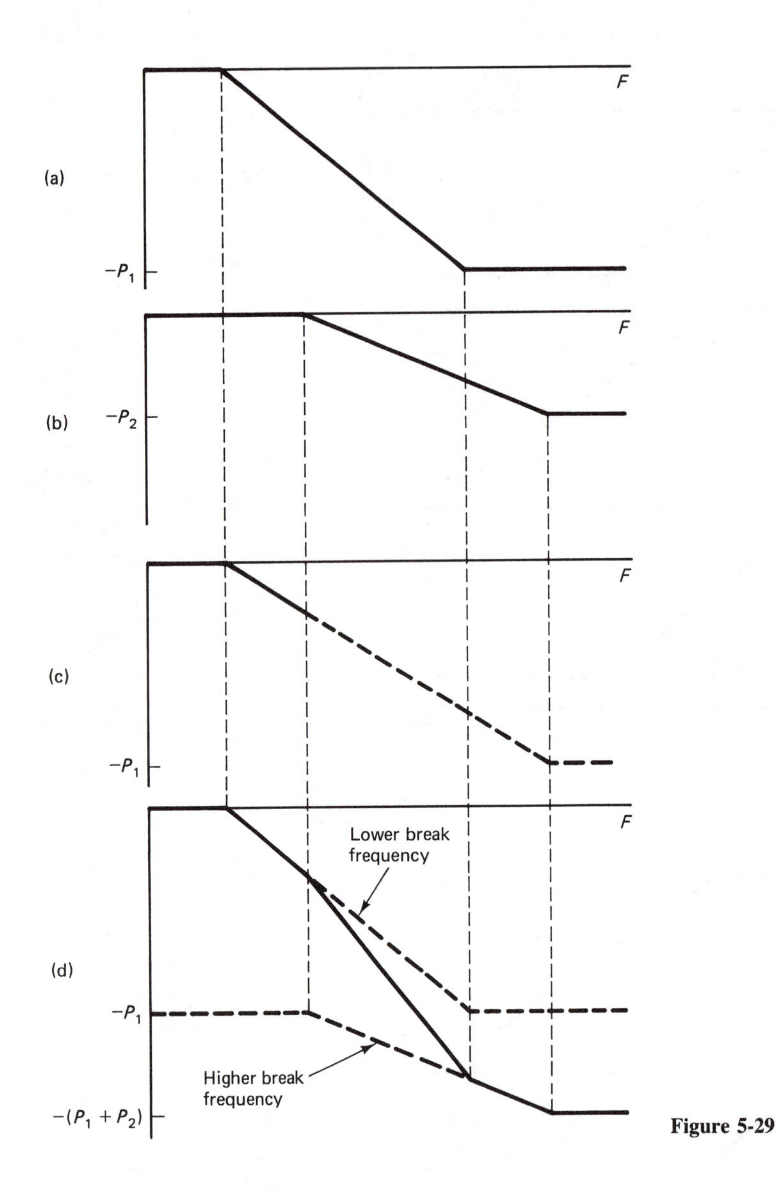

Figure 5-29

phase shift. Figure 5-28d shows this with the solid line as the part unaffected by the lower break frequency's phase shift. We notice that Fig. 5-28d is the same as Fig. 5-28b except shifted by the lower break frequency's final phase value. Figure 5-28e shows Figs. 5-28c and d combined on the same axis, and a line connecting the endpoints of the solid part of the first and second phase shift. Notice that we go from a slope of S_1 to a slope of $S_1 + S_2$ to the final slope of S_2.

We use Fig. 5-29 to summarize. Figures 5-29a and b show the two separate phase shifts.

1. Draw the lower break frequency in a solid line up to where the higher

break frequency starts affecting the lower break frequency, and then draw a dashed line the rest of the way. This is shown in Fig. 5-29c.

2. We then draw the second phase shift on Fig. 5-29c, which is shown in Fig. 5-29d. Assume that the higher break frequency is shifted by the amount of the final value of the lower break frequency. Draw the shifted higher break frequency with a solid line where the lower break frequency does not have a slope and a dashed line where the lower break frequency does have a slope.
3. Connect the two solid line endpoints with a straight line.

EXAMPLE 5-9

Draw the Bode plot for the transfer function shown. Calculate the approximate crossing of the 0-dB line.

$$\frac{-7.5 \times 10^6 s}{(s + 15)(s^2 + 400s + 1 \times 10^6)}$$

SOLUTION In the correct form for frequency analysis,

$$\frac{-0.5s}{\left(\dfrac{s}{15} + 1\right)\left[\dfrac{s^2}{(1 \times 10^3)^2} + \dfrac{0.4}{1 \times 10^3} s + 1\right]}$$

Listing the break frequencies and the gain constant in dB, we have

$$\omega_{C1Z} = 0 \text{ rad/sec}$$

$$\omega_{C2P} = 15 \text{ rad/sec}$$

$$\omega_{C3P} = 1 \times 10^3 \text{ rad/sec}$$

$$K_{\text{dB}} = 20 \log K$$

$$= 20 \log 0.5 = -6.0206 \text{ dB}$$

Notice that the gain is negative because the magnitude of the gain constant is less than 1—it is not caused by the "−" in front of the gain constant. The negative sign in front of the gain constant means that we have a $-180°$ constant phase angle shift. We also have a 90° constant phase shift due to the zero at a value of zero.

Figure 5-30a shows the magnitude function drawn from the break frequencies. The lowest is the break frequency from the zero, ω_{C1Z}. We know that this will pass through 1 rad/sec, so we can start there. This increases in magnitude by 20 dB/dec until we reach 15 rad/sec. The number of decades between these two points, using Eq. (D-2) in Appendix D, is

$$\text{ND} = \log\left(\frac{f_2}{f_1}\right) = \log\left(\frac{15}{1}\right) = 1.1761 \text{ decades}$$

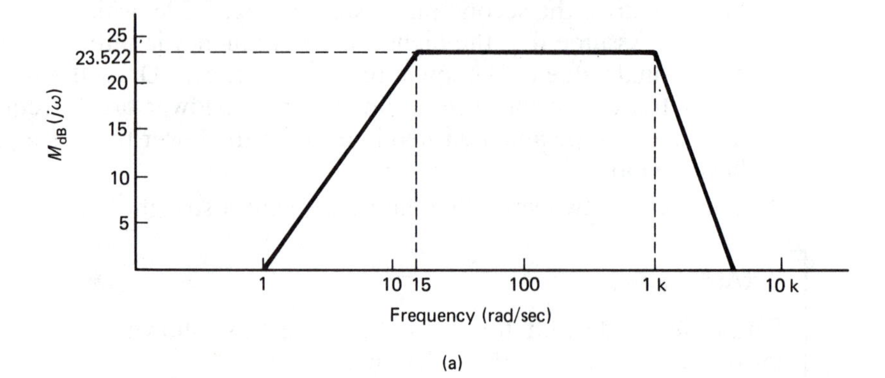

(a)

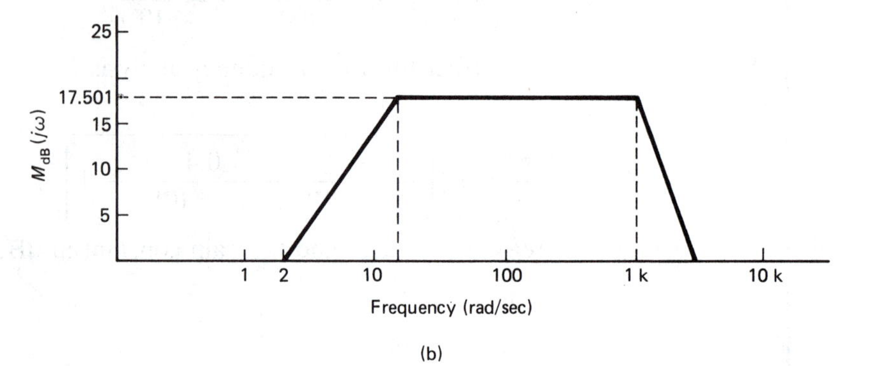

(b)

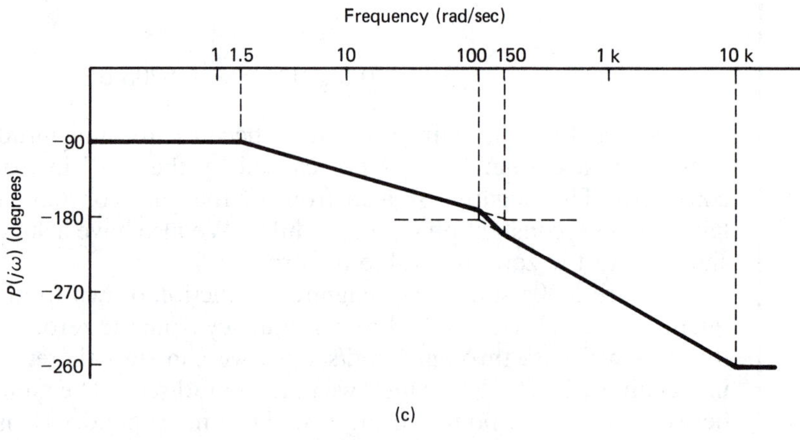

(c)

Figure 5-30

The change in magnitude between these two points using Eq. (D-5a) is then

$$\text{RR} = \frac{\Delta A}{\text{ND}}$$

$$\Delta A = (\text{ND})(\text{RR})$$

$$= (1.1761 \text{ decades})(20 \text{ dB/dec}) = 23.522 \text{ dB}$$

We can calculate the value this obtains at the point where ω_{C2P} starts by using Eq. (D-4).

$$\Delta A_{\text{dB}} = A_{2\text{dB}} - A_{1\text{dB}}$$

$$A_{2\text{dB}} = \Delta A_{\text{dB}} + A_{1\text{db}}$$

$$= 23.522 + 0 = 23.522 \text{ dB}$$

This second break has the same but opposite slope, and the result is therefore a constant value. The quadratic becomes active at 1×10^3 rad/sec and causes a roll-off of -40 dB/dec. Finally, we add in the gain constant, which subtracts 6.0206 dB from each value as shown in Fig. 5-30b.

Next we find the low-end x-axis crossing. Since at 15 rad/sec we are at 17.501 dB (23.522 − 6.0206) and at the x-axis crossing we are at 0 dB, we can use Eq. (D-4) to find the change in the magnitude.

$$\Delta A_{\text{dB}} = A_{2\text{dB}} - A_{1\text{dB}}$$

$$= 17.501 - 0$$

$$= 17.501 \text{ dB}$$

The slope to the left is 20 dB/dec, and the number of decades between the two frequencies can be found using Eq. (D-5a).

$$\text{RR} = \frac{\Delta A}{\text{ND}}$$

Solving for the number of decades gives

$$\text{ND} = \frac{\Delta A}{\text{RR}}$$

$$= \frac{17.501 \text{ dB}}{20 \text{ dB/dec}}$$

$$= 0.87506 \text{ decade}$$

Next we find this lower frequency by solving Eq. (D-2) for f_1.

$$\text{ND} = \log\frac{f_2}{f_1} \Rightarrow f_1 = \frac{f_2}{10^{\text{ND}}}$$

$$f_1 = \frac{15}{10^{0.87506}} = 2 \text{ rad/sec}$$

To find the x-axis crossing for the high end, we follow the same procedure except that the roll-off rate is -40 dB/dec. The change in the dB value using Eq. (D-4) is

$$\Delta A_{\text{dB}} = A_{2\text{dB}} - A_{1\text{dB}}$$
$$= 0 - 17.501$$
$$= -17.501 \text{ dB}$$

Next we find the number of decades using Eq. (D-5a).

$$\text{RR} = \frac{\Delta A}{\text{ND}}$$

$$\text{ND} = \frac{\Delta A}{\text{RR}}$$
$$= \frac{-17.501 \text{ dB}}{-40 \text{ dB/dec}}$$
$$= 0.43753 \text{ decade}$$

Solving for the high frequency f_2 in Eq. (D-2) yields

$$\text{ND} = \log\left(\frac{f_2}{f_1}\right) \Rightarrow f_2 = 10^{\text{ND}} f_1$$

$$f_2 = 10^{0.43753}(1 \times 10^3) = 2.7386 \text{ k rad/sec}$$

The phase function shown in Fig. 5-30c starts at $-90°$ due to the combined effect of the zero and the negative gain constant. A decade below 15 rad/sec, a slope of 45° per decade starts and goes to a decade above 15 rad/sec, but the last break frequency starts affecting this before it is completed. The construction lines are the dashed lines in Fig. 5-30c.

EXAMPLE 5-10

For the transfer function shown, draw the Bode plot.

$$\frac{2 \times 10^5(s + 1)}{s(s + 100)(s + 200)}$$

SOLUTION The correct form is

$$\frac{10\left(\frac{s}{1} + 1\right)}{s\left(\frac{s}{100} + 1\right)\left(\frac{s}{200} + 1\right)}$$

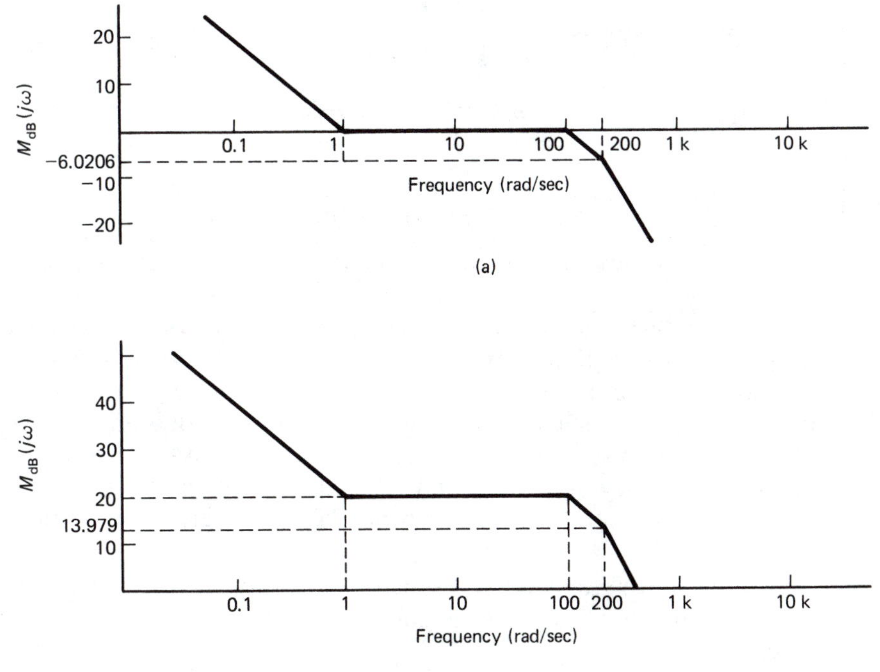

(b)

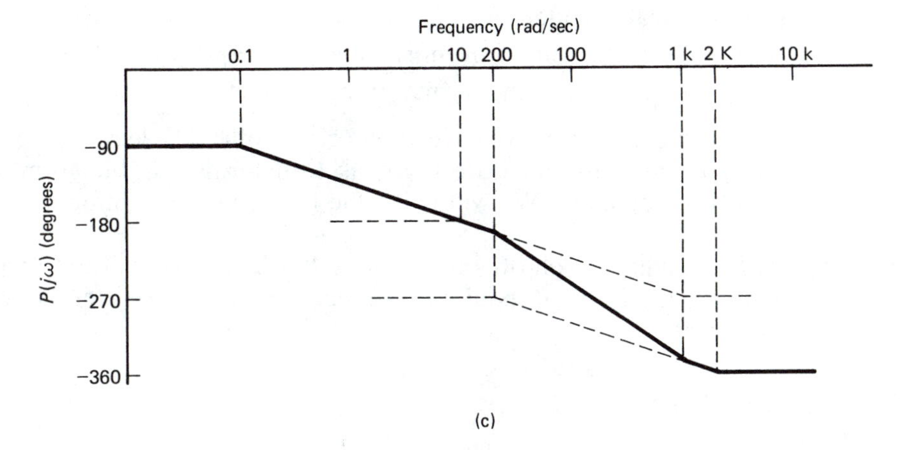

(c)

Figure 5-31

Listing the break frequencies and the dB gain, we have

$$\omega_{C1P} = 0 \text{ rad/sec}$$

$$\omega_{C2Z} = 1 \text{ rad/sec}$$

$$\omega_{C3P} = 100 \text{ rad/sec}$$

$$\omega_{C4P} = 200 \text{ rad/sec}$$

$$K_{\text{dB}} = 20 \log K$$

$$= 20 \log 10 = 20 \text{ dB}$$

Figure 5-31a shows the magnitude function drawn without the gain, and Fig. 5-31b shows the complete magnitude function.

When two real poles or zeros occur close to or at the same break frequency, they will add some more inaccuracy to the Bode plot. This is the case with the break at 100 rad/sec and 200 rad/sec. We do not try to include these effects in the drawing since the Bode plot is an approximation. However, we must remember this when we look at the drawing so that we know the accuracy of it. The phase function is shown in Fig. 5-31c. Notice the phase shift ω_{C3P} starts exactly when ω_{C2P} ends giving a continuous line.

After we have graphed a few functions, it becomes relatively easy to write the transfer function from the magnitude function. We simply reverse the process.

1. Shift the function to remove the gain factor. The amount of shift will be the gain constant in dB.
2. List the break frequencies.
3. Write the function in the form suitable for frequency analysis.
4. Write the function in the normal transfer function form.

One problem is where we have a roll-off rate of 40 dB/dec or more. From the exact plot we could measure the peak to determine ζ, but from a Bode plot this is not possible. We will make the following assumption.

RULE : When a roll-off rate exceeds 40 dB/sec, we will use quadratics with $\zeta = \sqrt{2}/2$ and, if required, one real pole or zero to meet the required roll-off rate.

EXAMPLE 5-11

Write the transfer function from the magnitude function shown in Fig. 5-32.

SOLUTION Using techniques shown in Appendix D, we find that the beginning slope of the function is 40 dB/dec. Since we have a starting slope of 40 dB/dec, we must have a double zero of $s = 0$. If there was no other

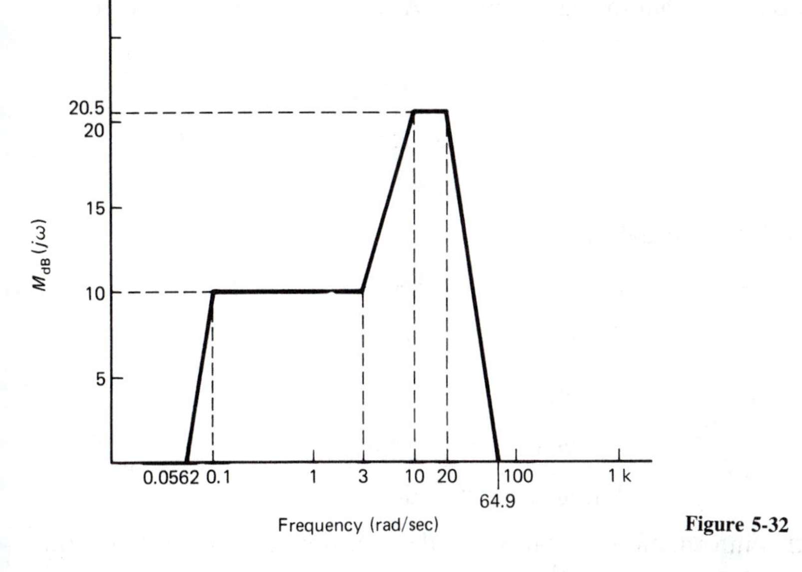

Figure 5-32

function present and no gain constant, the function would cross 0 dB at 1 rad/sec. To remove the gain constant we shift this graph so that the double zero by itself would cross 0 dB at 1 rad/sec. This is shown in Fig. 5-33. The dashed line shows this extension of the double zero to 1 rad/sec.

Since at 0.1 rad/sec there is a 50-dB difference between Figs. 5-32 and 5-33, each value in Fig. 5-32 has 50 dB subtracted from it to make Fig. 5-

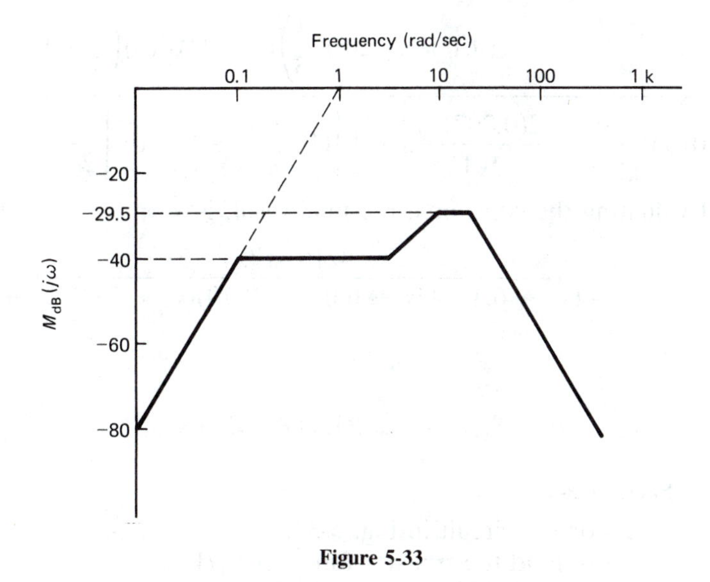

Figure 5-33

33. This also means that the gain constant K_{dB} is 50 dB. The gain constant is therefore

$$K_{dB} = 20 \log K$$

$$50 = 20 \log K$$

$$K = 316.23$$

Next we list the break frequencies.

$$2 \text{ at } \omega_{C1Z} = 0 \text{ rad/sec}$$

$$2 \text{ at } \omega_{C2P} = 0.1 \text{ rad/sec}$$

$$\omega_{C3Z} = 3$$

$$\omega_{C4P} = 10 \text{ rad/sec}$$

$$2 \text{ at } \omega_{C5P} = 20 \text{ rad/sec}$$

From this information we can write the transfer function in the form suitable for drawing the Bode plot from.

$$\frac{316.23\, s^2\left(\dfrac{s}{3} + 1\right)}{\left[\dfrac{s^2}{0.1^2} + \dfrac{2(0.70711)}{0.1}s + 1\right]\left(\dfrac{s}{10} + 1\right)\left[\dfrac{s^2}{20^2} + \dfrac{2(0.70711)}{20}s + 1\right]}$$

Next we put this in a proper transfer function form by making the coefficients of the highest power of s in each factor equal to 1.

$$\frac{(0.1^2)(10)(20^2)\left(\dfrac{1}{3}\right)(316.23)s^2(3)\left(\dfrac{s}{3} + 1\right)}{(0.1)^2\left[\dfrac{s^2}{0.1^2} + \dfrac{2(0.70711)}{0.1}s + 1\right](10)\left(\dfrac{s}{10} + 1\right)(20)^2\left[\dfrac{s^2}{20^2} + \dfrac{2(0.70711)}{20}s + 1\right]}$$

Evaluating the constants, the final result is

$$\frac{4216.4s^2(s + 3)}{(s^2 + 0.14142s + 0.01)(s + 10)(s^2 + 28.284s + 400)}$$

PROBLEMS

Section 5-1

1. For the circuit in Fig. 5-34:
   (a) Find the transfer function V_L/V_{in}.
   (b) Find the step and impulse response.

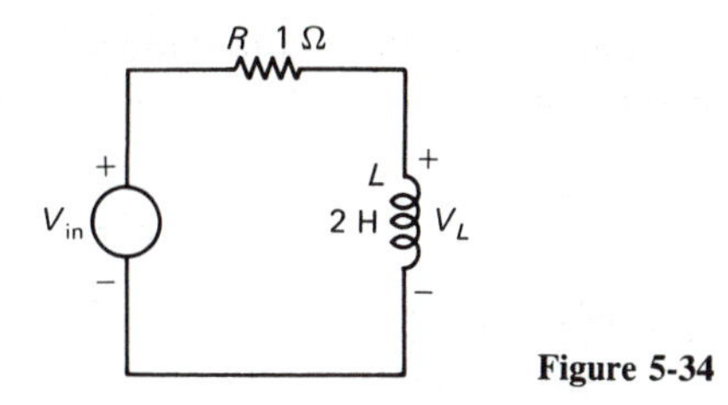

Figure 5-34

2. For the circuit in Fig. 5-35:
   (a) Find the transfer function I_L/V_{in}.
   (b) Using the transfer function, find $I_L(s)$ when $v_{\text{in}}(t) = 5e^{-2(t-1)}$ $u(t - 1)$.

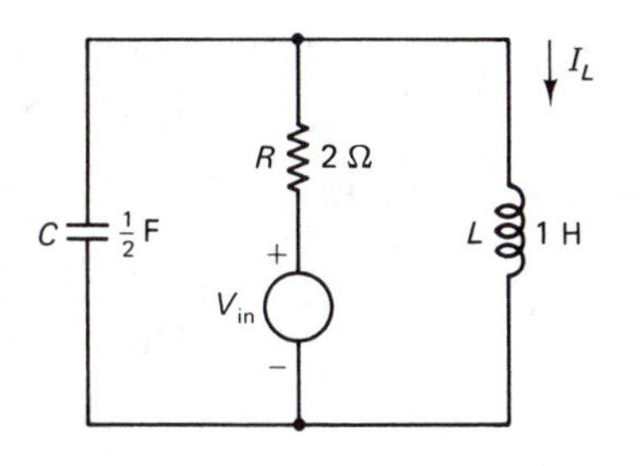

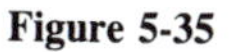

Figure 5-35

3. For the circuit in Fig. 5-36:
   (a) Find the transfer function V_L/V_{in}. (*Hint:* Find V_{in} in terms of I_1, and V_L in terms of I_1.)
   (b) Find the impulse response.

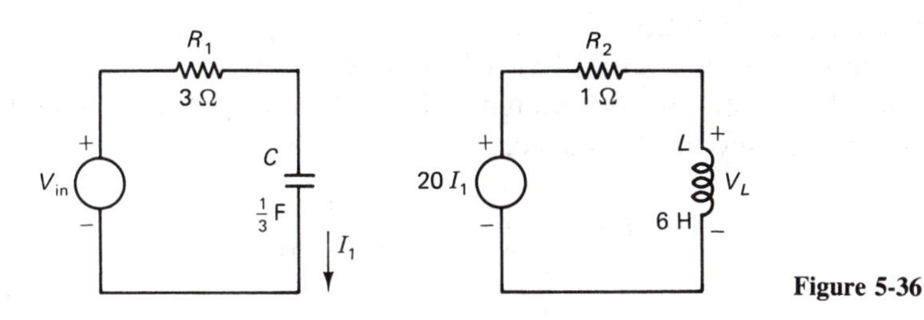

Figure 5-36

Section 5-2

4. For the transfer function shown:
   (a) Draw the pole–zero plot.
   (b) For the poles that are stable, find the transient and steady-state poles.
   (c) Is the system stable, unstable, or marginally stable?

(d) For the poles that are in the stable region, write them in order of their speed of decay, the fastest-decaying pole first.

$$G(s) = \frac{s + 2}{s(s + 3)[(s + 1)^2 + 36]}$$

5. Use the directions in Problem 4.

$$G(s) = \frac{s^2(s - 3)}{(s + 2)(s + 4)^2}$$

6. Use the directions in problem 4.

$$G(s) = \frac{(s - 2)^2 + 25}{(s + 1)(s - 3)(s + 4)^2}$$

7. Use the directions in Problem 4.

$$G(s) = \frac{(s - 3)(s + 4)}{(s^2 + 4)(s + 5)}$$

8. Use the directions in Problem 4.

$$G(s) = \frac{(s^2 + 9)[(s - 2)^2 + 1]}{s(s + 1)(s + 3)}$$

Section 5-3

Section 5-3-1

9. Write the transfer function in a form suitable for frequency analysis.

$$G(s) = \frac{5s^2(s + 10)}{(s + 2)[(s + 3)^2 + 9]}$$

10. For the transfer function shown:

(a) Write it in a form suitable for frequency analysis.

(b) Find the ac steady-state magnitude in dB and phase in degrees at a frequency of 10 rad/sec.

$$G(s) = \frac{15(s + 5)^2}{s^2(s^2 + 5s + 6)}$$

Section 5-3-2

11. Draw the magnitude function and phase function for:

(a) $K = 10$

(b) $K = -10$

12. Draw the magnitude function and phase function for:

(a) $K = 0.1$

(b) $K = -0.1$

Section 5-3-3

13. For the transfer function shown:
(a) Draw the Bode plot.
(b) Find the exact value of the magnitude in dB and phase in degrees at a frequency of 10 Hz.

$$G(s) = \frac{100}{s + 100}$$

14. Draw the Bode plot with the frequency scale in hertz.

$$G(s) = \frac{1}{s}$$

Section 5-3-4

15. Draw the Bode plot with the frequency scale in hertz.

$$G(s) = 0.1(s + 10)$$

16. For the transfer function shown:
(a) Draw the Bode plot.
(b) Calculate the exact magnitude in dB and phase in degrees at 12 rad/sec.

$$G(s) = s$$

Section 5-3-5

17. For the transfer function shown:
(a) Draw the Bode plot.
(b) Calculate ΔM_{dB} and the frequency of the peak.

$$G(s) = \frac{400}{s^2 + 16s + 400}$$

18. For the transfer function shown:
(a) Draw the Bode plot.
(b) Calculate ΔM_{dB} and the frequency of the peak.

$$G(s) = \frac{225}{s^2 + 225}$$

Section 5-3-6

19. For the transfer function shown:
(a) Draw the Bode plot.
(b) Calculate ΔM_{dB} and the frequency of the peak.

$$G(s) = 0.04(s^2 + 8s + 25)$$

20. For the transfer function shown:
(a) Draw the Bode plot.

(b) Calculate ΔM_{dB} and the frequency of the peak.

$$G(s) = 0.25(s^2 + 4)$$

Section 5-3-7

21. For the transfer function shown, draw the Bode plot when:
(a) $K = -10$
(b) $K = 0.1$

$$G(s) = Ks$$

22. Draw the Bode plot for the transfer functions and calculate the exact value of the magnitude in dB and phase in degrees at a frequency of 20 rad/sec.
(a) $\dfrac{120}{s + 150}$
(b) $12.5(s + 0.2)$

23. Draw the Bode plot for the transfer functions and calculate the magnitude at the peaking frequency. (This will be the maximum magnitude in one case and the minimum magnitude in the other case.)
(a) $\dfrac{5000}{s^2 + 4s + 1000}$
(b) $7.5 \times 10^{-8}(s^2 + 400s + 4 \times 10^6)$

24. Draw the Bode plot for the transfer function.

$$G(s) = \frac{-7.168 \times 10^7(s + 1)}{s(s + 800)^2}$$

25. With the frequency scale in hertz:
(a) Draw the Bode plot.
(b) Calculate the frequency of any x-axis crossing on the magnitude graph.
(c) Calculate the exact magnitude in dB and phase in degrees at 500 Hz.

$$G(s) = \frac{2 \times 10^4 s}{(s + 5)(s + 2000)}$$

26. For the transfer function:
(a) Draw the Bode plot.
(b) Find the exact value of magnitude in dB and phase in degrees at 0.3 rad/sec.

$$G(s) = \frac{(s + 0.3)^2}{s^2}$$

27. Draw the Bode plot for the transfer function.

$$G(s) = \frac{6.4 \times 10^7 s}{(s + 100)(s^2 + 800s + 6.4 \times 10^5)}$$

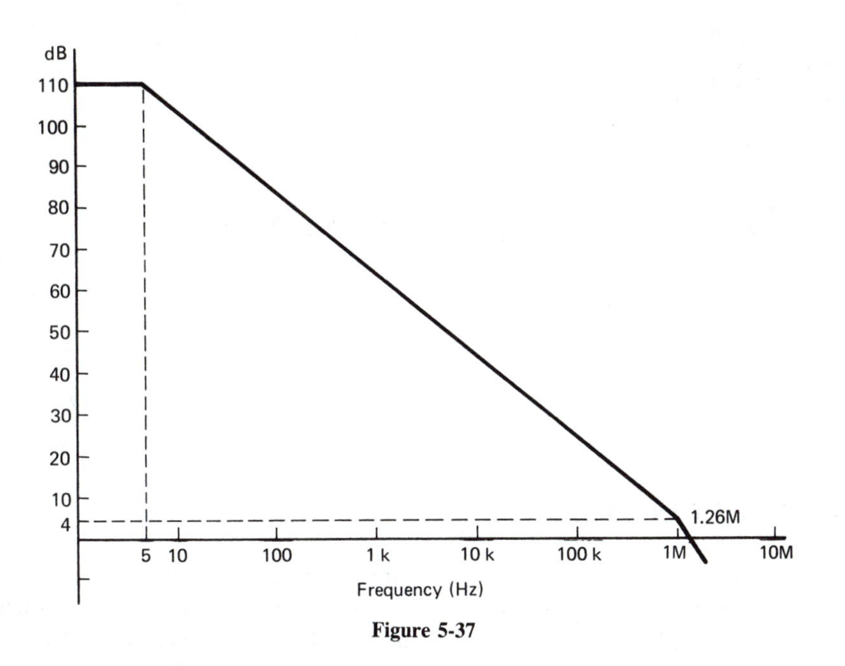

Figure 5-37

28. For the transfer function:

 (a) Draw the Bode plot.

 (b) Calculate any x-axis crossing on the magnitude graph.

$$\frac{4 \times 10^{12}s^2}{(s + 1 \times 10^3)(s + 4 \times 10^3)(s + 5 \times 10^4)(s + 2 \times 10^5)}$$

29. Figure 5-37 shows the magnitude plot of an op-amp. Write the transfer function for this op-amp.

30. From the magnitude plot shown in Fig. 5-38, write the transfer function.

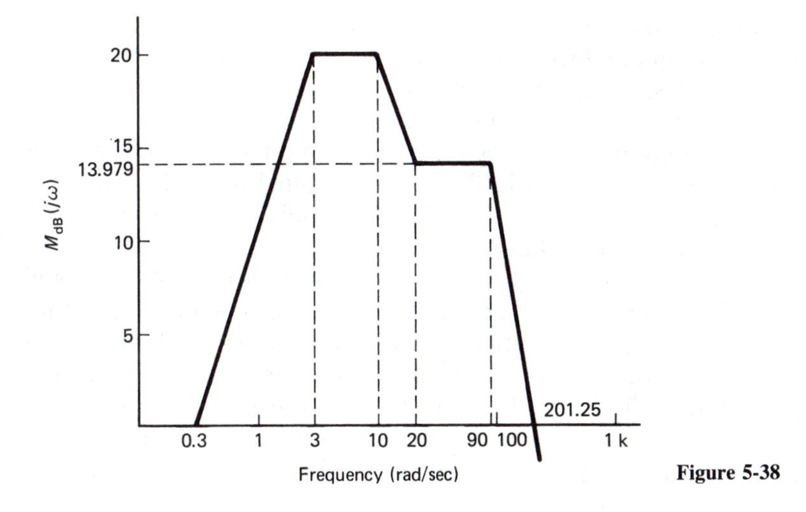

Figure 5-38

CHAPTER 6

Introduction to Filters

6-0 INTRODUCTION

In this chapter we define the terminology used when working with filters, and also typical assumptions used for the op-amp. This chapter is a transition from using Laplace to applying Laplace. The real power of Laplace can be demonstrated by its application in design work, and in this book we will use active filters.

6-1 FILTER GRAPHS

Filter graphs are graphs of transfer functions. We can plot the response of a filter or we can define what type of filter we want. For the type of filters used in this book, the phase is usually ignored. In advanced texts on filters, when phase is of concern, other types of filters are used.

6-1-1 Loss Function

When the transfer function is the ratio of the output voltage to the input voltage, it is called a gain function, and when the ratio is the input voltage to the output voltage, it is called a loss function. When working with a filter, we typically use the loss function instead of the gain function.

There are two reasons for using the loss function instead of the gain function. First we are concerned with how much a signal is attenuated. Second, since most of the function's response is due to poles, it is easier to work with the poles when they are in the numerator.

From this point on we must be careful about our use of the words "gain" and "loss," since they do not refer to the signal level. The gain can be either an increase (amplification) or a decrease (attenuation) in signal level. Similarly, the loss function does not mean a decrease in the signal. It means that if the loss function magnitude, in dB, is positive, there is a decrease (attenuation) in signal level, and if the loss function magnitude, in dB, is negative, there is an increase (amplification) in signal level. We must use "attenuation" and "amplification" to describe a decrease or increase in the signal level, and use "gain" and "loss" to indicate which type of function we are using.

The gain function, $G(j\omega)$, can be split into a magnitude function, $M(j\omega)$, and a phase function, $P(j\omega)$:

$$M(j\omega) \angle P(j\omega) = G(j\omega) \tag{6-1}$$

Since the loss function is the input over the output, it is the reciprocal of the gain function. The loss function, $L(j\omega)$, can also be split into a magnitude function, $ML(j\omega)$, and a phase function, $PL(j\omega)$:

$$ML(j\omega) \angle PL(j\omega) = L(j\omega) = \frac{1}{G(j\omega)} \tag{6-2}$$

In this book we will be concentrating on the magnitude function and ignore the phase shift when working with filters. The magnitude functions, in dB, for the gain function and loss function are

$$M_{\text{dB}}(j\omega) = 20 \log | G(j\omega) | \tag{6-3}$$

$$ML_{\text{dB}}(j\omega) = 20 \log | L(j\omega) | \tag{6-4}$$

Since the gain function, $G(j\omega)$, and loss function, $L(j\omega)$, are reciprocals,

$$ML_{\text{dB}}(j\omega) = -M_{\text{dB}}(j\omega) \tag{6-5}$$

When looking at a loss function graph we need to redesign our thinking. The larger the positive dB value, the more the signal is attenuated; and the smaller the dB value, the less the signal is attenuated. If the loss is a negative dB value, the signal is amplified. We could think in terms of stepping over the function. The higher the graphs go, the more the effort required to step over, or the more difficult it is for the signal to get through.

6-1-2 Normalized Graphs

The magnitude of the loss function typically has the frequency and the magnitude normalized. A gain function could also be normalized, but since we are concentrating on filters, we will only examine normalizing loss functions. We will start with frequency normalization.

To normalize the frequency:

1. Choose a frequency to normalize to. This is some significant frequency such as a break frequency or a center frequency.
2. Divide all frequencies by this chosen frequency.

When we normalize, the chosen frequency becomes the value 1. Since we are dividing a frequency by a frequency, the frequency axis will be unitless and will usually be indicated by Ω. Frequency normalization expressed in an equation form is

$$\Omega = \frac{\omega}{\omega_x} \tag{6-6a}$$

or expressed with the frequency in hertz:

$$\Omega = \frac{f}{f_x} \tag{6-6b}$$

where Ω = the normalized frequency (unitless)
ω or f = the frequency to be normalized
ω_x or f_x = the frequency we normalize to

EXAMPLE 6-1

Normalize the loss function shown in Fig. 6-1 to 3 krad/sec. What is the frequency of 25 krad/sec on the normalized graph?

SOLUTION Each frequency is divided by 3 krad/sec using Eq. (6-6) and plotted as shown in Fig. 6-2. Notice that this graph has the same shape as Fig. 6-1 but that the break occurs at 1.

Using Eq. (6-6), we can calculate the normalized value for 25 krad/sec.

$$\begin{aligned}\Omega &= \frac{\omega}{\omega_x} \\ &= \frac{25\text{k}}{3\text{k}} \\ &= 8.3333\end{aligned}$$

From the last example we see that the shape of the graphs is exactly the same. A big advantage of normalized graphs is that they will always have the

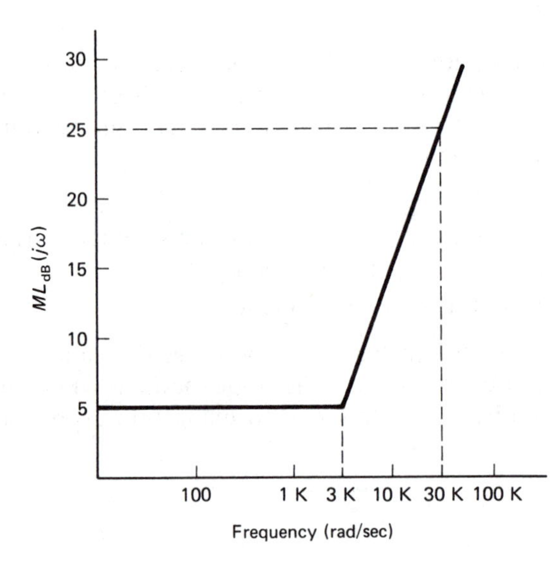

Figure 6-1

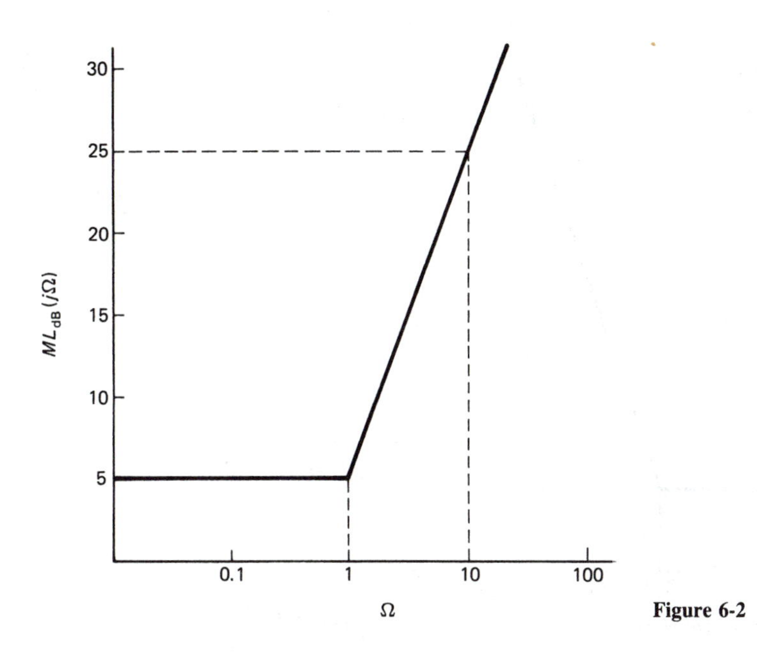

Figure 6-2

same shape when denormalized to any frequency. To denormalize a normalized graph:

1. Choose a frequency to denormalize the normalized frequency to.
2. Multiply each normalized frequency by this value.

This will change the unitless axis back to an axis with units. We can put this in equation form by solving Eq. (6-6) for ω or f.

$$\omega = \Omega\omega_x \tag{6-7a}$$

$$f = \Omega f_x \tag{6-7b}$$

where Ω = the normalized frequency (unitless)
ω or f = the denormalized frequency
ω_x or f_x = the frequency we are denormalizing to

The frequency ω_x does not have to be the same value as that used in Eq. (6-6). This allows us to shift the function to any frequency we want. Since the new graph must still have major divisions marked 1×10 to some integer power, the normalized locations on major divisions may or may not be on the major divisions of the denormalized graph.

EXAMPLE 6-2

Denormalize Fig. 6-2 to a frequency of 250 rad/sec. Find the denormalized frequency of 0.13.

SOLUTION We multiply each frequency by 250 rad/sec using Eq. (6-7). The denormalized graph is shown in Fig. 6-3.

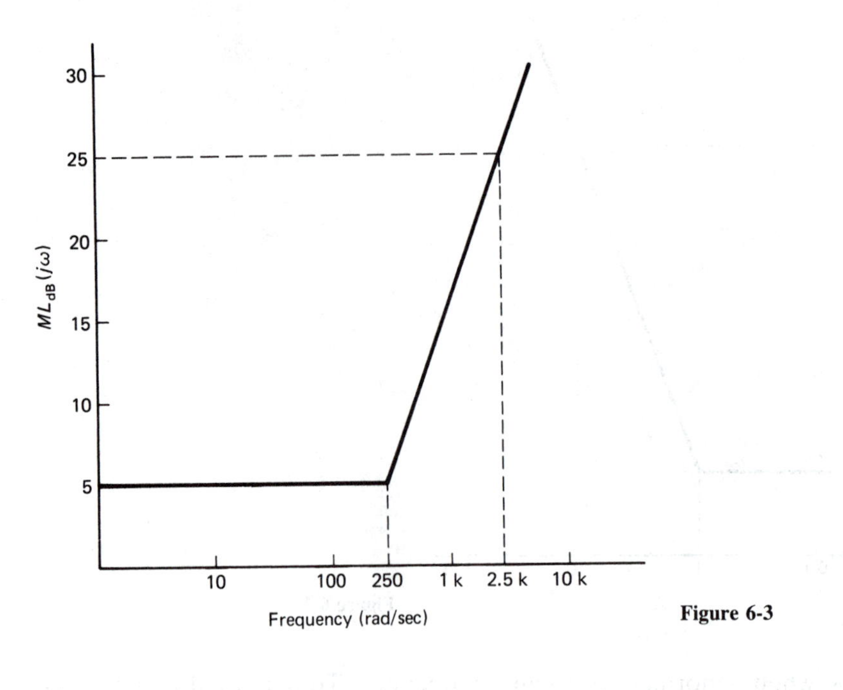

Figure 6-3

Using Eq. (6-7) gives

$$\begin{aligned}\omega &= \Omega\omega_x \\ &= (0.13)(250) \\ &= 32.5 \text{ rad/sec}\end{aligned}$$

A loss function can also have its magnitude normalized. To normalize a loss function's magnitude:

1. Choose a dB value to normalize to. On a loss function this is typically the lowest dB value of the graph.
2. Subtract it from each magnitude value.

We will notice that when this is done, the shape of the graph will remain the same, and the bottom of the graph will be 0 dB. In equation form

$$ML_{NdB}(j\omega) = ML_{dB}(j\omega) - M_{xdB} \tag{6-8}$$

where $ML_{NdB}(j\omega)$ = the normalized dB value
$ML_{dB}(j\omega)$ = the value to be normalized (ω may be the normalized frequency Ω)
M_{xdB} = the loss we are normalizing to

Since substraction of logarithms corresponds to a division of the values, we could express Eq. (6-8) as

$$ML_N(j\omega) = \frac{ML(j\omega)}{M_x(j\omega)}$$

Since this equation is the same as Eq. (6-6), we see that magnitude normalization is the same as frequency normalization. Therefore, magnitude denormalization is also the same as frequency denormalization.

To denormalize a loss function:

1. Choose a gain to denormalize the normalized gain to.
2. Add this value to each normalized value.

We can state this in equation form by solving Eq. (6-8) for $ML_{dB}(j\omega)$.

$$ML_{dB}(j\omega) = ML_{NdB}(j\omega) + M_{xdB} \tag{6-9}$$

where $ML_{NdB}(j\omega)$ = the normalized dB value
$ML_{dB}(j\omega)$ = the denormalized value (ω may be the normalized frequency Ω)
M_{xdB} = the loss we are denormalizing to

This M_{xdB} can be a different value than what was used to normalize the function in Eq. (6-8). This will shift the entire function up or down without modifying the shape of the curve.

EXAMPLE 6-3

Normalize the 5-dB level in Fig. 6-4 to 0 dB. Then denormalize the function to 3 dB with the frequency denormalized to 4 krad/sec.

SOLUTION Using Eq. (6-8) at significant points on the graph.

$$ML_{NdB}(j\omega) = ML_{dB}(j\omega) - M_{xdB}$$
$$= 5 - 5 = 0 \text{ dB}$$
$$ML_{NdB}(j\omega) = ML_{dB}(j\omega) - M_{xdB}$$
$$= 25 - 5 = 20 \text{ dB}$$

The function normalized in loss and frequency is shown in Fig. 6-5a.

When denormalizing, we may denormalize the loss and the frequency in any order. We will denormalize the loss first.

To denormalize the loss to 3 dB, we use Eq. (6-9) on the significant point to the normalized graph

$$ML_{dB}(j\omega) = ML_{NdB}(j\omega) + M_{xdB}$$
$$= 0 + 3 = 3 \text{ dB}$$
$$ML_{dB}(j\omega) = ML_{NdB}(j\omega) + M_{xdB}$$
$$= 20 + 3 = 23 \text{ dB}$$

This will shift Fig. 6-5a up by 3 dB.

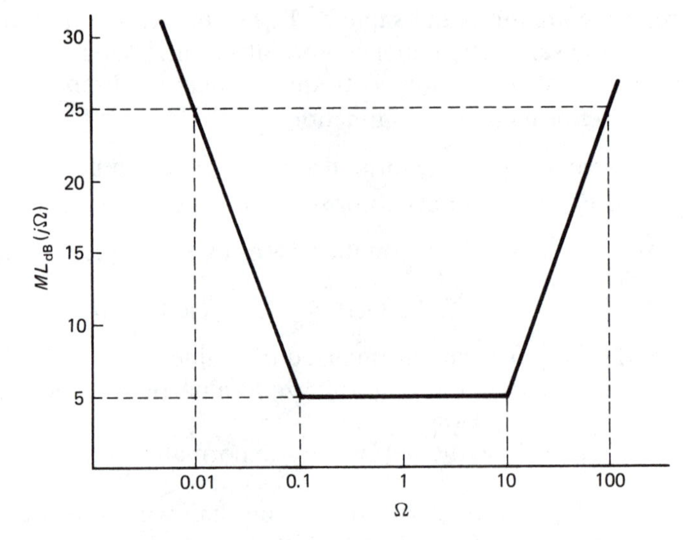

Figure 6-4

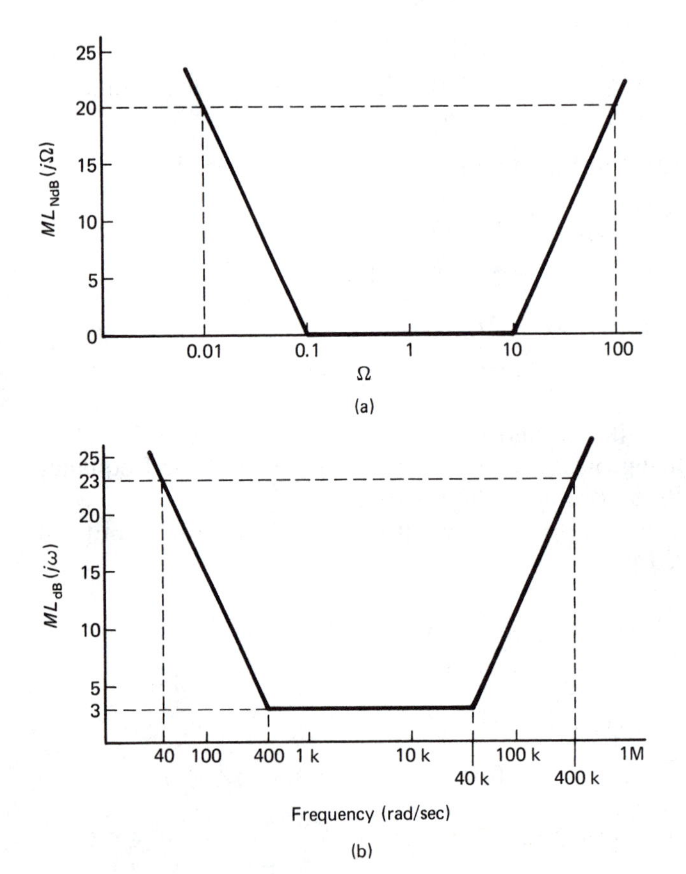

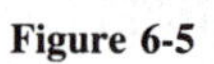
Figure 6-5

We denormalize the frequency using Eq. (6-7) at significant points on the graph.

$$\omega = \Omega\omega_x$$
$$= (10)(4 \text{ krad/sec}) = 40 \text{ krad/sec}$$
$$\omega = \Omega\omega_x$$
$$= (0.1)(4 \text{ krad/sec}) = 400 \text{ rad/sec}$$

This will shift Fig. 6-5a to the right by 4 krad/sec. Figure 6-5b shows the denormalized function.

6-1-3 Graph Specifications for Filters

To design a filter, it must first be specified. This is done on a graph. The graph is made up of bands of frequencies with loss areas blocked out. We want certain bands of frequencies passed with a limit on the maximum amount they can be attenuated, and a band of frequencies stopped by a minimum amount of attenuation.

An ideal filter would be one where, at one frequency, the loss would go from no attenuation to complete attenuation. Currently, this is impossible. We must therefore have a band of frequencies to give the filter a transition from the pass band to the stop band.

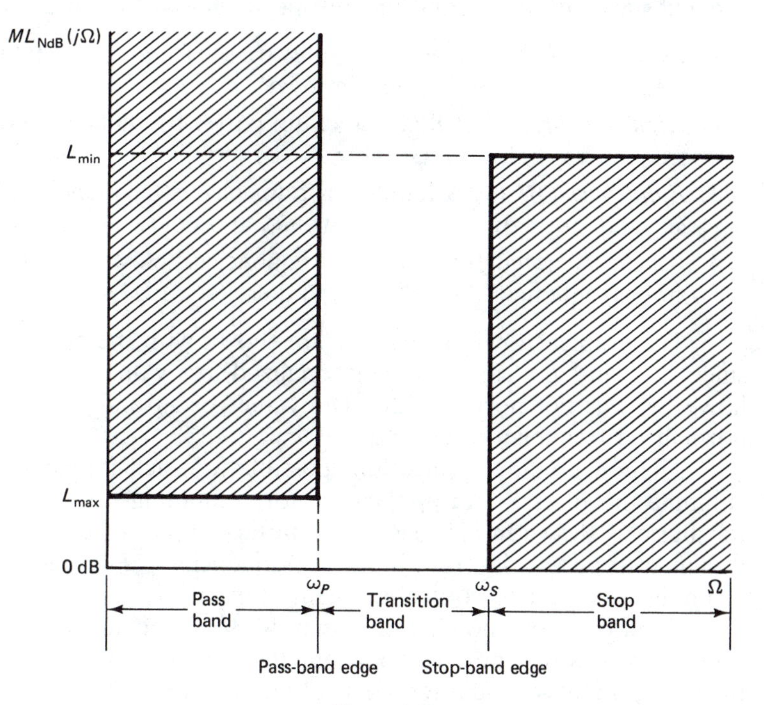

Figure 6-6

Let's define these terms and associate a variable name to them.

ω_s: the frequency of the edge of a stop band

ω_p: the frequency of the edge of a pass band

Transition band: the band of frequencies between ω_s and ω_p

L_{max}: the normalized maximum dB loss in the pass band

L_{min}: the normalized minimum dB loss in the stop band

Figure 6-6 shows a typical filter specification. A transition band will always exist between a pass-band edge frequency, ω_p, and a stop-band edge, ω_s. The filter to meet this specification must remain under L_{max} up to ω_p, and the filter must be above L_{min} in the stop band.

As ω_p and ω_s get closer, and as the difference between L_{min} and L_{max} gets larger, we approach the ideal filter. The closer we get to the ideal filter, the more complex and expensive the filter becomes.

6-2 FILTER DEFINITIONS

There are many types of filters. We will concentrate on a few. These are the basic building blocks that more complex filters are defined by.

6-2-1 Types of Filters: LP, HP, BP, Notch

We will start with the definition and specification of the four basic filter types.

1. *Low-pass* (*LP*). A low-pass filter passes frequencies below ω_P and stops frequencies above ω_s (see Fig. 6-7a).
2. *High-pass* (*HP*). A high-pass filter passes frequencies above ω_p and stops frequencies below ω_s (see Fig. 6-7b).
3. *Band-pass* (*BP*). A band-pass filter passes frequencies between ω_{p1} and ω_{p2} and stops frequencies above ω_{s2} and below ω_{s1} (see Fig. 6-7c).
4. *Notch.* A notch filter passes frequencies below ω_{p1} and above ω_{p2} and stops frequencies between ω_{s1} and ω_{s2} (see Fig. 6-7d).

The graphs shown in Fig. 6-7 are shown with the magnitude normalized since L_{max} and L_{min} are always normalized loss values. The frequency axis, however, was not normalized. The values of ω_p and ω_s may also appear as normalized frequencies, Ω_p and Ω_s.

There are many applications possible for each filter type. We can also combine the filters. An application for combining the filters is an equalizer for music as shown in Fig. 6-8. The music is input to several band-pass filters designed to pass a different band of frequencies. The output is applied to a summing amplifier to control the gain of each filter.

Band-pass and notch filters may be specified unsymmetrical. For this book we will force these to be symmetrical. This will be done by increasing the requirements of the filter. The object is to revise the requirements to meet or exceed the original specification so that the filter is symmetrical.

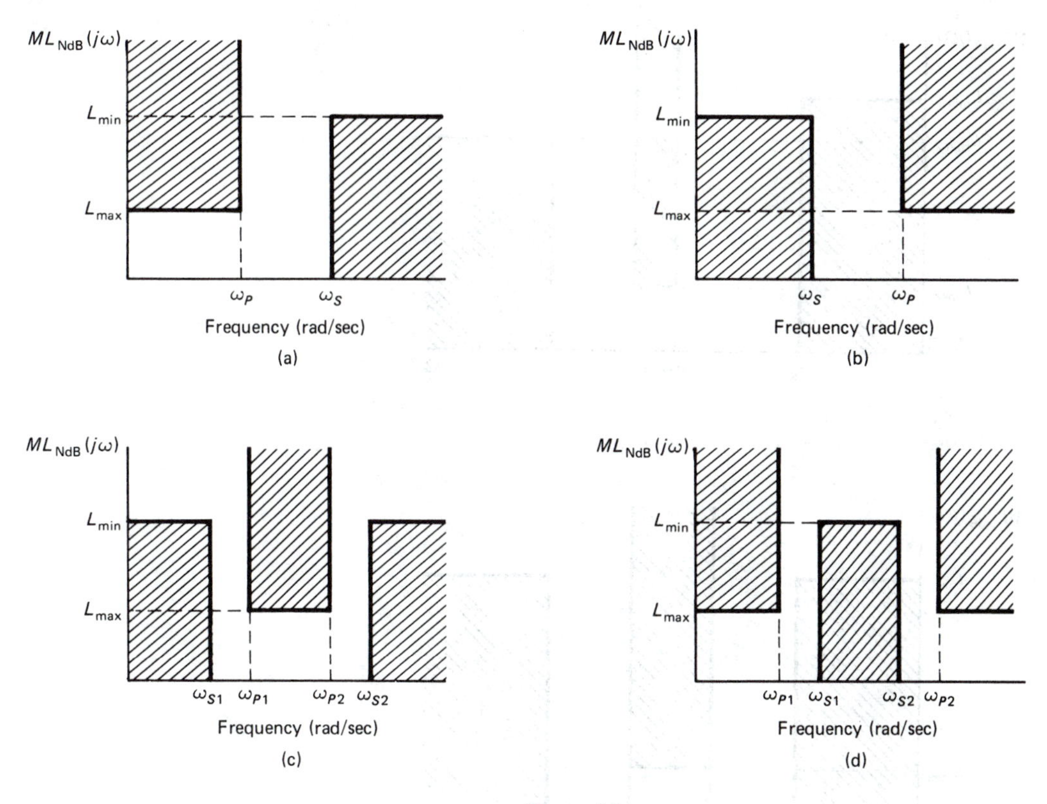

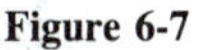

Figure 6-7

Figure 6-9a shows a band-pass filter that is unsymmetrical. First we will increase L_{min2}'s value to L_{min1}'s value, making both stop bands equal. Next we compare the number of decades between ω_{s1} and ω_{p1} to the number of decades between ω_{p2} and ω_{s2}. If they are not equal, we make them both equal to the smaller value. In this case we need to decrease ω_{p1}. We could have

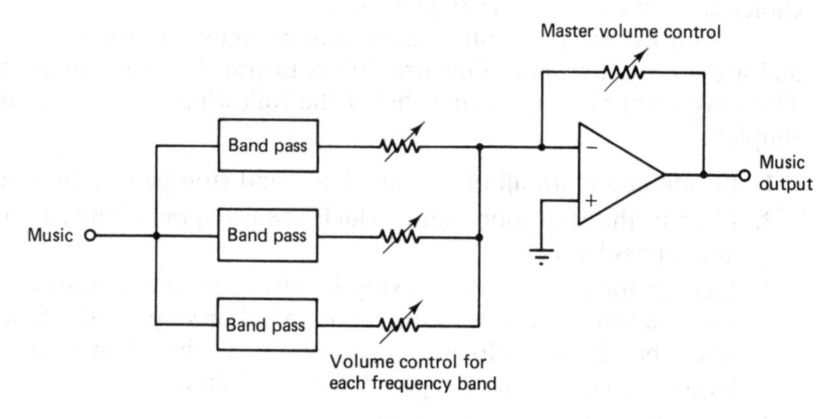

Figure 6-8

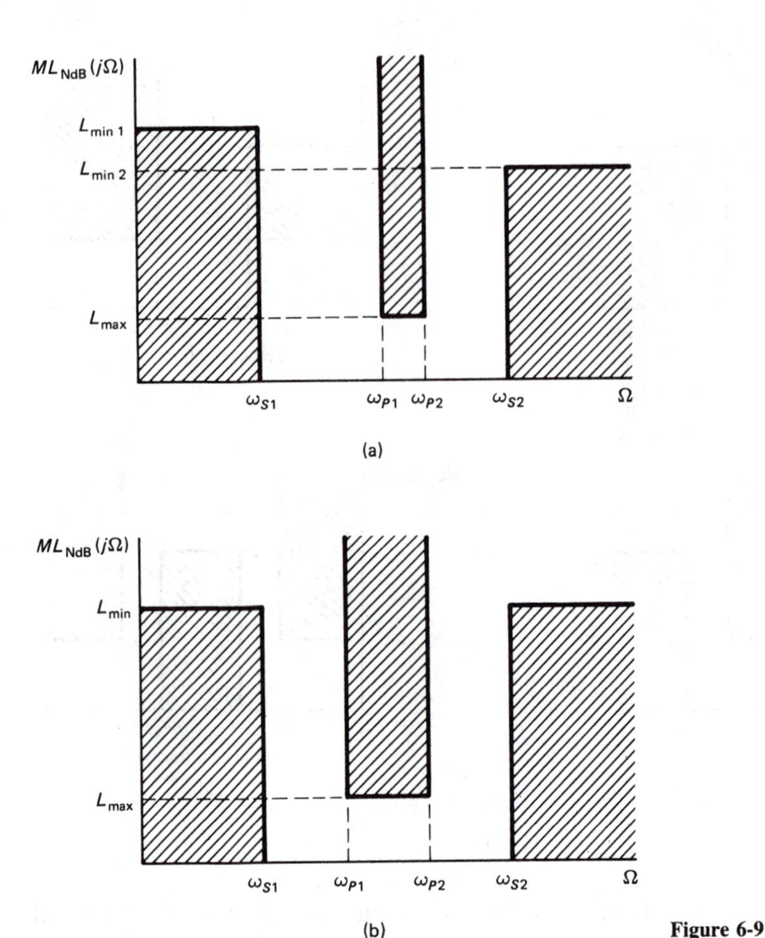

Figure 6-9

increased ω_{s1} instead of decreasing ω_{p1}, but this would also increase the Q of the circuit. Normally, the lower Q is preferred since it will usually lead to a less complex circuit. Unless the center frequency must remain the same, the choice that gives us the lower Q is used.

When we are given filter specifications, they are usually just the values and are not on a graph. Our first job is to transfer these values to a graph. This may seem difficult at first, but if the following steps are used, it is very simple:

1. Locate and mark all of the pass-band and stop-band edge frequencies.
2. Identify the transition bands which always appear between a stop band and a pass band.
3. Identify the pass bands and stop bands. The frequencies ω_p and ω_s will divide a transition band from a pass band or stop band. Since the transition bands have already been identified, the other side of the edge frequency ω_P/ω_s will be a pass band/stop band.
4. Draw the blocks.

EXAMPLE 6-4

Draw the following specification, and identify the type of filter.

$$\omega_{s1} = 50 \text{ rad/sec} \quad \omega_{s2} = 200 \text{ rad/sec}$$

$$\omega_{p1} = 10 \text{ rad/sec} \quad \omega_{p2} = 2 \text{ krad/sec}$$

$$L_{\min} = 40 \text{ dB} \qquad L_{\max 1} = 2 \text{ dB} \qquad L_{\max 2} = 5 \text{ dB}$$

SOLUTION First we mark the stop-band edge frequencies and the pass-band edge frequencies as shown in Fig. 6-10a. The transition bands are

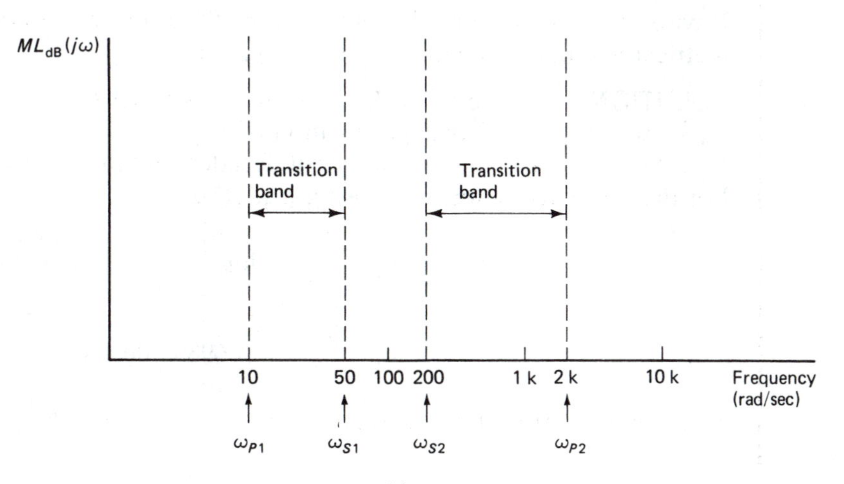

(a)

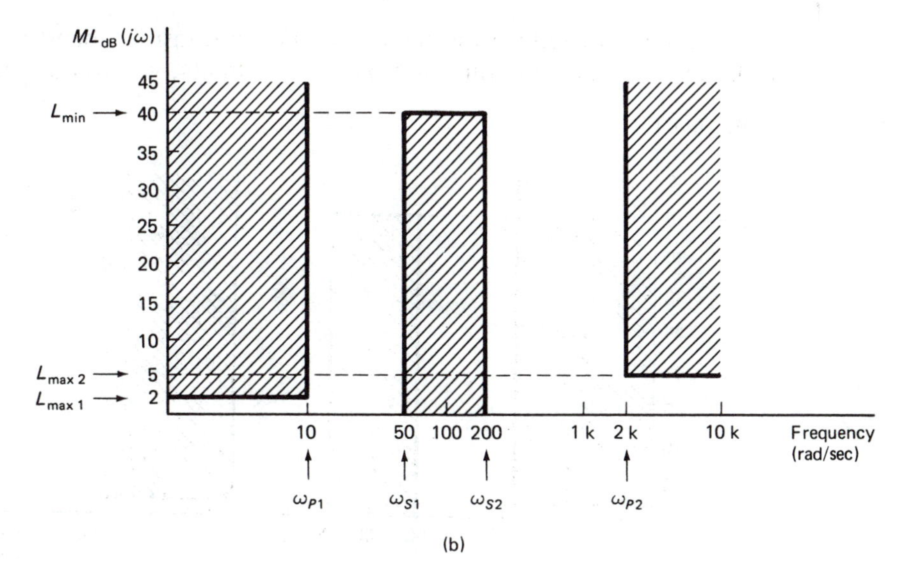

(b)

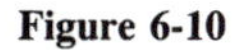
Figure 6-10

then identified between each pass-band edge frequency and stop-band edge frequency. Once the transition bands are identified, we see there is a stop band between ω_{s1} and ω_{s2}, and there are two pass bands.

Next, the blocks are drawn as shown in Fig. 6-10b. $L_{\min}$ goes in the stop band. $L_{\max 1}$ is used for the pass band associated with ω_{p1}, and $L_{\max 2}$ is used for the pass band associated with ω_{p2}. This filter is a notch filter since it stops a band of frequencies while allowing frequencies above and below the stop band to pass.

EXAMPLE 6-5

Revise the specifications shown in Fig. 6-10b to make the notch filter symmetrical without changing the center frequency.

SOLUTION First we move $L_{\max 2}$ so that it is equal $L_{\max 1}$, since if we meet $L_{\max 1}$, we will exceed the specification of $L_{\max 2}$.

Next we calculate the number of decades with the two transition bands. For the lower transition band using Eq. (D-2),

$$\text{ND} = \log\left(\frac{f_2}{f_1}\right) = \log\left(\frac{\omega_2}{\omega_1}\right)$$

$$= \log\left(\frac{50}{10}\right) = 0.69897 \text{ decade}$$

and the upper transition band is

$$= \log\left(\frac{2\text{ k}}{200}\right) = 1 \text{ decade}$$

We need to reduce the higher-frequency transition band to be equal to the lower-frequency transition band. Normally, we would increase ω_{s2},

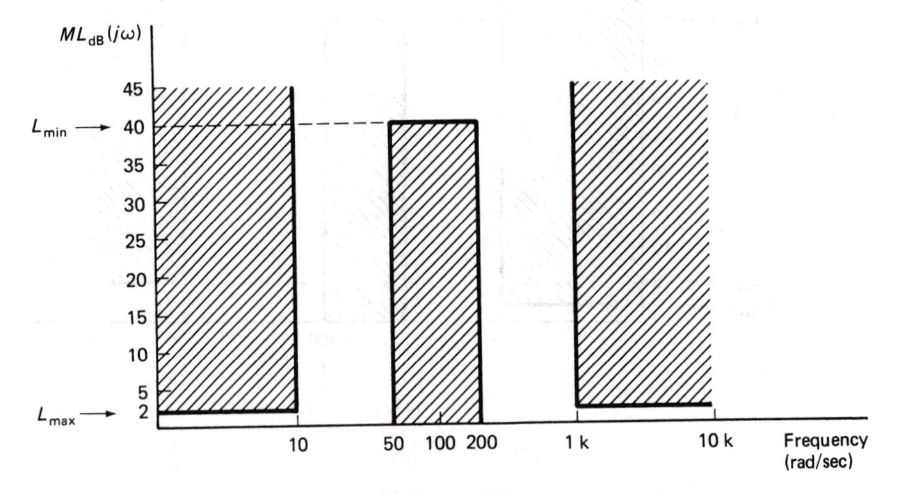

Figure 6-11

but this would change the center frequency. We must change ω_{p2}. Solving Eq. (D-2) for ω_2 gives

$$\omega_2 = \omega_1 \times 10^{ND}$$
$$= 200 \times 10^{0.69897}$$
$$= 1 \text{ krad/sec}$$

The revised filter specifications are shown in Fig. 6-11.

6-2-2 Passive and Active Filters

Filters can be put into two categories: passive and active. Active filters require external power, and passive filters do not. When the term "passive filter" is used, typically a filter using only *R*, *L*, and *C* is meant, but there are other types of passive filters available, such as mechanical and crystal filters.

A mechanical filter uses a transducer to convert an electrical signal to a mechanical vibration. The signal vibrates a material that is often a piece of metal cut in a particular shape. This material vibrates well at some frequencies, but will not at other frequencies. This allows only selected frequencies to pass. The vibration, after passing through the element, is then converted back to an electrical signal.

A crystal filter works very much like a mechanical filter except that it uses the piezoelectric effect. Crystals are well known for the ability to be resonant at a particular frequency, making them good as a very selective band-pass filter. Frequently, several crystals are used to obtain different filtering characteristics.

There are two types of active filters: digital and analog. Both types require external power supplies, and both can amplify and filter simultaneously.

The digital filter converts the signal to digital numbers that can be manipulated by a computer. The computer can perform calculations on these numbers to filter the signal. The resulting numbers are then converted back to an analog signal.

The analog filter uses an analog amplifier such as an op-amp. These analog op-amp filters are used at audio frequencies (<30 kHz) due to the frequency limit of the op-amp. Since we are at audio frequencies, only *R* and *C* elements are used. Inductors designed for these frequencies usually are large, expensive, and of poor quality (low *Q*). Since we have an active element, the op-amp, it is relatively easy to avoid using an inductor.

The rest of this book will use the analog op-amp active filter. When the term "active filter" is used, we will specifically be speaking of the analog op-amp active filter.

6-3 OP-AMPS

Currently, the upper limit of active filters is due to the op-amp's frequency response and to a lesser degree on the capacitors. With improvements in op-amps and capacitors, this upper limit will continue to increase. We would like an analytical method that will remain the same as technology improves. To do this we will analyze our circuits using a perfect op-amp. The op-amp will look perfect up to some limit, and as this limit increases, our analysis will continue to be appropriate.

6-3-1 Approximate Op-Amp

We must first define a current-technology op-amp. Then we will approximate the op-amp as a perfect op-amp with the limits required to make this assumption.

Using our knowledge of Bode plots and Laplace transforms, we can make an ac model of an op-amp. Let's first review the main characteristics of an op-amp.

1. The input impedance, r_i, is in the MΩ range.
2. The output impedance, r_o, is in the range 50 to 100 Ω.
3. The open-loop voltage gain below the lowest break frequency is greater than 100,000 (100 dB).
4. The lowest break frequency is about 10 Hz. There is a second break frequency at about 1 MHz which is close to the unity-gain point and is sometimes ignored.

We can simulate the op-amp's break frequencies by using either a series *RC* network driven with a voltage source or a parallel *RC* network driven with a current source. Since an op-amp is primarily a voltage amplifier and the gain is expressed as voltage gain, we will use the voltage source driving a series network. Figure 6-12 shows this series network.

The transfer function for the series network shown in Fig. 6-12 in a proper form to draw the Bode plot is

$$\frac{V_o}{V_{\text{in}}} = \frac{1}{RCs + 1}$$

The break frequency, as shown in Chapter 5, is calculated by

$$\omega = \frac{1}{RC}$$

Knowing the break frequency required, we may choose a value for either R or C and solve for the other component. Since we are using this circuit to simulate a break frequency, it is unimportant what value is initially chosen for R or C.

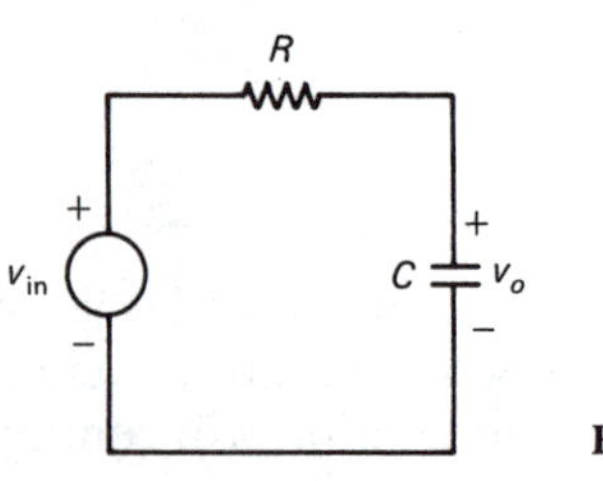

Figure 6-12

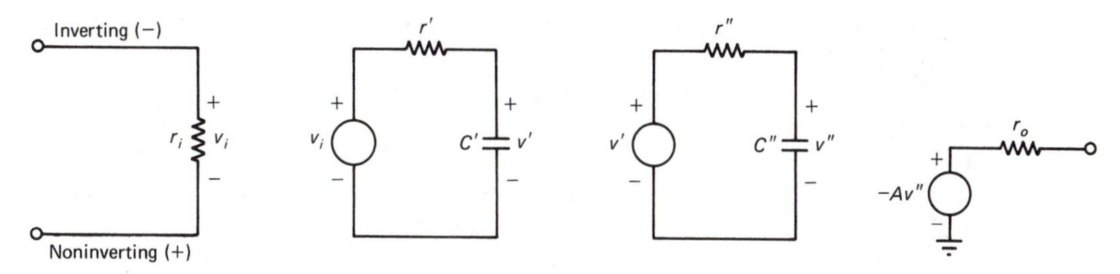

Figure 6-13

A convenient value to choose would be $R = 1$. This makes for an easy source conversion, using the resistor and the voltage source, if a current source is required.

To make another break frequency, the output of this circuit can be used to drive another circuit using a VCVS. We can cascade several stages; however, we typically need only two break frequencies.

To complete our model, we need to simulate the input and the output circuit. The input is a very high resistance, r_i, between the inverting and noninverting terminal as shown in Fig. 6-13. The voltage across this impedance is then amplified by the open-loop gain and passed through the break-frequency circuits. (Since this is an equivalent circuit, it does not matter where the actual open-loop gain is taken into account.) This is then passed to the output circuit, which is a voltage source with a series resistance, r_o.

It may be of some interest to examine the transfer function of this model of the op-amp. When we use a computer to evaluate a design based on an approximate model of the op-amp, we should use this accurate model to check the validity of our approximations. The transfer function, in this case, is a series of voltage divider equations multiplied together. Assuming an output load Z_L, we obtain

$$\left(\frac{v_i \dfrac{1}{r'C'}}{s + \dfrac{1}{r'C'}}\right)\left(\frac{\dfrac{1}{r''C''}}{s + \dfrac{1}{r''C''}}\right)\left(\frac{-AZ_L}{r_o + Z_L}\right)$$

If we use this model for the op-amp, we will have an accurate representation of an active filter. The analysis using this model will generate some very complex equations where a number of terms will have little affect on the filter's transfer function.

If we design the circuit gain such that its gain is much less than the op-amp's open-loop gain up to the highest frequency of use, we can ignore the op-amp's break frequencies. The highest frequency of use for a low-pass filter is at the pass-band edge frequency. After we are past the break frequency, the roll-off of the op-amp can only help us. However, a high-pass filter ideally passes frequencies from the pass-band edge frequency to infinity. In reality there is a maximum frequency above which we are no longer interested in passing signals.

Figure 6-14a shows the op-amp's open-loop gain and a low-pass filter frequency response. (This is drawn as a gain function, not a loss function.) Notice that if we have a minimum difference between the pass-band edge frequency of the circuit and the op-amp's open-loop gain, the op-amp's open-loop gain will always be larger than the circuit's gain.

Figure 6-14b shows the op-amp's open-loop gain and a high-pass filter frequency response. Here we need to know what is the highest frequency of use. At this point we want a minimum difference to make sure that the frequency is passed unaffected by the op-amp's roll-off.

We will recall that a typical assumption is the open-loop gain is infinity. Often in electronics a 10% error would be acceptable. For active filters this is often not enough. In this book we will use a 5% error. A 5% error will correspond to the open-loop gain being 20 times the circuit gain at the highest frequency of use. To increase the accuracy, this difference would be increased. In equation form,

$$A_{\text{op-amp}} \geqslant 20A_{\text{ckt}} \tag{6-10}$$

By meeting the condition of this equation, we can approximate the op-amp as shown in Fig. 6-15. This simplifies the analysis tremendously, but the analytical results are the same as using the full model. With this model we do the analysis assuming that the op-amp open-loop gain is infinity for all frequencies of interest.

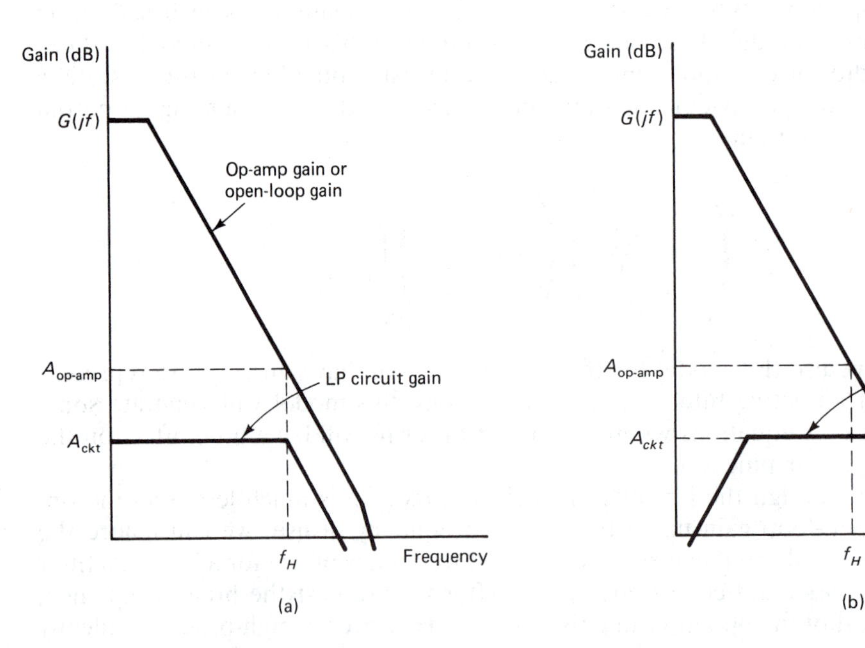

Figure 6-14

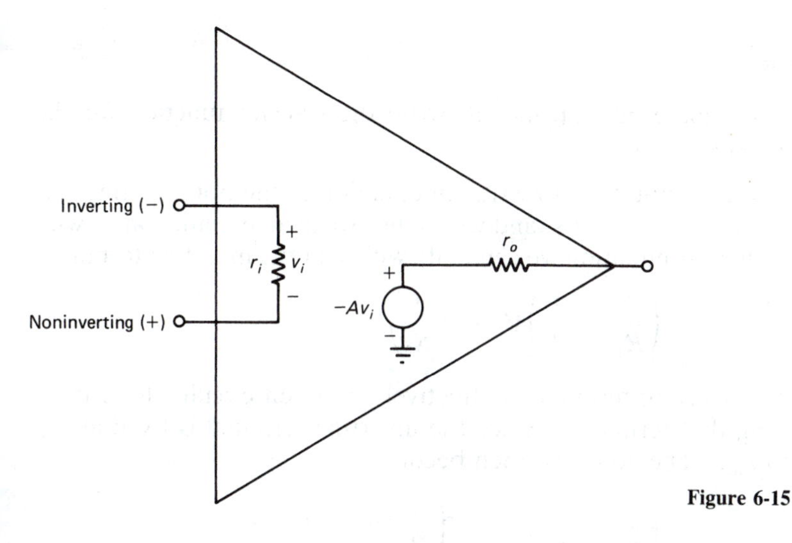

Figure 6-15

We can further simplify the op-amp by assuming that the input impedance, r_i, is much larger than the impedances attached to it, and that the output impedance, r_o, is zero if the output circuitry is much greater than r_o. This condition is easy to meet if we keep our resistive values in the kΩ range.

The approximate op-amp model we will use is shown in Fig. 6-16. There are some tricks to using this model successfully. Since this is an op-amp, the inverting and noninverting terminals will have the same voltage value. When writing node equations, do not write an equation at the output node since it will generate an indeterminate equation. Even though the input terminals have the same voltage value, they cannot be considered connected since the input impedance must be infinity; consequently, both terminals will not have a current entering or leaving.

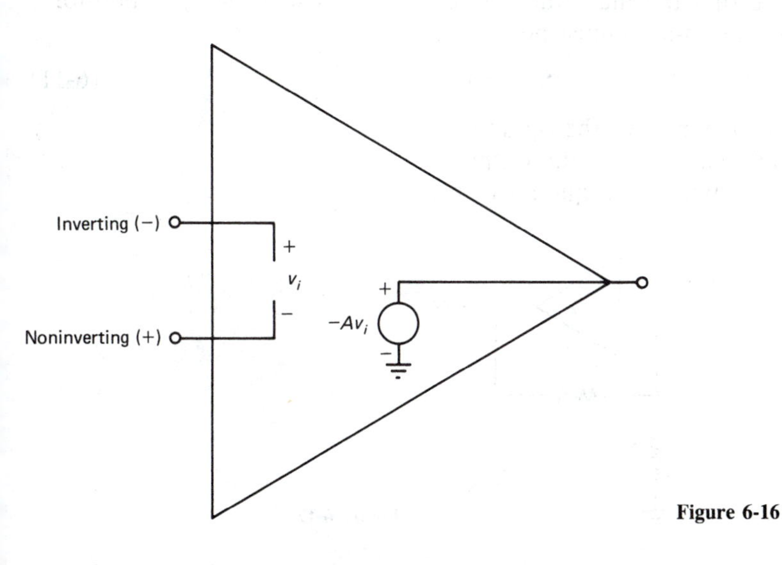

Figure 6-16

EXAMPLE 6-6

Using the approximate op-amp model, write the transfer function for the circuit in Fig. 6-17.

SOLUTION We do not need to write an equation at the noninverting terminal since it is fixed at v_{in} volts, and we cannot write an equation at v_o with this model. The only equation we can write will be at the inverting terminal.

$$\left(\frac{1}{R_1} + \frac{1}{R_2}\right)V_1 - \left(\frac{1}{R_2}\right)V_o = 0$$

Since the inverting terminal is effectively an open circuit, there is no current entering that terminal. Since the inverting terminal is fixed at v_{in}, V_1 is equal to v_{in}. The equation then becomes

$$\left(\frac{1}{R_1} + \frac{1}{R_2}\right)v_{in} - \left(\frac{1}{R_2}\right)V_o = 0$$

Solving for V_o/v_{in}, the transfer function is

$$\frac{V_o}{v_{in}} = 1 + \frac{R_2}{R_1}$$

We must also consider the slew rate so that the output will not become distorted. The slew rate tells us the rate at which the output can change. A sinusoidal function has a constantly changing slope. If we take the derivative of the sine function, we can define the maximum slope:

$$\frac{d}{dt} V_o \sin(\omega t) = V_o\omega \cos(\omega t)$$

From this we see that the maximum slope is when $\cos(\omega t) = 1$. Therefore, the slew rate of the op-amp must be

$$\text{SR} \geq V_o\omega \tag{6-11}$$

where SR = the slew rate of the op-amp
V_o = the peak value of the output
ω = the maximum frequency of use

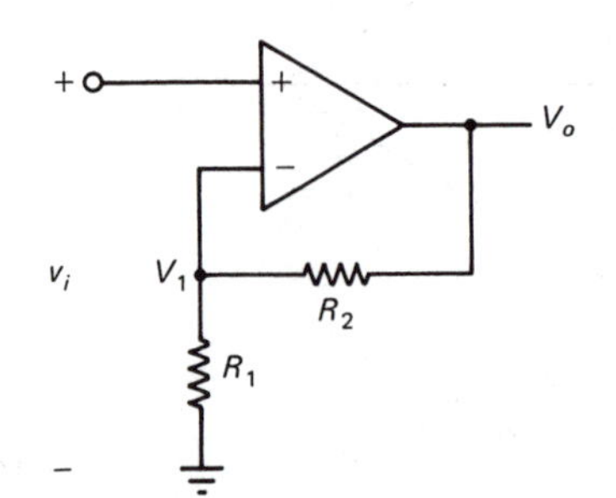

Figure 6-17

This equation shows that for a fixed slew rate, as we increase the frequency of use, we must reduce the peak output voltage.

As with all our analysis, we must remember that we are working with limits based on the sinusoidal function. This means that if we want the slew-rate limit for a square wave, we must convert the square wave to its sinusoidal components, determine the highest sinusoidal frequency required, and then apply Eq. (6-11).

In Eq. (6-11) there is a limit to the peak value due to the power supply. The output voltage will only be able to get to within a few volts of the power supply. This saturation voltage can be found in the data sheets. If this data value is unavailable, we will have to make an assumption. We will use a 2-V margin since typically we will be able to get within 2 V of the power supply voltage on most any op-amp.

Since there is a limit for the output, there is a limit for the maximum input voltage. Expressed in equation form,

$$V_{\text{in max}} = \frac{V_o}{A_{\text{ckt}}} \tag{6-12}$$

Once the maximum output voltage is determined, we can calculate the maximum limit for the applied input voltage.

Summarizing the approximate op-amp shown in Fig. 6-16:

1. The inverting and noninverting terminals are open circuits, but the voltage of these terminals will be equal.
2. An equation cannot be written at the op-amp's output voltage node.
3. The highest frequency of use must have an open-loop gain defined by Eq. (6-10):

$$A_{\text{op-amp}} \geqslant 20A_{\text{ckt}}$$

4. The slew rate must have a value defined by Eq. (6-11):

$$\text{SR} \geqslant V_o\omega$$

5. V_o must be less than the saturation level of the op-amp with the particular power supply voltage being used.

EXAMPLE 6-7

An active filter is to be designed using an op-amp with a slew rate of 0.5 V/μsec. The highest frequency of interest is 10 kHz, and a gain of 5 dB is required here. Find the op-amp's open-loop gain at the highest frequency of use. Find the maximum output voltage possible, to the nearest volt. Find what power supply voltage is required.

SOLUTION

(a) First we must find the voltage gain of the circuit.

$$A_{\text{ckt dB}} = 20 \log (A_{\text{ckt}})$$

$$A_{\text{ckt}} = 10^{A_{\text{ckt dB}}/20}$$

$$= 10^{5/20}$$

$$= 1.7783$$

From Eq. (6-10)

$$A_{\text{op-amp}} \geqslant 20A_{\text{ckt}}$$

$$\geqslant 20(1.7783)$$

$$\geqslant 35.566$$

(b) From Eq. (6-11),

$$\text{SR} \geqslant V_o\omega$$

$$V_o \leqslant \frac{\text{SR}}{\omega}$$

$$\leqslant \frac{0.5\ \text{V}/\mu\text{sec}}{2\pi \times 10\text{k}}$$

$$\leqslant 7.9577\ \text{V peak}$$

(Don't forget the 10^{-6} from the μ in the slew rate.)

The maximum input voltage using Eq. (6-12) is

$$V_{\text{in max}} = \frac{V_o}{A_{\text{ckt}}}$$

$$= \frac{7.9577}{1.7783}$$

$$= 4.4750\ \text{V peak}$$

(c) The power supply voltage should be about 2 V greater than the peak output voltage.

$$2 + 7.9577 = 9.9577\ \text{V}$$

And to the nearest volt, the supply should be

$$\pm 10$$

6-3-2 Filter Component Considerations

There are two basic component considerations. One is the component itself, and the other is the component's effect on the overall performance of the filter.

When a component ages and when it is exposed to temperature and humidity variations, its value will change. For the filter to maintain its design characteristics, we must use components that have small variations to temperature, humidity, and aging. The cost usually increases with an increase in temperature and humidity stability. We must choose the most stable components that we can afford.

A fixed-value component typically has a nomimal value and a tolerance. For a capacitor the range of this tolerance can be astonishingly large. Therefore, ceramic or electrolytic capacitors should be avoided, and Mylar or polystyrene capacitors should be used. The tolerance for the capacitor should be $\pm 5\%$ or less for good results.

Resistor tolerances of $\pm 1\%$ are available. These precision resistors are usually expensive. Another alternative is the precision potentiometer, especially for experimental work. These are only slightly larger than a resistor and can be soldered directly into a PC board. When a precision potentiometer is used, it allows us some "tweaking" to compensate for capacitors that are not exact values.

When a component is not the exact calculated value or when it drifts due to aging or environment, the overall response of the filter is affected. Some components will affect the transfer function greatly, but others only slightly. The degree to which the components affect the transfer function is called sensitivity. The study of sensitivity is beyond the scope of this book, but the basic concept should be understood.

The sensitivity of a component is not due to the component. It is due to how the component is used in the transfer function and its magnitude relationship to the other components. With experience, we will notice that in some designs, when a component deviates from its calculated value only slightly, the filter's response deviates considerably. Sometimes a relatively large change in a component, in a different design, will change the response slightly. The difference is the sensitivity.

PROBLEMS

Section 6-1

1. **(a)** When the loss function is -5 dB, is the signal amplified or attenuated?
   (b) Find this value expressed in terms of a gain function.
2. Write the transfer function shown in the form of a loss function.

$$\frac{(s + 3)[(s + 2)^2 + 15]}{s^2(s + 4)^2(s + 7)}$$

3. For Fig. 6-18a:

(a) Draw the loss function with the frequency axis normalized to 3 krad/sec and the magnitude of 6 dB normalized to 0 dB.

(b) Find the normalized magnitude and normalized frequency of the 10 krad/sec location.

(c) Denormalize the graph drawn in part (a) to a frequency of 15 rad/sec and the 0-dB level to −2 dB.

4. For Fig. 6-18b:

(a) Draw the loss function with the frequency axis normalized to 50 rad/sec (notice that this is not at the break frequency) and the magnitude of −5 dB normalized to 0 dB.

(b) What are the normalized magnitude and normalized frequency of the 5 krad/sec location?

(c) Denormalize the graph drawn in part (a) to a frequency of 1 krad/sec and the 0-dB level to 4 dB.

5. For Fig. 6-18c:

(a) Draw the loss function with the frequency axis normalized to 800 rad/sec and the magnitude of −10 dB normalized to 0 dB.

(b) Find the normalized magnitude and normalized frequency of the 5 krad/sec location.

(c) Denormalize the graph drawn in part (a) to a frequency of 1 krad/sec and the 0-dB level to −2 dB.

Section 6-2

6. Draw the graph of the following specifications and determine if the function is an HP, LP, BP, or notch.

$$\omega_s = 25 \text{ rad/sec} \qquad \omega_p = 200 \text{ rad/sec}$$

$$L_{min} = 20 \text{ dB} \qquad L_{max} = 3 \text{ dB}$$

7. Draw the graph of the following specifications and determine if the function is an HP, LP, BP, or notch.

$$\omega_s = 10 \text{ krad/sec} \qquad \omega_{\mathrm{p}} = 2 \text{ krad/sec}$$

$$L_{min} = 40 \text{ dB} \qquad L_{max} = 0.5 \text{ dB}$$

8. Draw the graph of the following specifications and determine if the function is an HP, LP, BP, or notch.

$$\omega_{s1} = 2 \text{ krad/sec} \qquad \omega_{s2} = 12 \text{ krad/sec}$$

$$\omega_{p1} = 5 \text{ krad/sec} \qquad \omega_{p2} = 10 \text{ krad/sec}$$

$$L_{min1} = 40 \text{ dB} \qquad L_{min2} = 20 \text{ dB} \qquad L_{max} = 3 \text{ dB}$$

9. Revise the specifications in Problem 8 so that the filter is symmetrical.

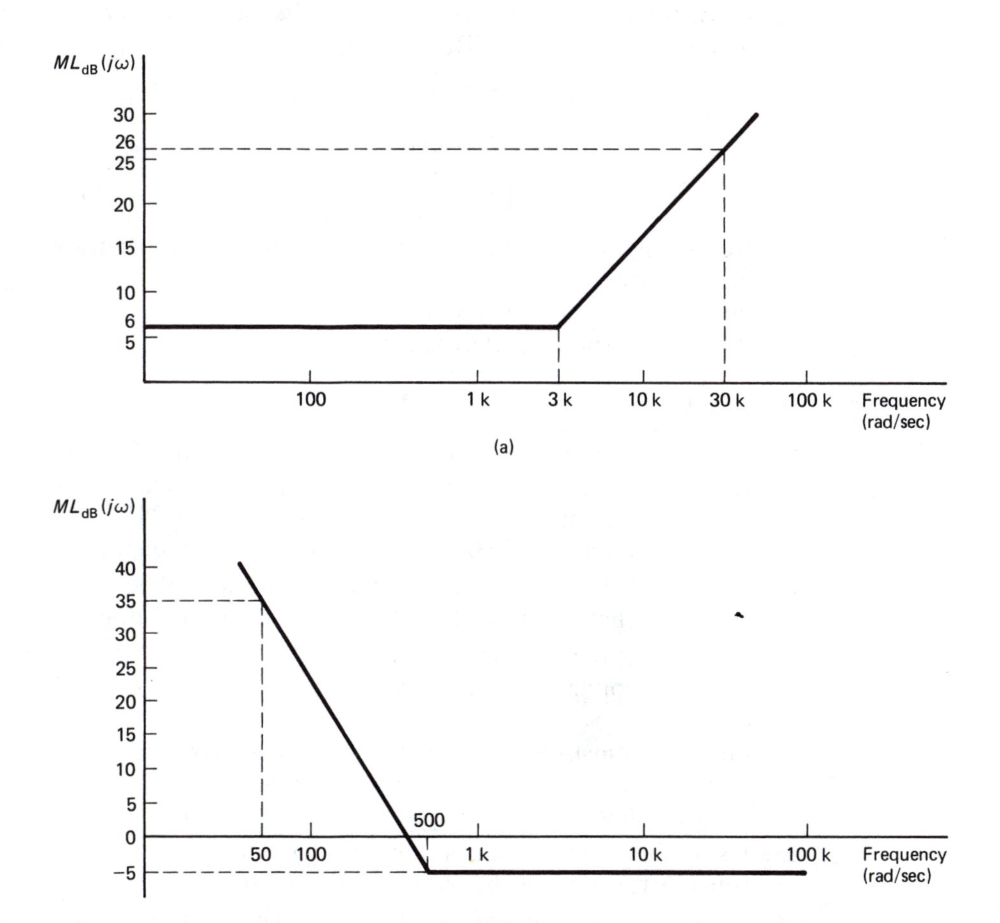

$ML_{dB}(j\omega)$
30
26
25
20
15
10
6
5
100
1 k
3 k
10 k
30 k
100 k
Frequency
(rad/sec)
(a)
$ML_{dB}(j\omega)$
40
35
30
25
20
15
10
5
0
−5
50
100
500
1 k
10 k
100 k
Frequency
(rad/sec)
(b)

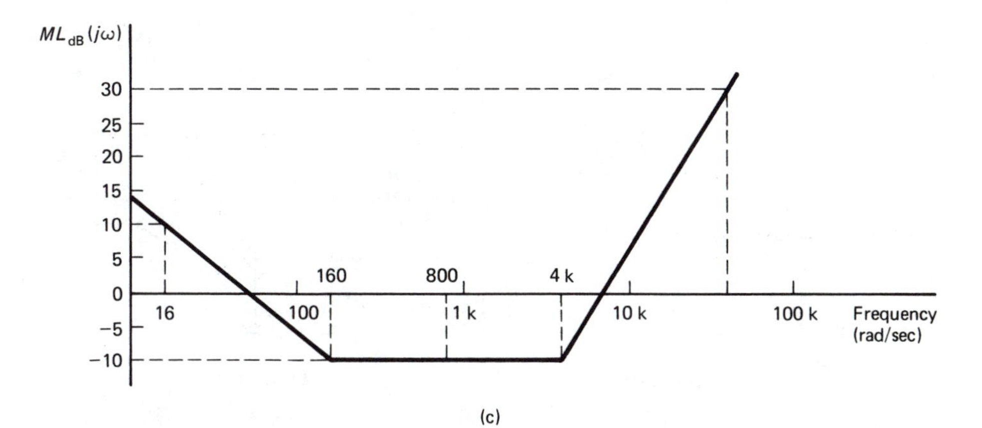

$ML_{dB}(j\omega)$
30
25
20
15
10
5
0
−5
−10
16
100
160
800
1 k
4 k
10 k
100 k
Frequency
(rad/sec)
(c)

Figure 6-18

10. Draw the graph of the following specifications and determine if the function is an HP, LP, BP, or notch.

$$\omega_{s1} = 5 \text{ krad/sec} \qquad \omega_{s2} = 10 \text{ krad/sec}$$

$$\omega_{p1} = 2 \text{ krad/sec} \qquad \omega_{p2} = 12 \text{ krad/sec}$$

$$L_{\min} = 40 \text{ dB} \qquad L_{\max 1} = 2 \text{ dB} \qquad L_{\max 2} = 5 \text{ dB}$$

11. Revise the specifications in Problem 10 so that the filter is symmetrical when:
    (a) The center frequency is not changed.
    (b) The center frequency is changed.

Section 6-3

12. Using the op-amp model shown in Fig. 6-13, draw the equivalent circuit of an op-amp that has an input impedance of 1.5 MΩ, an output impedance of 70 Ω, an open-loop gain of 150,000, a lower break frequency at 8 Hz, and an upper break frequency of 1 MHz.
13. Write the transfer function for the op-amp described in Problem 12 assuming that a load of 10 kΩ is attached to the output.
14. In the pass band of an op-amp filter, V_o must be 9.5 V peak with an input voltage of 1 V peak. The highest frequency of use is 4 kHz.
    (a) What must the open-loop gain be at 4 kHz?
    (b) What must the op-amp's slew rate be in volts/μsec?
    (c) Calculate power supply voltage to the nearest volt.
15. Using the approximate op-amp model shown in Fig. 6-16, find the transfer function of the circuit in Fig. 6-19.
16. Using the approximate op-amp model shown in Fig. 6-16, find the transfer function of the circuit in Fig. 6-20 using the component values.

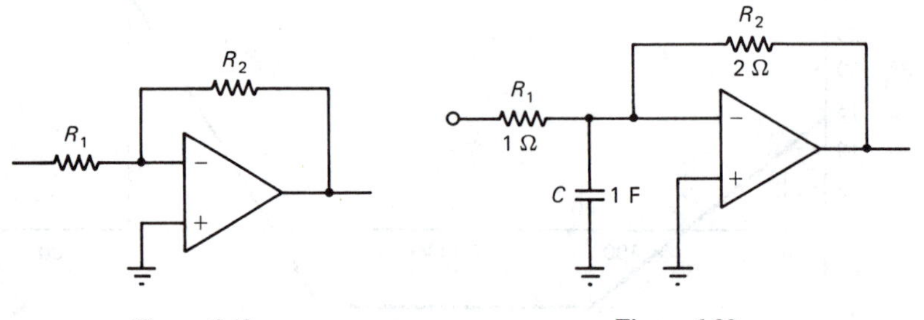

Figure 6-19 **Figure 6-20**

CHAPTER 7

Normalized Low-Pass Filter

7-0 INTRODUCTION

In this chapter we will learn how to use the generic active filter equations to make a normalized low-pass filter. In Chapter 8 we will use these equations to make practical low-pass, high-pass, band-pass, and notch active filters.

A normalized filter will have unrealistic values of resistance and capacitance, usually a few ohms or farads. Generic equations are low-pass by nature and have their break or center frequency normalized. Therefore, an active filter may be normalized in both component value and frequency.

The generic equations are actually found in the last section of this chapter. In the first few sections we define how to apply any filter equation to a circuit. These first few sections apply to low-pass, high-pass, band-pass, and notch filters, but we specifically apply them to a normalized low-pass filter in this chapter.

From this chapter we will have a good working knowledge of three basic normalized low-pass filters. There are a lot of mathematical ways to look at these filters, but we will look at them in the context of previous chapters. Most active-filter texts examine the mathematical wonders of these filters. We approach them from a how-to-design approach.

7-1 TOPOLOGY

Topology is the form of the circuit. It is a map of the locations of components. We will be unable to look at all the possible topologies, but we will focus our attention to five basic topologies: noninverting single feedback, noninverting dual feedback, inverting single feedback, inverting dual feedback, and twin-T.

We use nodal analysis to determine the loss function of a topology. It will be easier if we use admittance, $Y(s)$, for each component. We will also drop the s in the function notation.

$$Y = \frac{1}{Z(s)} \tag{7-1}$$

Later, when we have the loss function expressed in terms of admittances, we can determine the type of component, R or C, to use. The type and value of the components chosen to match a desired loss function will be determined by coefficient matching shown in the next section.

Figure 7-1a shows the noninverting single-feedback topology. Since components R_1 and R_2 are almost always resistors, we will leave them as resistors. Writing the equations at voltage nodes V_1 and V_2, we have

$$(Y_A + Y_B)V_1 - Y_A V_{\text{in}} = 0$$

$$\left(\frac{1}{R_1} + \frac{1}{R_2}\right)V_2 - \left(\frac{1}{R_1}\right)V_o = 0$$

We can combine all the resistors in the second equation into a single coefficient

$$\left(1 + \frac{R_1}{R_2}\right)V_2 - V_o = 0$$

Notice that this coefficient is the same as the gain for a noninverting op-amp amplifier. We will therefore define this coefficient as K:

$$K = 1 + \frac{R_1}{R_2} \tag{7-2}$$

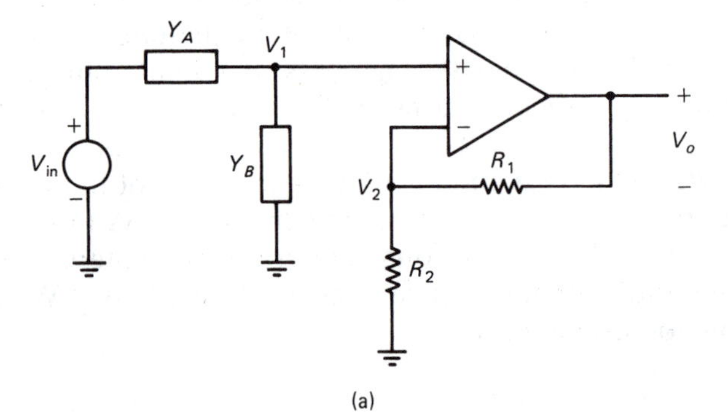

(a)

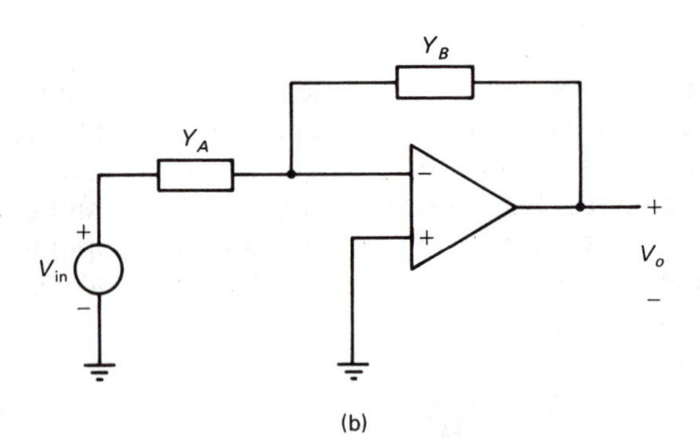

(b)

Figure 7-1

The second equation then becomes

$$KV_2 - V_o = 0$$

Since V_2 is the inverting terminal, it must be equal to V_1. Substituting V_1 into the second equation and solving for V_1, we have

$$V_1 = \left(\frac{1}{K}\right)V_o$$

This is substituted back into the first equation for the node voltage V_1:

$$(Y_A + Y_B)\left(\frac{1}{K}\right)V_o - Y_A V_{\text{in}} = 0$$

Solving for the loss function, we have

$$\frac{V_{\text{in}}}{V_o} = \frac{Y_A + Y_B}{KY_A} \tag{7-3}$$

By a similar procedure we can find the loss function for the inverting single-feedback topology shown in Fig. 7-1b.

$$\frac{V_{\text{in}}}{V_o} = \frac{-Y_B}{Y_A} \tag{7-4}$$

Notice that this loss function does not have a gain constant, and the noninverting single-feedback topology does. This is a characteristic difference between the inverting and noninverting topologies.

The single-feedback topologies are used for first-order filters. The dual-feedback topologies are used for second-order filters.

Figure 7-2 shows the dual-feedback topologies. The same procedure is followed, but with these topologies, solving by substitution is not practical. Instead, we will use determinants to solve the equations.

Let's derive the noninverting dual-feedback topology in Fig. 7-2a. We write the equations at the three nodes, substituting in V_2 for V_3 and using K as defined in Eq. (7-2).

$$(Y_A + Y_B + Y_C + Y_E)V_1 - Y_C V_2 - Y_B V_o = Y_A V_{\text{in}}$$

$$-Y_C V_1 + (Y_C + Y_D)V_2 + 0V_o = 0$$

$$0V_1 + KV_2 - V_o = 0$$

The voltage V_o is

$$V_o = \frac{\begin{vmatrix} Y_A + Y_B + Y_C + Y_E & -Y_C & Y_A V_{\text{in}} \\ -Y_C & Y_C + Y_D & 0 \\ 0 & K & 0 \end{vmatrix}}{\begin{vmatrix} Y_A + Y_B + Y_C + Y_E & -Y_C & -Y_B \\ -Y_C & Y_C + Y_D & 0 \\ 0 & K & -1 \end{vmatrix}}$$

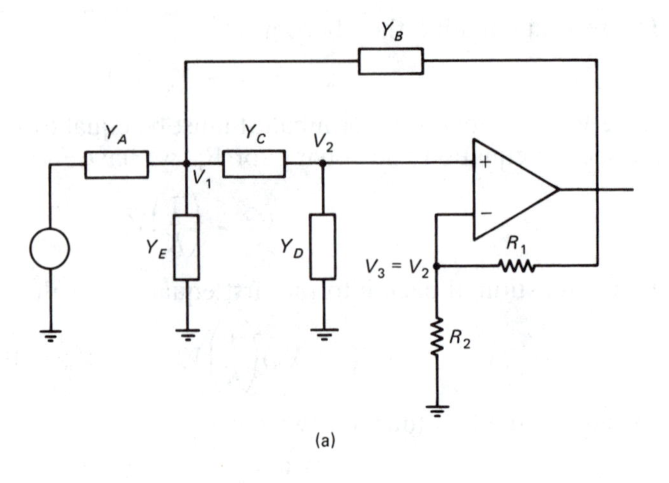

(a)

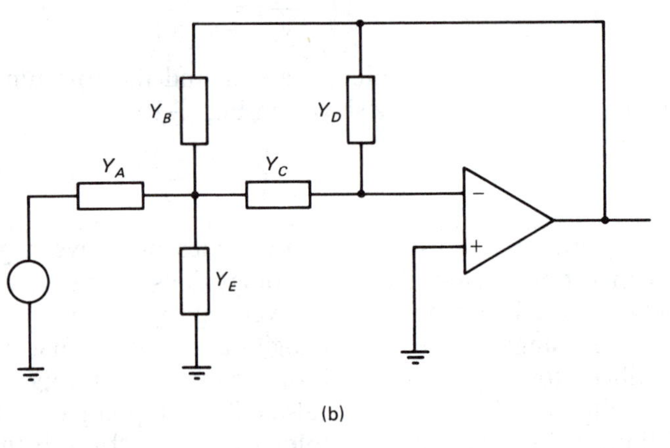

(b)

Figure 7-2

Solving this for V_{in}/V_o yields

$$\frac{V_{in}}{V_o} = \frac{(Y_A + Y_B + Y_E)(Y_C + Y_D) + Y_C(Y_D - Y_B K)}{Y_A Y_C K} \tag{7-5}$$

Using the same procedure, the inverting dual-feedback topology in Fig. (7-2b) can be found.

$$\frac{V_{in}}{V_o} = \frac{Y_D(Y_A + Y_B + Y_C + Y_E) + Y_B Y_C}{-Y_A Y_C} \tag{7-6}$$

The last topology we will use is called twin-T and is shown in Fig. 7-3. This is called twin-T since Y_B–Y_C–Y_F and Y_A–Y_D–Y_E each form the shape of a "T." The loss function for this topology is

$$\frac{V_{in}}{V_o} = \frac{Y_G Y_1 Y_2 + Y_F Y_1(Y_B + Y_C - KY_C) + Y_E Y_2(Y_A + Y_D)}{K(Y_A Y_E Y_2 + Y_B Y_F Y_1)} \tag{7-7}$$

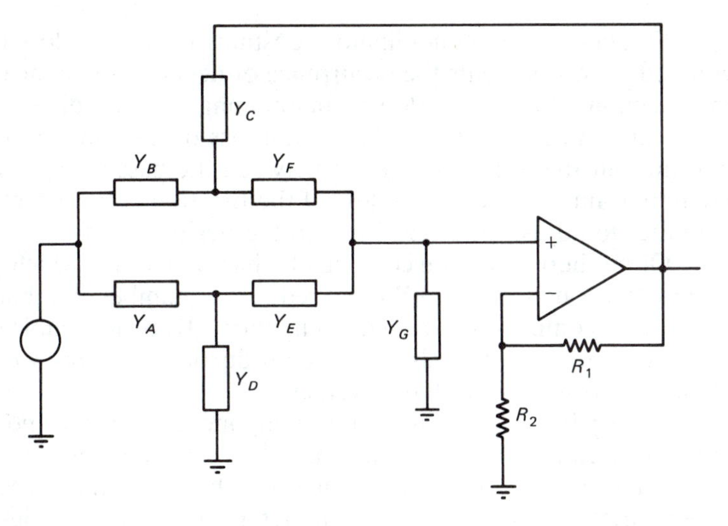

Figure 7-3

where $K = 1 + R_1/R_2$ [defined in Eq. (7-2)]
$Y_1 = Y_A + Y_D + Y_E$
$Y_2 = Y_B + Y_C + Y_F$

The twin-T topology is sometimes modified so that Y_D is tied to the op-amp output instead of reference. When this is done, the loss function changes only slightly. In this case, the last term of the numerator becomes

$$Y_E Y_2(Y_A + Y_D - KY_D)$$

Appendix E contains a summary of the topologies we will use.

7-2 COEFFICIENT MATCHING

When designing a filter, we normally start with a desired loss function equation, choose an appropriate topology, determine what components to use for the admittances, and match the desired loss function's coefficients to the topology loss function's coefficients. This last step is called coefficient matching. The desired loss function comes either from a Bode plot (as shown in Chapter 5) or from a defined function (which we will look at in Section 7-4).

Coefficient matching is as follows:

1. From the topology loss function, determine which admittances should be resistors and which should be capacitors. Choose the admittances such that the desired loss function and the topology loss function have the same number of constants and powers of *s* present in both numerator and denominator.
2. Make the highest power of *s* in the topology loss function equal to 1.
3. Form a set of simultaneous equations by equating the coefficients of the desired loss function to the coefficients of the topology's loss function.

Frequently, the denominator constants of the two loss functions are not equated. This prevents the occurrence of negative resistance values but shifts the function's loss up or down, maintaining the loss difference between the stop band and pass band. The overall loss of the function is typically unimportant, but the differences in dB between the pass bands and the stop bands are important. If the overall loss of the function is important, we may add an amplifier to adjust the overall loss to the desired level.

Often there are more coefficients than equations, which means that there is more than one solution. We can reduce the number of unknowns by allowing some of the components to equal each other. By proper choice this will simplify the equation. This will, also, reduce production cost since we can buy in larger quantities with fewer different values.

Even when we equate some components, we may not have the same number of equations and unknowns. We will frequently have to select some values and use the equation to solve for the rest. Since resistors come in a larger variety of values and potentiometers are smaller and less expensive than variable capacitors, we desire to select capacitor values and solve for resistor values.

EXAMPLE 7-1

From a Bode plot of a desired frequency response we found the transfer function shown. Use coefficient matching to determine a suitable filter.

$$\frac{V_o}{V_{\text{in}}} = \frac{80}{s + 10}$$

SOLUTION First we need to write the desired transfer function as a loss function.

$$L(s) = \frac{s + 10}{80}$$

Since this is a first-order circuit, we will start with the noninverting single-feedback topology in Fig. 7-1a. This topology also has the same sign (positive). From Eq. (7-3),

$$L(s) = \frac{Y_A + Y_B}{KY_A}$$

where $K = 1 + R_1/R_2$ [defined in Eq. (7-2)]

We must choose what type of components to use for Y_A and Y_B. Since the desired loss function has an s in it, one of our admittances must be a capacitor. If we choose Y_A as a capacitor and Y_B as a resistor, the resulting loss function is

$$L(s) = \frac{sC_A + \dfrac{1}{R_B}}{KsC_A}$$

This will not work since the desired loss function does not have an s in the denominator. If we choose Y_A as a resistor and Y_B as a capacitor, the resulting loss function is

$$L(s) = \frac{\frac{1}{R_A} + sC_B}{K\frac{1}{R_A}}$$

This will work since we have the same powers of s in the correct locations.

We then make the highest powers of s equal to 1.

$$L(s) = \frac{s + \frac{1}{R_A C_B}}{K\left(\frac{1}{R_A C_B}\right)}$$

Finally, we write the simultaneous equations by coefficient matching.

$$\frac{1}{R_A C_B} = 10$$

$$\left(1 + \frac{R_1}{R_2}\right)\frac{1}{R_A C_B} = 80$$

Here we have two equations and four unknowns.

When we choose a value for C_B we will still have one more equation than unknowns. We could select a value for R_1 or R_2, but we may not choose a value for R_A because we have chosen a value of C_B in the first equation. To have the most components of the same value, it would be better to let R_1 or R_2 equal R_A.

Let's select $C_B = 1\ \mu\text{F}$ and $R_1 = R_A$. Solving the first equation,

$$R_A = \frac{1}{10C_B} = \frac{1}{10(1\ \mu\text{F})} = 100\ \text{k}\Omega$$

Substituting the known values into the second equation gives

$$\left(1 + \frac{100\ \text{k}\Omega}{R_2}\right)\frac{1}{(100\ \text{k}\Omega)(1\ \mu\text{F})} = 80$$

$$R_2 = 14.286\ \text{k}\Omega$$

Figure 7-4 shows the final circuit.

Most of the time we are not concerned with the phase of the response since we are interested only in the voltage magnitude. When this is the case, we may ignore the sign of the desired loss function and the sign of the topology's loss function. We therefore will have two choices for the design. Even when we must maintain the sign of the desired loss function, we still have two choices by adding an inverting circuit to the output of the filter.

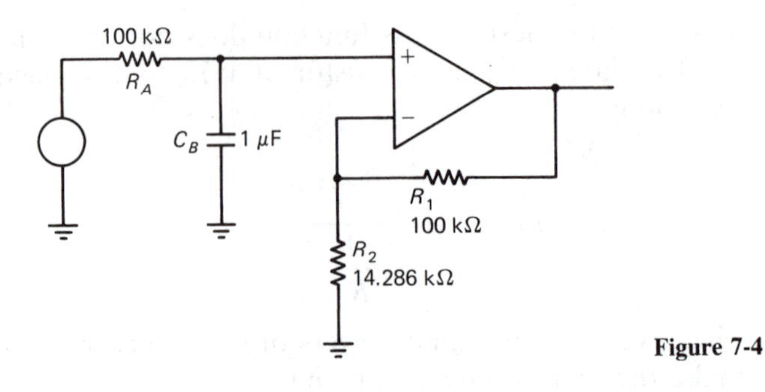

Figure 7-4

In Example 7-1 we could have used the inverting single-feedback topology. The loss function for this from Eq. (7-4) was

$$\frac{V_{in}}{V_o} = \frac{-Y_B}{Y_A}$$

which does not appear to match the example's loss function. If we make Y_B a resistor in parallel with a capacitor and Y_A a resistor, we will be in the correct form.

$$L(s) = \frac{\dfrac{1}{R_B} + sC_B}{-\left(\dfrac{1}{R_A}\right)}$$

In this case we would either ignore the minus sign, or add an inverting amplifier to the output of the filter.

This leads us to conclude that admittances do not have to be single components, but could be complex networks. We will use single-component values for admittances on our designs as much as possible so that the process of the designs will be easy to follow.

In Example 7-1 we selected a practical capacitor value, but we could have selected any value. In Chapter 8 we will learn how to change the values of the components after the design. We may therefore select any convenient value. For calculation purposes it will be easier to select values around 1. When this is done, we call the circuit a normalized circuit. If we had chosen $C_B = 1$ F, the resulting resistor values would have been calculated: $R_A = R_1 = 0.1\ \Omega$, $R_2 = 0.014286\ \Omega$.

We can pre-solve the coefficient matching equations by using variables instead of numbers in the desired loss function. There are a number of ways to solve these, but let's define some general guidelines.

Since resistors have a larger assortment of values, we desire to choose capacitor values. We would also like to have as many component values equal as practical. We may ignore the desired function's multiplying factor since it may be too restrictive, as shown in the following example.

EXAMPLE 7-2

Using coefficient matching, write the general equations to solve the loss function:

$$\frac{s^2 + as + b}{-h}$$

SOLUTION This requires an inverting dual-feedback topology if we wish to match the minus sign. By the process shown in Example 7-1, we can determine the component selection as shown in Fig. 7-5. Making the highest power of s equal to 1, we therefore have

$$\frac{s^2 + \dfrac{1}{C_E}\left(\dfrac{1}{R_A} + \dfrac{1}{R_B} + \dfrac{1}{R_C}\right)s + \dfrac{1}{C_D C_E R_B R_C}}{-\dfrac{1}{C_D C_E R_B R_C}}$$

If we are required to match the h in the desired loss function, this will not work since h may not equal b. This is often the case, and why we usually do not try to match the multiplying factor h.

We will start by allowing $R_B = R_C = R$. The loss function reduces to

$$\frac{s^2 + \dfrac{1}{C_E}\left(\dfrac{1}{R_A} + \dfrac{2}{R}\right)s + \dfrac{1}{C_D C_E R^2}}{-\dfrac{1}{C_D C_E R^2}}$$

Matching the coefficients gives

$$\frac{1}{C_E}\left(\frac{1}{R_A} + \frac{2}{R}\right) = a$$

$$\frac{1}{C_D C_E R^2} = b$$

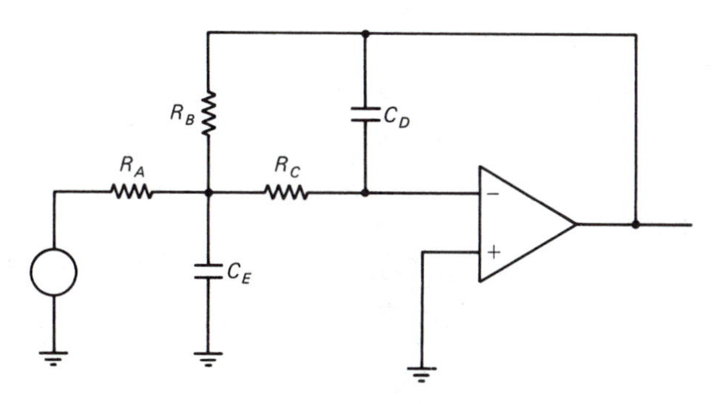

Figure 7-5

Since we will select capacitor values, we will solve for R in the first equation.

$$R = \frac{1}{\sqrt{bC_DC_E}}$$

Also, since C_D and C_E will be selected, we will be able to solve for a particular value of R. Solving the first equation for the last unknown value, R_A, we have

$$R_A = \frac{1}{C_Ea - \frac{2}{R}}$$

Notice that in this equation, for R_A to be positive, C_Ea must be greater than $2/R$. Therefore,

$$R > \frac{2}{C_Ea}$$

This restricts the value of R previously calculated, or we may say that

$$\frac{2}{C_Ea} < R = \frac{1}{\sqrt{bC_DC_E}}$$

$$< \frac{1}{\sqrt{bC_DC_E}}$$

If we choose C_E first we can solve this equation for C_D. We will then be able to determine an appropriate value for C_D so that R_A will be positive.

$$C_D < \frac{C_Ea^2}{4b}$$

We could have assumed that we choose C_D first, and then solve this equation for C_E. Either way it allows us to choose capacitor values and solve for resistor values. The order in which to use these equations is as follows:

1. Choose C_E.

2. Choose C_D such that

$$C_D < \frac{C_Ea^2}{4b}$$

3. Solve for R_B and R_C.

$$R = \frac{1}{\sqrt{bC_DC_E}}$$

4. Solve for R_A.

$$R_A = \frac{1}{C_E a - \dfrac{2}{R}}$$

5. Calculate the resulting multiplying factor.

$$h = \frac{1}{C_D C_E R^2}$$

The topologies shown in Appendix E have a list of pre-solved equations which were found by the process shown in Example 7-2.

7-3 BIQUADS

From Bode plots in Chapter 5 and filter specifications in Chapter 6, we can easily determine some general loss equations for low-pass, high-pass, band-pass, and notch filters.

$$\text{First-order low-pass} = \frac{s + b}{\pm h}$$

$$\text{First-order high-pass} = \frac{s + b}{\pm hs}$$

$$\text{Second-order low-pass} = \frac{s^2 + as + b}{\pm h} \qquad \zeta < 1$$

$$\text{Second-order high-pass} = \frac{s^2 + as + b}{\pm hs^2} \qquad \zeta < 1$$

$$\text{Band pass} = \frac{s^2 + as + b}{\pm hs} \qquad \zeta > 1$$

$$\text{Notch} = \frac{s^2 + as + b}{\pm h(s^2 + e)} \qquad \zeta > 1$$

These are not the only forms, but they are a subset of a general form called a biquad. A biquad is the ratio of two quadratic functions. The loss function, expressed in biquads, is

$$L(s) = \prod_{i=1}^{\text{int}[(N+1)/2]} \frac{m_i s^2 + a_i s + b_i}{\pm h(n_i s^2 + d_i s + e_i)} \qquad \text{(7-8)}$$

where int means the integer part
N is the filter order
m_i and n_i may be either 1 or 0
h, a_i, b_i, and e_i are any real number
d_i will typically be 0 for the filters we will use

A system may require the multiplication of several biquads. All of our topologies can only make one biquad function since they are capable of up to a second-order circuit. We could develop more complex topologies, or we could cascade the topologies we have.

For circuits above second order, we will cascade first- and second-order filters to obtain the desired order. Cascading is possible since op-amp circuits have high input impedance and low output impedance. This allows us to split up a loss function and design individual parts since each op-amp circuit is relatively isolated. Normally when splitting a function into biquads, we ignore the gain constant. Since we often have a resulting multiplying factor, we cannot always match the multiplying factor.

EXAMPLE 7-3

Split the loss function into biquads.

$$\frac{(s + 0.47558)(s^2 + 0.22958s + 1.0863)(s^2 + 0.69854s + 0.59202)}{6.6305 \times 10^{-3}(s^2 + 10.568)(s^2 + 4.3650)}$$

SOLUTION

$$\frac{s + 0.47558}{h_1}$$

$$\frac{s^2 + 0.22958s + 1.0863}{h_2(s^2 + 10.568)}$$

$$\frac{s^2 + 0.69854s + 0.59202}{h_3(s^2 + 4.3650)}$$

There will be a resulting multiplying constant equal to

$$h = h_1 h_2 h_3$$

From Example 7-3 we would continue by using coefficient matching or the pre-solved equations in Appendix E. The first equation could be a non-inverting single-feedback topology, and the last two equations could be inverting dual-feedback topologies. We could use all noninverting forms. If we were not required to match the sign of the multiplying constant, there are several more possibilities.

We notice there is an option of which denominator factor to associate with a numerator factor. Different pairing will yield different component values, but the overall loss function will be the same. We will pair the lowest numerator quadratic constant with the lowest denominator quadratic constant, the next-to-lowest numerator quadratic constant with the next-to-lowest denominator quadratic constant, and so on. This will tend to keep the resistors of a stage close in value to each other.

The order of which stage to cascade first, second, and so on, is determined by the Q of the numerator quadratic. We previously expressed the quadratic in the general form

$$s^2 + 2\zeta\omega_n s + \omega_n^2 \tag{7-9}$$

In Chapter 5 we examined the peaking in the Bode plot, which was in proportion to ζ. The quadratic peaking can also be expressed as a Q where

$$Q = \frac{1}{2\zeta} \tag{7-10}$$

The quadratic of Eq. (7-9) can then be expressed as

$$s^2 + \frac{\omega_n}{Q} s + \omega_n^2 \tag{7-11}$$

Since we may express the quadratic as powers of s with numerical coefficients, a and b,

$$s^2 + as + b$$

we can express the Q as

$$Q = \frac{\sqrt{b}}{a} \tag{7-12}$$

Once each biquad is determined, we can examine the numerator Q of each biquad loss function. We put the circuit with the lowest numerator Q first and the highest numerator Q last. This is done since a high-Q circuit at the peak frequency will cause a large signal that may overdrive the next circuit. With the low-Q circuit first, the signal input to the high-Q circuit will be at a lower level.

An odd-order function will contain a first-order factor. We will still refer to this as a biquad, even though it is not a quadratic. When the loss function contains a first-order biquad, we will always use this circuit first since there is no possibility of it having a peak frequency.

EXAMPLE 7-4

The biquads from Example 7-3 are listed below. What is the numerator Q of each biquad, and in what order should they be put?

$$\frac{s + 0.47558}{h_1}$$

$$\frac{s^2 + 0.22958s + 1.0863}{h_2(s^2 + 10.568)}$$

$$\frac{s^2 + 0.69854s + 0.59202}{h_1(s^2 + 4.3650)}$$

SOLUTION Since the first term has a first-order numerator, it will be the first stage in the filter. The other two biquads' Q can be calculated by Eq. (7-12).

$$Q = \frac{\sqrt{b}}{a}$$

$$Q = \frac{\sqrt{1.0863}}{0.22958} = 4.5398$$

$$Q = \frac{\sqrt{0.59202}}{0.69854} = 1.1015$$

Therefore, the third term will be for the second stage, and the second term will be for the last stage.

7-4 LOW-PASS FILTER APPROXIMATIONS

More often than not, we are given the filter specifications as shown in Chapter 6. We cannot design a filter that is able to form blocks like a filter specification. Instead, we approximate the filter specification by staying within the limits of L_{max} and L_{min}.

There are a number of defined low-pass functions that will approximate a filter specification. We will look at three approximations: Butterworth, Chebyshev, and elliptic. These approximations are based on the magnitude of the loss. The phase is not restricted since a large number of filter applications are not concerned with the phase.

These equations can be transformed from low-pass to high-pass, band-pass, and notch filters. We will, therefore, call these "generic equations." Since they are used to approximate a filter specification, we may also refer to these as "approximation equations."

In this section we concentrate on how to use these generic equations to make circuits that approximate a filter specification. Appendix F contains information on how the generic or approximation equations are formed, and how tables of these equations can be calculated.

In this section we will also restrict ourselves to a normalized low-pass filter. This means that we will find the low-pass filter with the pass-band edge frequency at 1 and choose our capacitor values at 1 (when possible). In Chapter 8 we will convert the normalized circuit to practical values of resistance and capacitance, and the circuit to any frequency.

7-4-1 General Form

Each of the three functions are based on the loss function form

$$L(s) = \sqrt{1 + \varepsilon^2 F^2(s)} \tag{7-13}$$

$F(s)$ is the specific function that determines whether the loss function will be Butterworth, Chebyshev, or elliptic. ε is a constant that is determined by the maximum pass-band loss, L_{max}. In all three cases this constant is calculated by

$$\varepsilon = \sqrt{10^{0.1 L_{max}} - 1} \tag{7-14}$$

TABLE 7-1 Butterworth: $L_{max} = 3.0103$ dB

N	Denominator constant	Numerator
1	1.0000	$s + 1.0000$
2	1.0000	$s^2 + 1.4142s + 1.0000$
3	1.0000	$(s + 1.0000)(s^2 + 1.0000s + 1.0000)$
4	1.0000	$(s^2 + 1.8478s + 1.0000)(s^2 + 0.76537s + 1.0000)$
5	1.0000	$(s + 1.0000)(s^2 + 1.6180s + 1.0000)(s^2 + 0.61803s + 1.0000)$

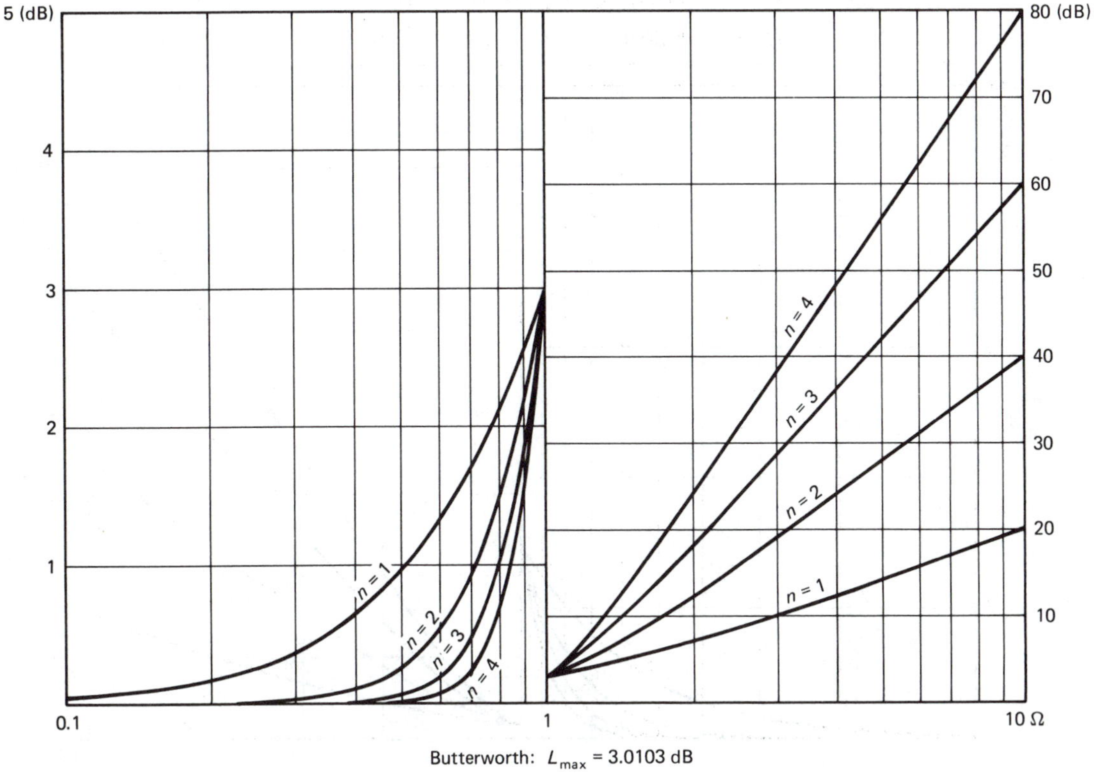

Butterworth: $L_{max} = 3.0103$ dB

Figure 7-6

7-4-2 Butterworth

The response of a Butterworth filter is shown in Figs. 7-6, 7-7, and 7-8. The Butterworth equations are shown in Tables 7-1, 7-2, and 7-3. Notice that the graphs shown in the figures are split at the frequency 1. The graph to the left of 1 has a different vertical scale than it does on the right, but the horizontal axis is the same for both sections. This allows us to see more clearly the shape of the function before the break frequency since we have magnified it.

Each figure has a different L_{max} value. We see that the curve is a smooth curve. This is why it is sometimes called the maximally flat approximation.

The order of filter required for a particular filter specification can be

TABLE 7-2 Butterworth: L_{max} = 1.5000 dB

N	Denominator constant	Numerator
1	1.5569	$s + 1.5569$
2	1.5569	$s^2 + 1.7646s + 1.5569$
3	1.5569	$(s + 1.1590)(s^2 + 1.1590s + 1.3433)$
4	1.5569	$(s^2 + 2.0640s + 1.2478)(s^2 + 0.85494s + 1.2478)$
5	1.5569	$(s + 1.0926)(s^2 + 1.7678s + 1.1937)(s^2 + 0.67525s + 1.1937)$

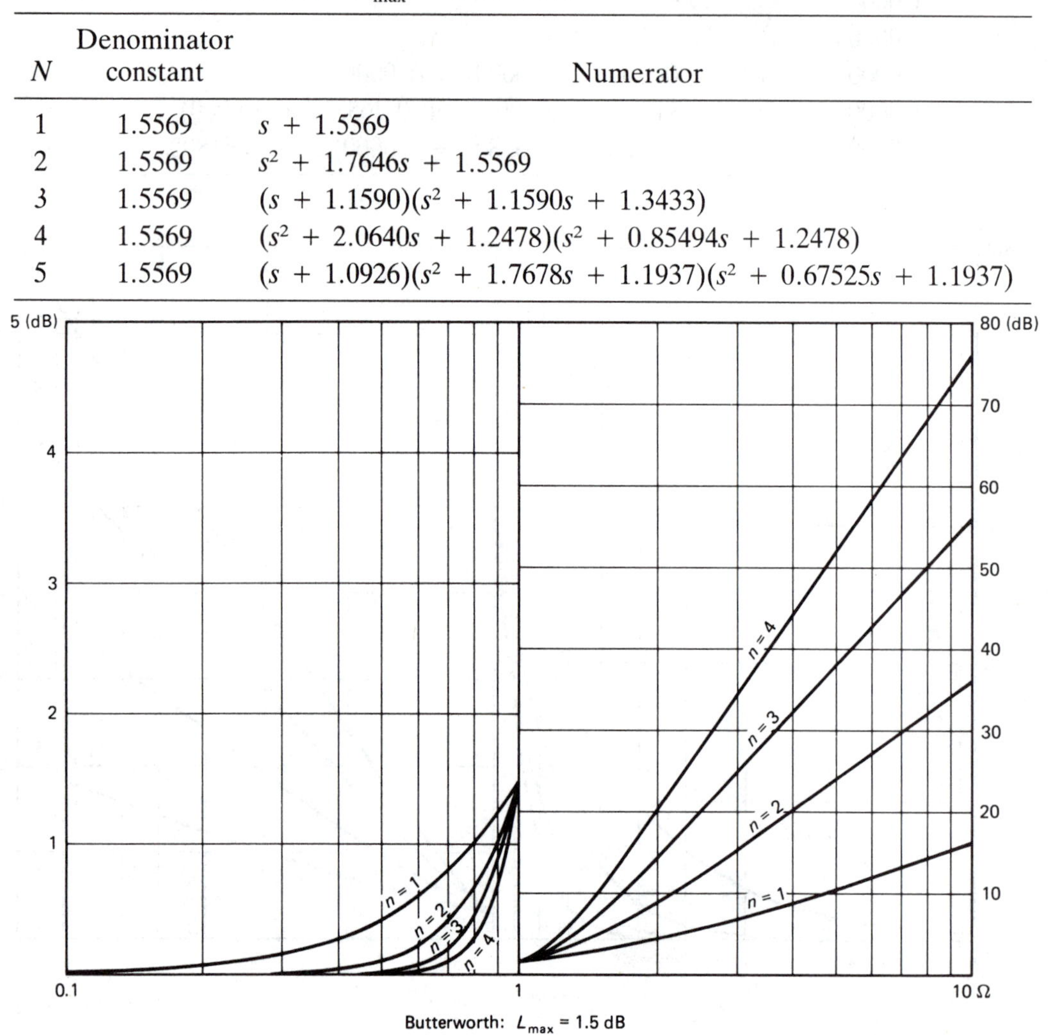

Figure 7-7

found in two different ways. One method is graphic and the other is by calculation. In both cases we find the normalized stop-band frequency first.

$$\Omega_s = \frac{\omega_s}{\omega_p} \tag{7-15}$$

Using this equation will make our resulting filter have a break frequency at 1 rad/sec, not at the originally desired break frequency. In Chapter 8 we will learn how to shift this filter back up to the desired break frequency.

In the graphic method we determine which figure of Figs. 7-6, 7-7, and

TABLE 7-3 Butterworth: $L_{\max} = 0.2500$ dB

N	Denominator constant	Numerator
1	4.1081	$s + 4.1081$
2	4.1081	$s^2 + 2.8664s + 4.1081$
3	4.1081	$(s + 1.6016)(s^2 + 1.6016s + 2.5650)$
4	4.1081	$(s^2 + 2.6306s + 2.0268)(s^2 + 1.0896s + 2.0268)$
5	4.1081	$(s + 1.3266)(s^2 + 2.1464s + 1.7598)(s^2 + 0.81986s + 1.7598)$

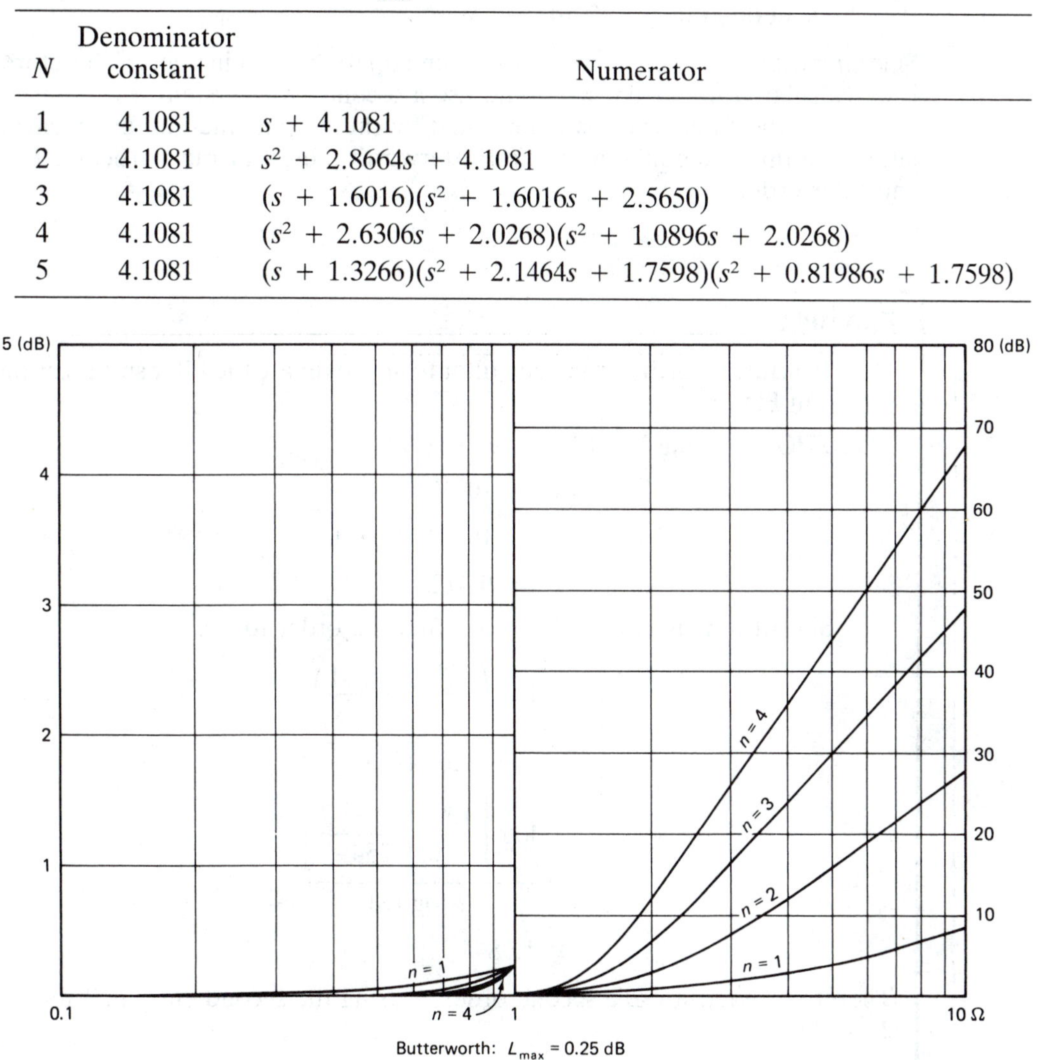

Figure 7-8

7-8 to use by our required L_{max}. We then locate our L_{min} at our Ω_s on this graph. The filter order that just touches or is above this point will be the filter order to use. The disadvantage of this method is that we must have a graph with our particular L_{max}. The calculation method does not require a graph.

In the calculation method we find the filter order by

$$n \geq \frac{\log\left(\dfrac{10^{0.1L_{min}} - 1}{\varepsilon^2}\right)}{2 \log(\Omega_s)} \tag{7-16}$$

where n is an integer
ε is defined in Eq. (7-14)

Since n must be an integer, we always round up to the next integer. Therefore, if we calculated $n = 1.03$, we would use a second-order system.

Once the filter order is determined by the graphic method or by calculation, we find the equation in the Butterworth tables for our particular L_{max} and filter order.

EXAMPLE 7-5

Design a Butterworth normalized filter to approximate the filter specification shown in Fig. 7-9.

SOLUTION Using Eq. (7-14), we find ε^2.

$$\begin{aligned}\varepsilon^2 &= 10^{0.1L_{max}} - 1 \\ &= 10^{(0.1)(1.5)} - 1 \\ &= 0.41254\end{aligned}$$

Substituting into Eq. (7-16), we find the order to use

$$\begin{aligned}n &\geq \frac{\log\left(\dfrac{10^{0.1L_{min}} - 1}{\varepsilon^2}\right)}{2 \log(\Omega_s)} \\ &\geq \frac{\log\left[\dfrac{10^{(0.1)(32)} - 1}{0.41254}\right]}{2 \log(8)} \\ &\geq 1.9844\end{aligned}$$

Therefore, we will use a second order. Using the second entry in Table 7-2, we have

$$\frac{s^2 + 1.7646s + 1.5569}{1.5569}$$

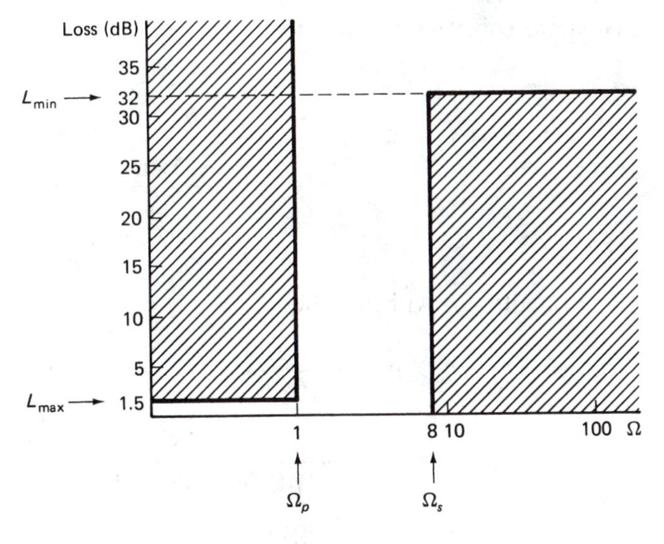

Figure 7-9

To design the circuit we use the topologies in Appendix E. Either the inverting or the noninverting low-pass dual-feedback topology will work. Let's use the noninverting topology since we will match the sign of the denominator. Using the general biquad equation (E-1) in Appendix E, we can determine

$$a = 1.7646 \qquad b = 1.5569 \qquad h = 1.5569$$

From the noninverting dual-feedback low-pass section in Appendix E, we must first determine if the topology will work.

$$0 \leq 2 - \frac{a}{\sqrt{b}}$$

$$\leq 2 - \frac{1.7646}{\sqrt{1.5569}}$$

$$\leq 0.58578$$

Since this is true, we may use this topology.

We then choose a value for $C_B = C_D = C$. Since we want a normalized circuit, we will choose $C = 1$ F. Therefore, R is

$$R = \frac{1}{C\sqrt{b}}$$

$$= \frac{1}{1\sqrt{1.5569}}$$

$$R_A = R_C = R = 0.80144\ \Omega$$

Next we solve for the K value.

$$K = 3 - \frac{a}{\sqrt{b}}$$

$$= 3 - \frac{1.7646}{\sqrt{1.5569}}$$

$$= 1.5858$$

The resulting h will then be

$$h = \frac{K}{C_B C_D R_A R_C}$$

$$= \frac{1.5858}{1(1)(0.80144)(0.80144)}$$

$$= 2.4689$$

We then calculate R_1 and R_2 using the relationship of R_1, R_2, and K shown in Eq. (E-2).

$$K = 1 + \frac{R_1}{R_2}$$

In this equation we can assume a value for one of the resistors and solve for the other. We will arbitrarily assume a value of R_2 and solve for R_1. For economic reasons we would like to have as many resistors equal in value; therefore, we will choose R_2 equal to 0.80144. Solving for R_1, we have

$$R_1 = R_2(K - 1)$$

$$= (0.80144)(1.5858 - 1)$$

$$= 0.46947\ \Omega$$

The final normalized circuit is shown in Fig. 7-10.

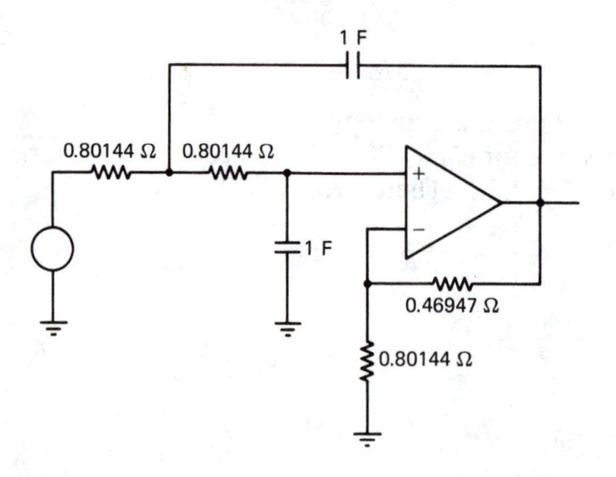

Figure 7-10

In Example 7-5 our resulting h was not equal to our desired h. If we had to match the h, we could have added another amplifier. This would require a noninverting amplifier, or we could change our design to an inverting low-pass dual feedback topology followed by an inverting amplifier.

TABLE 7-4 Chebyshev: $L_{max} = 3.0103$ dB

N	Denominator constant	Numerator
1	1.0000	$s + 1.0000$
2	0.5000	$s^2 + 0.64359s + 0.70711$
3	0.2500	$(s + 0.29804)(s^2 + 0.29804s + 0.83883)$
4	0.1250	$(s^2 + 0.41044s + 0.19579)(s^2 + 0.17001s + 0.90290)$
5	0.06250	$(s + 0.17719)(s^2 + 0.28670s + 0.37689)(s^2 + 0.10951s + 0.93590)$

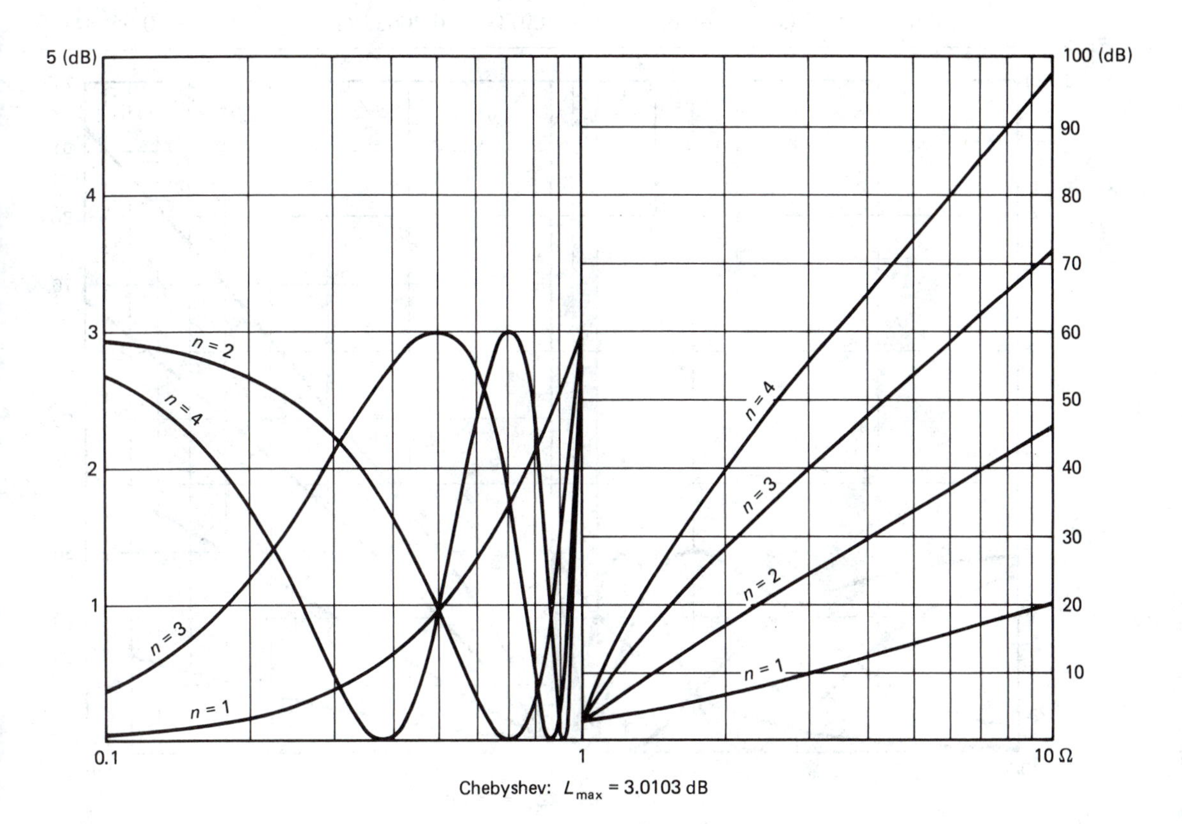

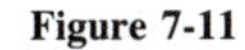
Figure 7-11

7-4-3 Chebyshev

The response of a Chebyshev filter is shown in Figs. 7-11, 7-12, and 7-13. The Chebyshev equations are shown in Tables 7-4, 7-5, and 7-6. The graphs shown in the figures are split at 1, the same as with the Butterworth graphs.

The main characteristic of this response is the equal height ripple in the pass band. This is why this response is sometimes called the equiripple approximation.

The equations for the Chebyshev response are similar to the Butterworth equations. We notice in the Chebyshev equations that the break frequencies occur at a lower frequency than the Butterworth equations. Looking at the

TABLE 7-5 Chebyshev: $L_{max} = 1.5000$ dB

N	Denominator constant	Numerator
1	1.5569	$s + 1.5569$
2	0.77846	$s^2 + 0.92218s + 0.92521$
3	0.38923	$(s + 0.42011)(s^2 + 0.42011s + 0.92649)$
4	0.19462	$(s^2 + 0.57521s + 0.24336)(s^2 + 0.23826s + 0.95046)$
5	0.097308	$(s + 0.24765)(s^2 + 0.40071s + 0.40682)(s^2 + 0.15306s + 0.96584)$

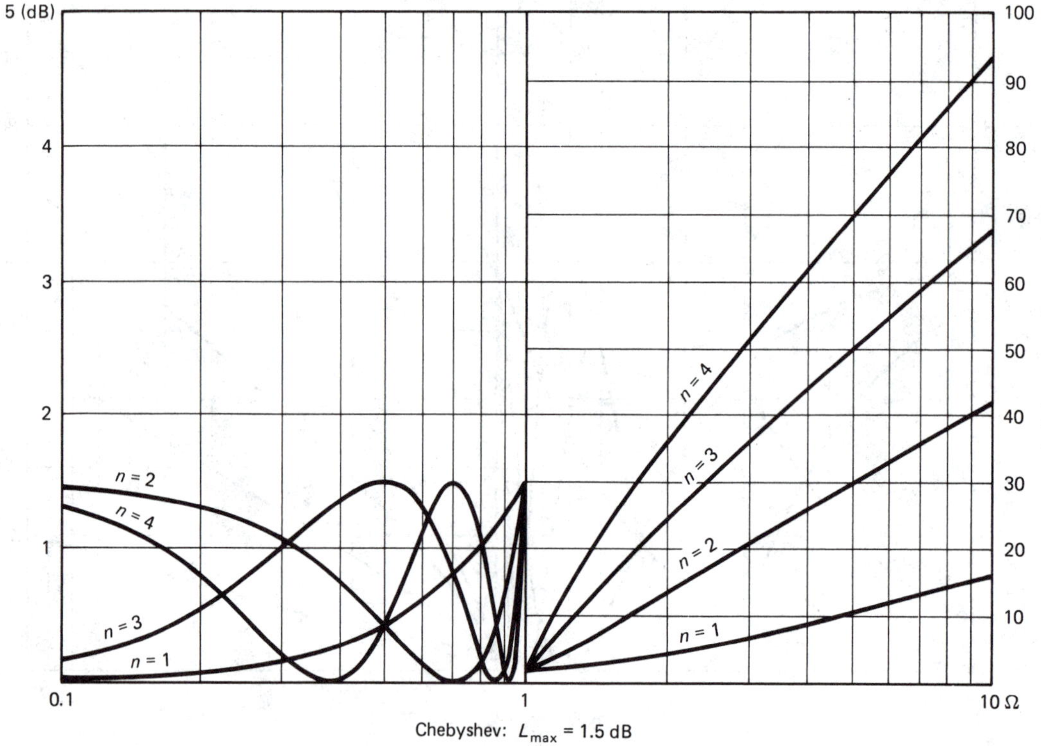

Figure 7-12

graphs, we see that to accomplish this, each quadratic has some peaking (due to ζ being small). Notice that the first-order Chebyshev equation is not a quadratic and will therefore be the same equation as the first-order Butterworth.

Since the Chebyshev has an earlier break frequency than the Butterworth, the roll-off rate of 20 dB/dec per filter order starts sooner, and as a result, the Chebyshev will have a greater attenuation at any stop-band frequency than the Butterworth. Both types of filters eventually have the same roll-off rate, but the Chebyshev has a greater loss in the stop band.

Using the Chebyshev equations is similar to using the Butterworth equations. First you normalize the stop-band frequency using Eq. (7-15). Then we may apply the graphic method described in the Butterworth section to the

TABLE 7-6 Chebyshev: L_{max} = 0.2500 dB

N	Denominator constant	Numerator
1	4.1081	$s + 4.1081$
2	2.0541	$s^2 + 1.7967s + 2.1140$
3	1.0270	$(s + 0.76722)(s^2 + 0.76722s + 1.3386)$
4	0.51351	$(s^2 + 1.0261s + 0.45485)(s^2 + 0.42504s + 1.1620)$
5	0.25676	$(s + 0.43695)(s^2 + 0.70700s + 0.53642)(s^2 + 0.27005s + 1.0954)$

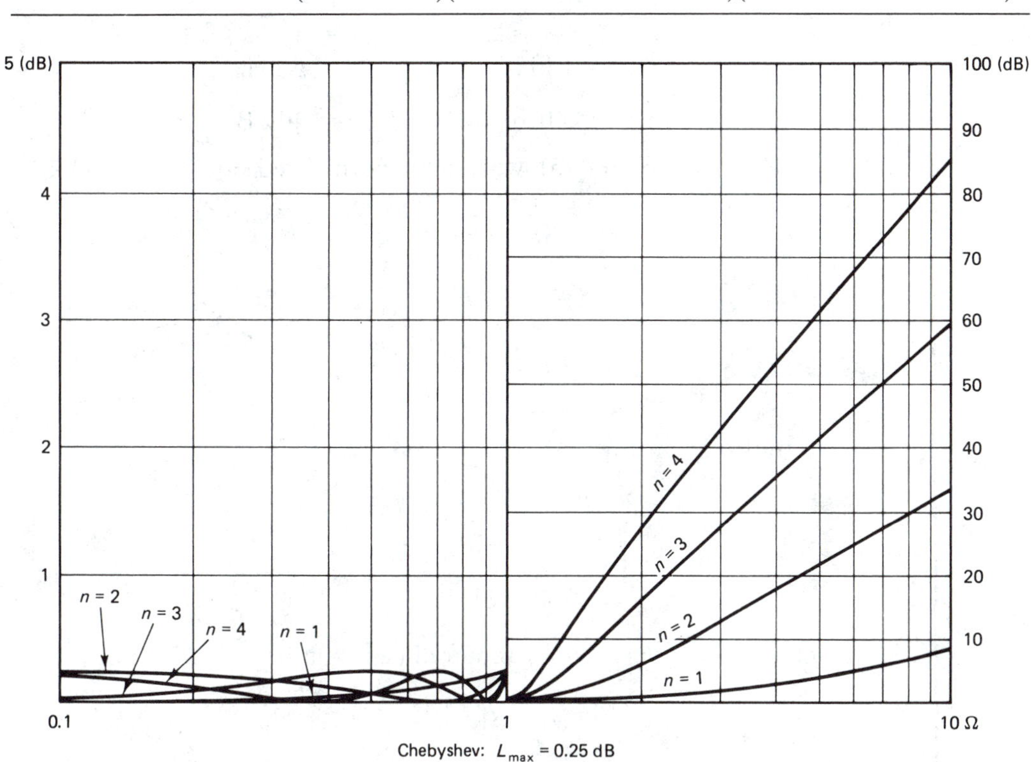

Chebyshev: L_{max} = 0.25 dB

Figure 7-13

Chebyshev graphs, or we may calculate the filter order by

$$n \geq \frac{\cosh^{-1}\left(\dfrac{10^{0.1L_{min}} - 1}{\varepsilon^2}\right)^{1/2}}{\cosh^{-1}(\Omega_s)} \tag{7-17}$$

where n is an integer

ε is defined in Eq. (7-14)

$\cosh^{-1}(x) = \ln\left(x + \sqrt{x^2 - 1}\right)$

Since n must be an integer, we always round up to the next integer value.

Once the order is determined, we use the table containing our particular L_{max} and order.

Since the Chebyshev equations have peaking, we must be careful to cascade the lowest Q circuits first as pointed out in Section 7-3. These tables are written with the lowest Q to highest Q when read left to right.

EXAMPLE 7-6

Design a normalized Chebyshev filter to approximate the following specifications:

$$f_p = 1 \text{ kHz} \qquad f_s = 5 \text{ kHz}$$

$$L_{max} = 3.0103 \text{ dB} \qquad L_{min} = 40 \text{ dB}$$

SOLUTION Using Eq. (7-15), we find the normalized stop band frequency.

$$\Omega_s = \frac{\omega_s}{\omega_p} = \frac{2\pi(5 \text{ kHz})}{2\pi(1 \text{ kHz})} = 5$$

We find ε^2 from Eq. (7-14).

$$\varepsilon^2 = 10^{0.1L_{max}} - 1 = 10^{(0.1)(3.0103)} - 1 = 1.0000$$

The correct order of filter required can then be found by Eq. (7-17).

$$n \geq \frac{\cosh^{-1}\left(\dfrac{10^{0.1L_{min}} - 1}{\varepsilon^2}\right)^{1/2}}{\cosh^{-1}(\Omega_s)}$$

$$\geq \frac{\cosh^{-1}\left[\dfrac{10^{(0.1)(40)} - 1}{1}\right]^{1/2}}{\cosh^{-1}(5)}$$

$$\geq 2.3112$$

Therefore, we need a third-order filter.

From Table 7-4 we find the third-order equation.

$$\frac{(s + 0.29804)(s^2 + 0.29804s + 0.83883)}{0.2500}$$

Here we will use a low-pass inverting single-feedback topology followed by a low-pass inverting dual-feedback topology. Since in Appendix E we notice that the single-feedback topology allows us to set h to a particular value, we may design the second-order filter first and determine the gain required for the first-order circuit to give us an overall h of 0.2500.

We will first design the second-order circuit using the biquad.

$$\frac{s^2 + 0.29804s + 0.83883}{h_1}$$

Using the equations listed in Appendix E, we first choose $C_E = 1$ F, then choose C_D such that

$$C_D < \frac{C_E a^2}{4b}$$

$$< \frac{1(0.29804)^2}{4(0.83883)}$$

$$< 0.026474$$

Since values of capacitors have the same magnitudes with different powers of 10, we will choose $C_D = 0.01$ F.

Solving for the resistors gives

$$R = \frac{1}{\sqrt{bC_D C_E}}$$

$$= \frac{1}{\sqrt{(0.83883)(0.01)(1)}}$$

$$R_B = R_C = R = 10.919\ \Omega$$

$$R_A = \frac{R}{RC_E a - 2}$$

$$= \frac{10.919}{(10.919)(1)(0.29804) - 2}$$

$$= 8.7059\ \Omega$$

The resulting h for the second-order circuit is

$$h_1 = \frac{1}{C_D C_E R_B R_C}$$

$$= \frac{1}{(0.01)(1)(10.919)^2}$$

$$= 0.83883$$

Since we want to make the overall denominator constant equal to 0.25, we need to solve for the h_2 required for the first-order circuit.

$$h_1 h_2 = 0.25000$$

$$h_2 = \frac{0.25000}{h_1}$$

$$= 0.29803$$

We find the procedure for the inverting single-feedback topology in Appendix E. Choose $C_B = 1$ F. Solving for R_B yields

$$R_B = \frac{1}{bC_B}$$

$$= \frac{1}{(0.29804)(1)}$$

$$= 3.3553\ \Omega$$

Since we want a particular h, we will solve for R_A.

$$R_A = \frac{1}{hC_B}$$

$$= \frac{1}{(0.29803)(1)}$$

$$= 3.3553\ \Omega$$

R_A and R_B do not calculate to the same exact value, but they are so close that we will use the same resistor. The resulting circuit is shown in Fig. 7-14.

7-4-4 Elliptic

The response of the elliptic filter is shown in Figs. 7-15, 7-16, and 7-17. The elliptic equations are shown in Tables 7-7, 7-8, and 7-9. The graphs shown in the figures are split at 1, the same as the Butterworth and Chebyshev graphs. This type of filter is sometimes referred to as the Cauer approximation.

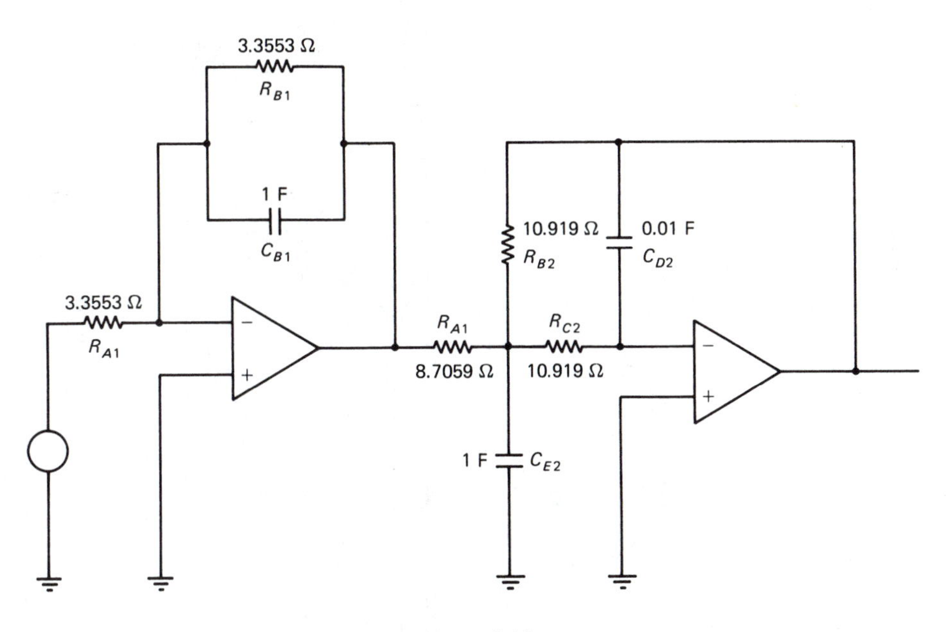

Figure 7-14

These equations are similar to the Chebyshev equations except that a denominator quadratic is added. The quadratic in the denominator has a ζ value of zero, which causes an infinite peak at a frequency in the stop band. This will pull the loss response such that the roll-off will be more than 20 dB/dec per order. Because of this denominator quadratic, the equations are calculated using a particular stop band, Ω_s. The denominator quadratic does have a disadvantage.

The response in the pass band will not continue to attenuate for even-order equations. For odd-order equations the response will eventually increase in attenuation, but at a rate of only 20 dB/dec.

The most efficient use of this filter is when the desired Ω_s and the equation's Ω_s are the same. With the availability of computers it does not make sense to design the filter any other way. Therefore, in this book we will only have stop band frequencies for the elliptic at 2 or 5, which matches the tables available in this book.

Since these equations have large peaks in the loss response, we need to make sure we have the lowest-Q biquad first. These tables are organized in biquads with the lowest Q in the top entry and the highest Q in the bottom entry for each order. The appropriate denominator quadratic to use with the numerator quadratic is listed directly to the right.

To find the order of filter required, we use the table with the appropriate $L_{\max}$ and Ω_s, then find the entry that has an $L_{\min}$ equal to or greater than our desired $L_{\min}$.

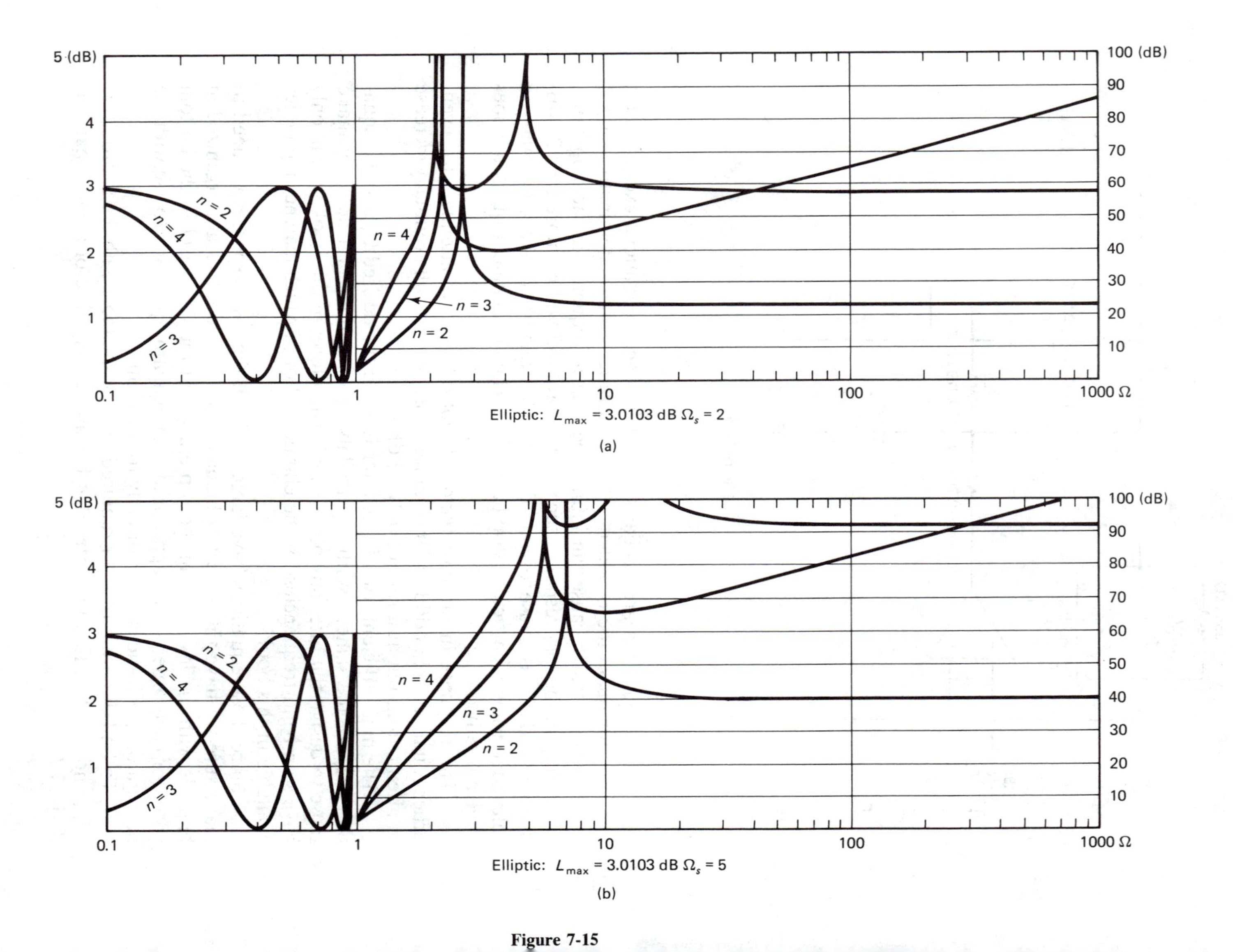

5 (dB)
4
3
2
1
0.1
1
10
100
1000 Ω
100 (dB)
90
80
70
60
50
40
30
20
10
n = 2
n = 4
n = 3
n = 4
n = 3
n = 2
Elliptic: L_{max} = 3.0103 dB Ω_s = 2
(a)
5 (dB)
4
3
2
1
0.1
1
10
100
1000 Ω
100 (dB)
90
80
70
60
50
40
30
20
10
n = 2
n = 4
n = 3
n = 4
n = 3
n = 2
Elliptic: L_{max} = 3.0103 dB Ω_s = 5
(b)

Figure 7-15

TABLE 7-7a Elliptic: $L_{max} = 3.0103$ dB

N	L_{min} (dB)	Denominator constant	Numerator	Denominator
			$\Omega_s = 2$	
2	22.900	7.1612×10^{-2}	$s^2 + 0.60745s + 0.75593$	$s^2 + 7.4641$
3	40.321	5.3882×10^{-2}	$s + 0.32193$	
			$s^2 + 0.26805s + 0.86251$	$s^2 + 5.1532$
4	57.775	1.2920×10^{-3}	$s^2 + 0.42709s + 0.22182$	$s^2 + 4.5933$
			$s^2 + 0.15023s + 0.91666$	$s^2 + 24.227$
5	75.229	1.6140×10^{-3}	$s + 0.19059$	
			$s^2 + 0.28496s + 0.41342$	$s^2 + 4.3650$
			$s^2 + 0.095977s + 0.94485$	$s^2 + 10.568$

TABLE 7-7b Elliptic: $L_{max} = 3.0103$ dB

N	L_{min} (dB)	Denominator constant	Numerator	Denominator
			$\Omega_s = 5$	
2	39.824	1.0205×10^{-2}	$s^2 + 0.63888s + 0.71429$	$s^2 + 49.495$
3	65.756	7.6540×10^{-3}	$s + 0.30136$	
			$s^2 + 0.29371s + 0.84232$	$s^2 + 33.165$
4	91.688	2.6038×10^{-5}	$s^2 + 0.41287s + 0.19936$	$s^2 + 29.203$
			$s^2 + 0.16711s + 0.90494$	$s^2 + 167.77$
5	117.62	3.2547×10^{-5}	$s + 0.17906$	
			$s^2 + 0.28654s + 0.38201$	$s^2 + 27.586$
			$s^2 + 0.10751s + 0.93723$	$s^2 + 71.405$

EXAMPLE 7-7

Design an elliptic filter to approximate the following filter specification.

$$\omega_p = 2 \text{ kHz} \qquad \omega_s = 10 \text{ kHz}$$

$$L_{max} = 0.2500 \text{ dB} \qquad L_{min} = 25 \text{ dB}$$

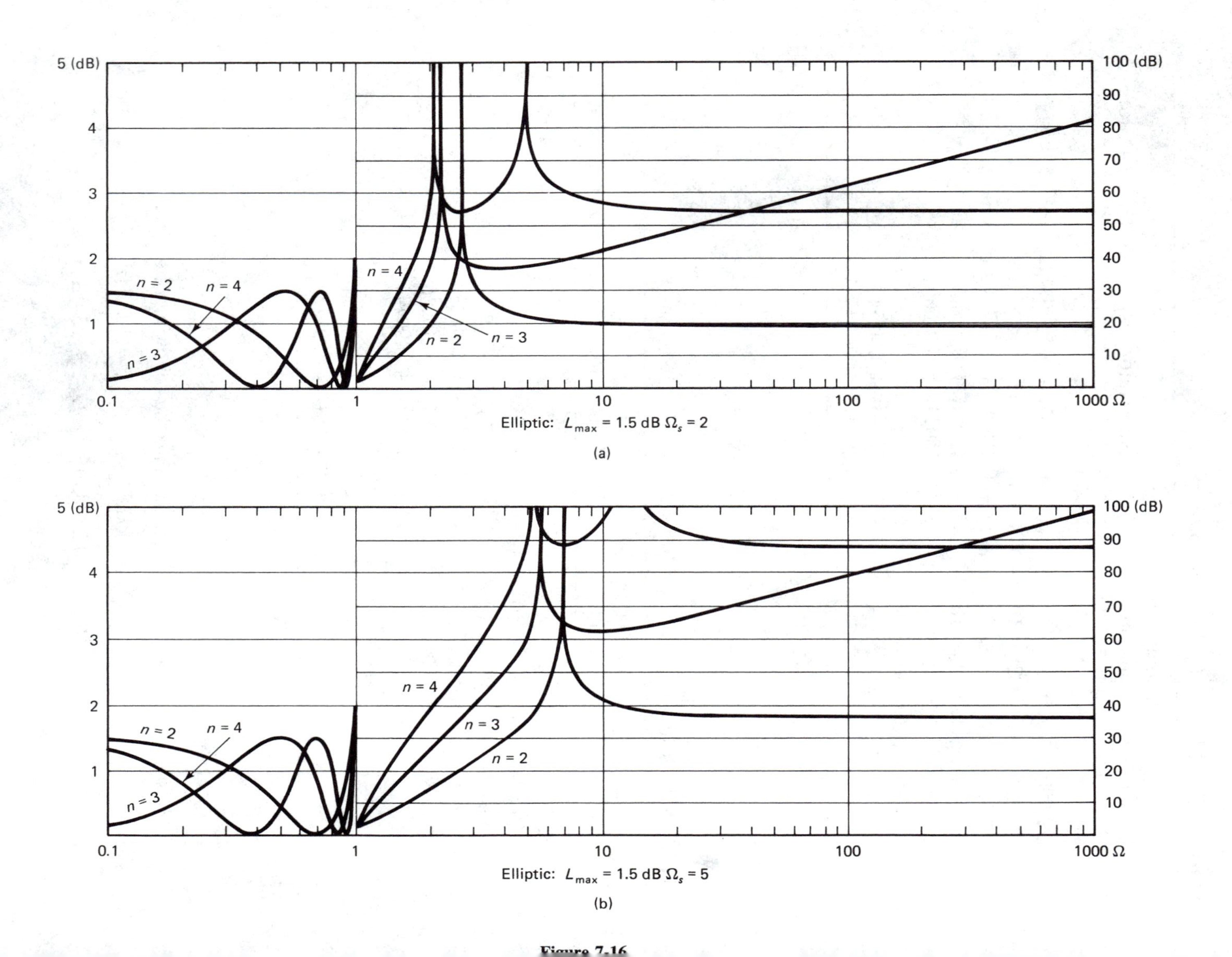

Elliptic: $L_{max} = 1.5$ dB $\Omega_s = 2$

(a)

Elliptic: $L_{max} = 1.5$ dB $\Omega_s = 5$

(b)

Figure 7.16

TABLE 7-8a Elliptic: $L_{max} = 1.5000$ dB

N	L_{min} (dB)	Denominator constant	Numerator	Denominator
			$\Omega_s = 2$	
2	19.086	1.1109×10^{-1}	$s^2 + 0.85334s + 0.98549$	$s^2 + 7.4641$
3	36.476	8.3891×10^{-2}	$s + 0.45674$	
			$s^2 + 0.37285s + 0.94651$	$s^2 + 5.1532$
4	53.929	2.0116×10^{-3}	$s^2 + 0.59927s + 0.27702$	$s^2 + 4.5933$
			$s^2 + 0.20876s + 0.96038$	$s^2 + 24.227$
5	71.383	2.5129×10^{-3}	$s + 0.26697$	
			$s^2 + 0.39784s + 0.44677$	$s^2 + 4.3650$
			$s^2 + 0.13338s + 0.97180$	$s^2 + 10.568$

TABLE 7-8b Elliptic: $L_{max} = 1.5000$ dB

N	L_{min} (dB)	Denominator constant	Numerator	Denominator
			$\Omega_s = 5$	
2	35.979	1.5887×10^{-2}	$s^2 + 0.91330s + 0.93453$	$s^2 + 49.495$
3	61.910	1.1917×10^{-2}	$s + 0.42518$	
			$s^2 + 0.41326s + 0.92952$	$s^2 + 33.165$
4	87.843	4.0539×10^{-5}	$s^2 + 0.57873s + 0.24796$	$s^2 + 29.203$
			$s^2 + 0.23392s + 0.95196$	$s^2 + 167.77$
5	113.77	5.0673×10^{-5}	$s + 0.25034$	
			$s^2 + 0.40044s + 0.41243$	$s^2 + 27.586$
			$s^2 + 0.15015s + 0.96673$	$s^2 + 71.405$

SOLUTION Finding the normalized stop band using Eq. (7-15), we obtain

$$\Omega_s = \frac{\omega_s}{\omega_p}$$

$$= \frac{10 \text{ kHz}}{2 \text{ kHz}}$$

$$= 5$$

Therefore, we use Table 7-9b since this is for $L_{max} = 0.25$ dB and $\Omega_s = 5$. The first entry has an $L_{min} = 27.558$ dB, which exceeds our require-

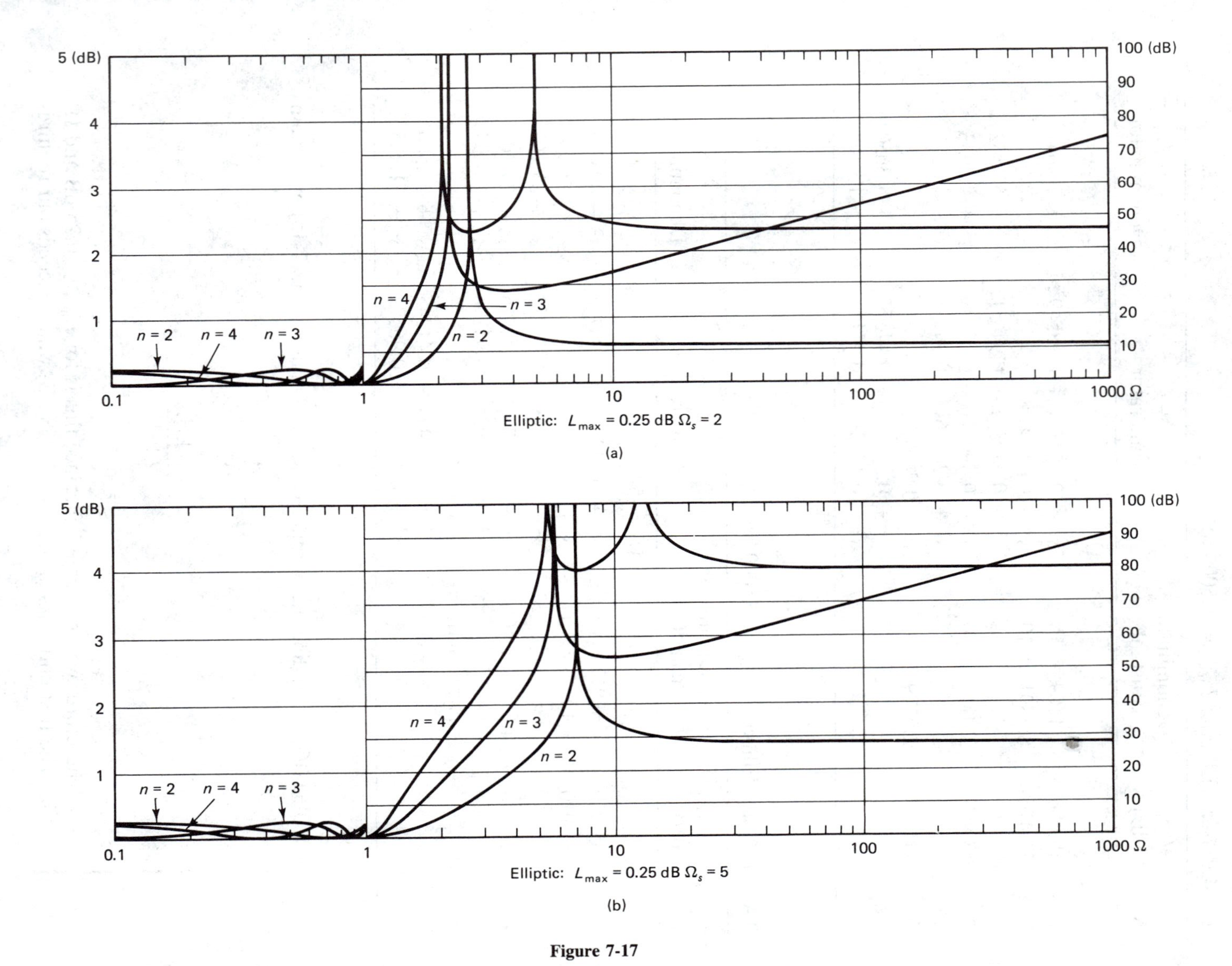
5 (dB)
4
3
2
1
100 (dB)
90
80
70
60
50
40
30
20
10
n = 2
n = 4
n = 3
n = 4
n = 3
n = 2
0.1
1
10
100
1000 Ω
Elliptic: L_{max} = 0.25 dB Ω_s = 2
(a)
5 (dB)
4
3
2
1
100 (dB)
90
80
70
60
50
40
30
20
10
n = 2
n = 4
n = 3
n = 4
n = 3
n = 2
0.1
1
10
100
1000 Ω
Elliptic: L_{max} = 0.25 dB Ω_s = 5
(b)

Figure 7-17

TABLE 7-9a Elliptic: $L_{max} = 0.2500$ dB

N	L_{min} (dB)	Denominator constant	Numerator	Denominator
			$\Omega_s = 2$	
2	10.967	2.8290×10^{-1}	$s^2 + 1.4716s + 2.1733$	$s^2 + 7.4641$
3	28.055	2.2135×10^{-1}	$s + 0.86031$	
			$s^2 + 0.63895s + 1.3259$	$s^2 + 5.1532$
4	45.502	5.3077×10^{-3}	$s^2 + 1.0746s + 0.52880$	$s^2 + 4.5933$
			$s^2 + 0.35862s + 1.1496$	$s^2 + 24.227$
5	62.956	6.6305×10^{-3}	$s + 0.47558$	
			$s^2 + 0.69854s + 0.59202$	$s^2 + 4.3650$
			$s^2 + 0.22958s + 1.0863$	$s^2 + 10.568$

TABLE 7-9b Elliptic: $L_{max} = 0.2500$ dB

N	L_{min} (dB)	Denominator constant	Numerator	Denominator
			$\Omega_s = 5$	
2	27.558	4.1887×10^{-2}	$s^2 + 1.7565s + 2.1337$	$s^2 + 49.495$
3	53.483	3.1443×10^{-2}	$s + 0.77977$	
			$s^2 + 0.74833s + 1.3373$	$s^2 + 33.165$
4	79.415	1.0697×10^{-4}	$s^2 + 1.0333s + 0.46487$	$s^2 + 29.203$
			$s^2 + 0.41512s + 1.1602$	$s^2 + 167.77$
5	105.35	1.3370×10^{-4}	$s + 0.44229$	
			$s^2 + 0.70615s + 0.54424$	$s^2 + 27.586$
			$s^2 + 0.26400s + 1.0941$	$s^2 + 71.405$

ment; therefore, we may use a second-order equation

$$\frac{s^2 + 1.7565s + 2.1337}{4.1887 \times 10^{-2}(s^2 + 49.495)}$$

Referring to Appendix E, we see that the only appropriate topology is the twin-T. We first test to see if this topology works.

$$\begin{aligned} 0 &\le e + b - a\sqrt{e} \\ &\le 49.495 + 2.1337 - 1.7565\sqrt{49.495} \\ &\le 39.271 \end{aligned}$$

If this was not true, we would have to search for another topology which is not in this appendix.

Next we choose $C = 1$ F since we want a normalized circuit. Therefore, $C_A = C_E = 1$ F, and $C_C = 2C = 2$ F. Solving for R gives

$$R = \frac{1}{C\sqrt{e}}$$

$$= \frac{1}{1\sqrt{49.495}}$$

$$R_B = R_F = R = 0.14214\ \Omega$$

$$R_D = \frac{R}{2} = 0.071070\ \Omega$$

Next we choose C_G such that

$$C_G \geq \frac{C(e - b)}{2b}$$

$$\geq \frac{(1)(49.495 - 2.1337)}{(2)(2.1337)}$$

$$\geq 11.098$$

We will choose C_G equal to 20 F. Sometimes C_G needs to be $\geq$ a negative number. In this case we may choose $C_G = 0$, but it must be at least zero.

Finding R_G, we obtain

$$R_G = \frac{2\sqrt{e}}{C(b - e) + 2C_G b}$$

$$= \frac{2\sqrt{49.495}}{1(2.1337 - 49.495) + 2(20)(2.1337)}$$

$$= 0.37041\ \Omega$$

In a case where $b = e$ in this equation and C_G is chosen as 0, R_G will be infinity. This indicates using an open circuit for both R_G and C_G.

We then calculate K.

$$K = \frac{\dfrac{2}{R_G} + C(4\sqrt{e} - a) + 2C_G(\sqrt{e} - a)}{2C\sqrt{e}}$$

$$= \frac{\dfrac{2}{0.37041} + 1(4\sqrt{49.495} - 1.7565) + 2(20)(\sqrt{49.495} - 1.7565)}{(2)(1)\sqrt{49.495}}$$

$$= 17.265$$

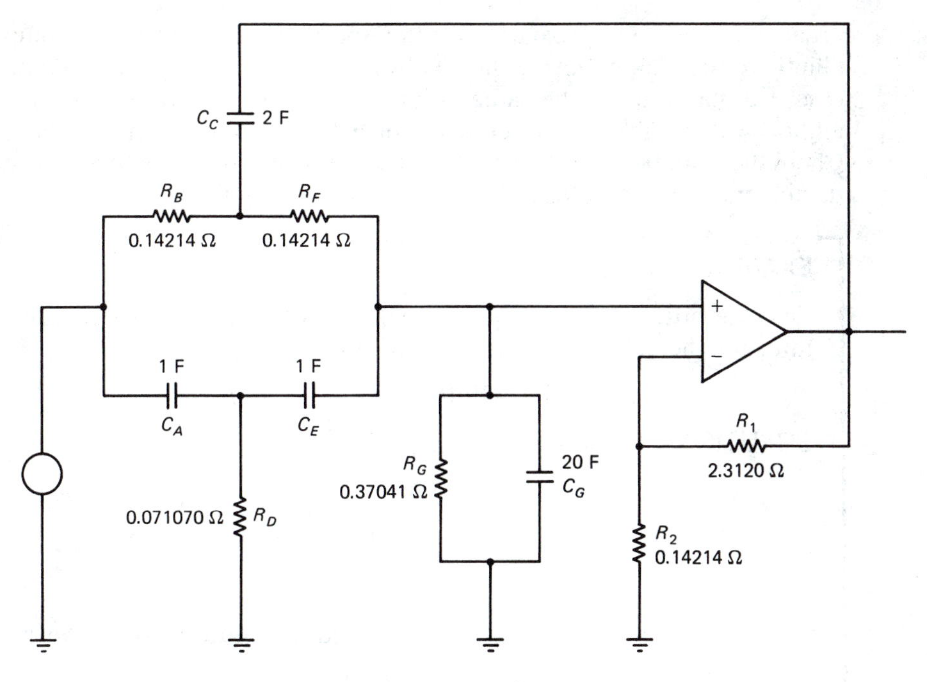

Figure 7-18

The resulting h is

$$h = K\frac{C}{C + 2C_G}$$

$$= 17.265\left[\frac{1}{1 + 2(20)}\right]$$

$$= 0.42111$$

Since this is not the original h desired, we could add an amplifier stage with a gain of 9.9468×10^{-2} if it was necessary.

Finally, we choose $R_2 = 0.14214\ \Omega$ and calculate R_1.

$$R_1 = R_2(K - 1)$$

$$= (0.14214)(17.265 - 1)$$

$$= 2.3120\ \Omega$$

The completed circuit is shown in Fig. 7-18.

7-4-5 Comparison of Butterworth, Chebyshev, and Elliptic

The obvious difference between the three approximations is the shape of the responses. Another important difference is the filter order required to approximate a filter specification. The elliptic filter will always give us the lowest-order equation to approximate a filter specification. When we compare the elliptic to the Chebyshev or the Chebyshev to the Butterworth, in a particular case, we may require the same-order filter. When there is a difference, the elliptic will be a lower order than the Chebyshev or Butterworth, and the Chebyshev will be a lower order than the Butterworth. The following example demonstrates a case where there is a difference in all cases.

EXAMPLE 7-8

Find the order of filter required for a Butterworth, Chebyshev, and elliptic filter for the following filter specification.

$$L_{max} = 3.0103 \text{ dB} \qquad L_{min} = 20 \text{ dB} \qquad \Omega_s = 2$$

SOLUTION We find ε^2 from Eq. (7-14).

$$\begin{aligned}\varepsilon^2 &= 10^{0.1L_{max}} - 1 \\ &= 10^{(0.1)(3.0103)} - 1 \\ &= 1\end{aligned}$$

Substituting into Eq. (7-16), we find the Butterworth order required.

$$\begin{aligned}n &\geq \frac{\log\left(\dfrac{10^{0.1L_{min}} - 1}{\varepsilon^2}\right)}{2\log(\Omega_s)} \\ &\geq \frac{\log\left[\dfrac{10^{(0.1)(20)} - 1}{1}\right]}{2\log(2)} \\ &\geq 3.3147\end{aligned}$$

Therefore, we will require a fourth-order Butterworth.

The order of filter required for the Chebyshev is found by using Eq. (7-17).

$$\begin{aligned}n &\geq \frac{\cosh^{-1}\left(\dfrac{10^{0.1L_{min}} - 1}{\varepsilon^2}\right)^{1/2}}{\cosh^{-1}(\Omega_s)} \\ &\geq \frac{\cosh^{-1}\left[\dfrac{10^{(0.1)(20)} - 1}{1}\right]^{1/2}}{\cosh^{-1}(2)} \\ &\geq 2.2690\end{aligned}$$

Therefore we need a third-order Chebyshev filter, which is one order less than the order required for the Butterworth.

For the elliptic we use Table 7-7a since this is for $L_{max} = 3.0103$ dB and $\Omega_s = 2$. The first entry has an $L_{min} = 22.900$ dB, which exceeds our requirement; therefore, we may use the second-order elliptic.

In this example each filter required one less order to approximate the filter specification. This is not always the case, but should be considered before we decide whether to use the Butterworth, Chebyshev, or elliptic filter.

In general, the fewer components we use, the less expensive the filter is to manufacture, and lower-order filters typically require fewer components. Frequently, economics are not the only factor to consider. Each type of filter has different characteristics, which can enhance or deteriorate the overall performance. To choose the best filter, we must know the exact application. If using the least op-amps is the main goal, the clear choice in Example 7-8 is the elliptic filter.

PROBLEMS

Section 7-1

1. Derive the loss function for the inverting single-feedback topology shown in Fig. 7-1b.

2. Derive the loss function for the inverting dual-feedback topology shown in Fig. 7-2b.

3. Derive the loss function for the twin-T topology shown in Fig. 7-3.

Section 7-2

4. Using the noninverting single-feedback topology in Fig. 7-1a, state what type of components (R, C, or combination) should be used for the loss function.

(a) $\dfrac{s + b}{h}$

(b) $\dfrac{s + b}{hs}$

5. State what type of components should be used for the loss function

$$\frac{s^2 + bs + d}{hs^2}$$

when using the:

(a) Noninverting dual-feedback topology in Fig. 7-2a.

(b) Inverting dual-feedback topology in Fig. 7-2b.

6. Design a filter for the loss function

$$\frac{s + 5}{10}$$

with the capacitor set to 1 μF. It may not be possible to match the denominator constant. If not, what is the resulting denominator constant?

(a) Use the noninverting single-feedback topology in Fig. 7-1a. (Assume that $R_2 = R_A$)

(b) Use the inverting single-feedback topology in Fig. 7-1b.

7. Design a filter for the loss function

$$\frac{s^2 + 1.4142s + 1}{1}$$

with the capacitors set to 1 F (normalized circuit). It may not be possible to match the denominator constant. If not, what is the resulting denominator constant? Use the noninverting dual-feedback topology in Fig. 7-2a with Y_B and Y_D as capacitors and all resistors equal in value. (Assume that $R_2 = 1\ \Omega$.)

8. Using coefficient matching and assuming that C will be chosen first, write the general equations to solve the loss function.

$$\frac{s^2 + as + b}{-hs^2}$$

(a) Use the inverting dual-feedback topology in Fig. 7-2b. Assume that $Y_A = Y_B = Y_C = sC$, $Y_D = 1/R_D$, and $Y_E = 1/R_E$.

(b) Use the noninverting dual-feedback topology in Fig. 7-2a. Assume that $Y_A = Y_C = sC$, and $Y_B = Y_D = 1/R$.

Section 7-3

9. Identify the order and type (low pass, high pass, band pass, or notch) for the loss functions shown.

(a) $\dfrac{s^2 + 5s + 25}{5}$

(b) $\dfrac{(s + 10)(s + 1000)}{s}$

(c) $\dfrac{s + 1}{s}$

10. List the biquads in a form suitable to design a cascaded filter. List the biquad for the first stage first and the last stage last.

(a) $\dfrac{(s^2 + 1.6016s + 2.5650)(s + 1.6016)}{4.1081}$

(b) $\dfrac{(s^2 + 0.15306s + 0.96584)(s + 0.24765)(s^2 + 0.40071s + 0.40682)}{0.097308}$

(c) $\dfrac{(s^2 + 0.10751s + 0.93723)(s^2 + 0.28654s + 0.38201)(s + 0.17906)}{3.2547 \times 10^{-5}(s^2 + 27.586)(s^2 + 71.405)}$

Section 7-4

Section 7-4-2

11. For a Butterworth filter, find L_{min} if it is designed as a third-order filter with $\Omega_s = 3$ and $L_{max} = 0.25$ dB.

12. Design a normalized Butterworth filter to approximate the following specifications.

$$\omega_p = 1 \text{ krad/sec} \qquad \omega_s = 20 \text{ krad/sec}$$

$$L_{max} = 3.0103 \text{ dB} \qquad L_{min} = 25 \text{ dB}$$

(a) Use only inverting topologies that will match the overall h.
(b) Use only noninverting topologies that will match the overall h.

13. Design a normalized Butterworth filter to approximate the following specifications.

$$\Omega_s = 2 \qquad L_{max} = 1.5 \text{ dB} \qquad L_{min} = 18 \text{ dB}$$

(a) Use the inverting topologies that match the overall h.
(b) Use the noninverting topologies and one inverting amplifier to match the overall h. (Assume that $R_2 = 1\ \Omega$.)

14. Design a third-order normalized Butterworth filter with $L_{max} = 0.25$ dB (matching the magnitude of h in the overall equation).
(a) Use a noninverting first-order circuit (assume that $R_2 = 1\Omega$), an inverting second-order circuit, and an inverting amplifier.
(b) Use only two inverting stages.

Section 7-4-3

15. Design a normalized Chebyshev filter to approximate the following specifications.

$$\omega_p = 2 \text{ kHz} \qquad \omega_s = 20 \text{ kHz}$$

$$L_{max} = 3.0103 \text{ dB} \qquad L_{min} = 25 \text{ dB}$$

(a) Use inverting topologies only. Calculate the resulting h.
(b) Use noninverting topologies only. Calculate the resulting h. (Assume that $R_2 = 1\Omega$.)

16. Design a normalized Chebyshev filter to approximate the specifications shown in Problem 13.

$$\Omega_s = 2 \qquad L_{max} = 1.5 \text{ dB} \qquad L_{min} = 18 \text{ dB}$$

(a) Use only inverting topologies matching the overall h.
(b) Use only noninverting topologies. Calculate the resulting h. (Assume that $R_2 = 1\ \Omega$.)

17. Using a fourth-order Chebyshev equation with $L_{max} = 3.0103$ dB, calculate the exact magnitude and phase at a frequency of:
(a) 0.9 rad/sec
(b) 1.2 rad/sec

Section 7-4-4

18. Design a normalized elliptic filter to approximate the specifications shown in Problems 13 and 16. (Assume that $R_2 = 1\ \Omega$.)

$$\Omega_s = 2 \qquad L_{\max} = 1.5\ \text{dB} \qquad L_{\min} = 18\ \text{dB}$$

19. Design a normalized elliptic filter to approximate the following specifications.

$$\omega_p = 10\ \text{rad/sec} \qquad \omega_s = 50\ \text{rad/sec}$$

$$L_{\max} = 3.0103\ \text{dB} \qquad L_{\min} = 60\ \text{dB}$$

CHAPTER 8

Practical Filters from the Generic Equations

8-0 INTRODUCTION

In this chapter we will learn how to design practical active filters. First we will learn how to change the normalized low-pass filters designed in Chapter 7 into filters with practical break frequencies and appropriate component values. Then we will learn how to design high-pass, band-pass, and notch filters from the low-pass equations.

8-1 FREQUENCY SHIFTING

So far we have designed our circuits at a pass-band frequency of 1 rad/sec. Frequency shifting allows us to move the pass-band frequency to any frequency we desire. The technique shown in this section will work on any *RC* active filter, including the high-pass, band-pass, and notch. This process will change the filter equation, but it will be unnecessary to express this shifted equation.

Looking at the transfer function in Appendix E, we notice that the coefficients occur as an *RC* product. We also notice that the break frequencies always occur as the reciprocal of an *RC* product. Therefore, the frequency is proportional to the reciprocal of the *RC* combination.

$$f \propto \frac{1}{RC} \tag{8-1}$$

If we increase the *RC* product by some factor, F_f, we will decrease the frequency by a proportional factor, and if we decrease the *RC* product by some factor, we will increase the frequency by a proportional factor.

We typically know the frequency at which the circuit was designed and the frequency to which we want to shift. We need to know the factor that relates the new frequency, f_{new} or ω_{new}, to the old frequency, f_{old} or ω_{old}.

$$f_{\text{new}} = F_f f_{\text{old}} \tag{8-2a}$$

$$\omega_{\text{new}} = F_f \omega_{\text{old}} \tag{8-2b}$$

From this we can solve for the frequency shifting factor.

$$F_f = \frac{f_{new}}{f_{old}} \tag{8-3a}$$

$$F_f = \frac{\omega_{new}}{\omega_{old}} \tag{8-3b}$$

Once this factor is calculated, we divide either all the frequency-dependent R's or all the frequency-dependent C's. (The R's in the constant K are not frequency dependent.)

$$R_{new} = \frac{R_{old}}{F_f} \tag{8-4a}$$

$$C_{new} = \frac{C_{old}}{F_f} \tag{8-4b}$$

It does not matter whether we use the R's or the C's since when we use impedance shifting in the next section, we will arrive at the same final values.

Some of the transfer functions in Appendix E have a gain constant K. Since K is a ratio of resistors, it will not be affected by frequency shifting. The resistors associated with the K factor do not have to be changed with the other R's of the circuit, but they may be changed if it is convenient. We may say that the R's associated with the K factor are not frequency dependent.

EXAMPLE 8-1

A filter is designed at a frequency of 1 rad/sec. Find the values of the components when shifted to 500 Hz.

$$R_1 = 0.8\ \Omega \qquad R_2 = 1.2\ \Omega \qquad C_1 = 1\ \text{F} \qquad C_2 = 0.1\ \text{F}$$

SOLUTION First we find the frequency shifting factor, F_f, from Eq. (8-3a).

$$F_f = \frac{f_{new}}{f_{old}}$$

$$= \frac{500\ \text{Hz}}{1/2\pi}$$

$$= 3141.6$$

At this point there are two possible solutions to this problem. One would be to change the R's, and the other would be to change the C's.

If we change the R's using Eq. (8-4a), the solution is

$$R_{1\ new} = \frac{R_{1\ old}}{F_f} = \frac{0.8}{3141.6} = 2.5465 \times 10^{-4}\ \Omega$$

$$R_{2\ new} = \frac{R_{2\ old}}{\text{F}_f} = \frac{1.2}{3141.6} = 3.8197 \times 10^{-4}\ \Omega$$

The final values for this approach are

$$R_1 = 2.5465 \times 10^{-4}\ \Omega \qquad R_2 = 3.8197 \times 10^{-4}\ \Omega$$

$$C_1 = 1\ \text{F} \qquad C_2 = 0.1\ \text{F}$$

If we change the C's using Eq. (8-4b), the solution is

$$C_{1\ \text{new}} = \frac{C_{1\ \text{old}}}{F_f} = \frac{1}{3141.6} = 3.1831 \times 10^{-4}\ \text{F}$$

$$C_{2\ \text{new}} = \frac{C_{2\ \text{old}}}{F_f} = \frac{0.1}{3141.6} = 3.1831 \times 10^{-5}\ \text{F}$$

The final values for this approach are

$$R_1 = 0.8\ \Omega \qquad R_2 = 1.2\ \Omega$$

$$C_1 = 3.1831 \times 10^{-4}\ \text{F} \qquad C_2 = 3.1831 \times 10^{-5}\ \text{F}$$

8-2 IMPEDANCE SHIFTING

Even after we do frequency shifting, we still have impractical values of resistors and capacitors. In this section we will finally achieve a filter at a desired frequency with practical components using impedance shifting.

Impedance shifting is also based on Eq. (8-1). If we multiply the capacitors by a factor, F_z, and divide the resistors by the same factor, the frequency will remain the same.

$$f \propto \frac{1}{\left(\frac{R}{F_z}\right) F_z C} \tag{8-5}$$

In this way we may change the component values, or shift their impedance, without affecting the frequency of the filter.

Since there is not as wide of a selection of capacitor values as there are resistor values, we would like to select a final capacitor value. The new capacitor value will be

$$C_{\text{new}} = F_z C_{\text{old}} \tag{8-6a}$$

Normally, we choose a value for the new capacitor, and we know the current value of the capacitor, C_{old}. We should therefore solve for the impedance factor, F_z, in Eq. (8-6a).

$$F_z = \frac{C_{\text{new}}}{C_{\text{old}}} \tag{8-6b}$$

We then calculate the new resistor value which will keep the filter frequency the same.

$$R_{new} = \frac{R_{old}}{F_z} \tag{8-7}$$

EXAMPLE 8-2

Using the two solutions calculated in Example 8-1, determine the component values when C_1 is chosen to be 1 μF.

SOLUTION Using the first solution

$$R_1 = 2.5465 \times 10^{-4}\ \Omega \qquad R_2 = 3.8197 \times 10^{-4}\ \Omega$$

$$C_1 = 1\text{ F} \qquad C_2 = 0.1\text{ F}$$

We calculate the impedance shifting factor F_z using Eq. (8-6b).

$$F_z = \frac{C_{new}}{C_{old}} = \frac{1\ \mu\text{F}}{1\text{ F}} = 1 \times 10^{-6}$$

Substituting into Eqs. (8-6a) and (8-7), we obtain

$$C_{1\,new} = F_z C_{1\,old} = (1 \times 10^{-6})(1\text{ F}) = 1\ \mu\text{F}$$

$$C_{2\,new} = F_z C_{2\,old} = (1 \times 10^{-6})(0.1\text{ F}) = 0.1\ \mu\text{F}$$

$$R_{1\,new} = \frac{R_{1\,old}}{F_z} = \frac{2.5465 \times 10^{-4}}{1 \times 10^{-6}} = 254.65\ \Omega$$

$$R_{2\,new} = \frac{R_{2\,old}}{F_z} = \frac{3.8197 \times 10^{-4}}{1 \times 10^{-6}} = 381.97\ \Omega$$

Using the second solution

$$R_1 = 0.8\ \Omega \qquad R_2 = 1.2\ \Omega$$

$$C_1 = 3.1831 \times 10^{-4}\text{ F} \qquad C_2 = 3.1831 \times 10^{-5}\text{ F}$$

Again we calculate the impedance shifting factor F_z using Eq. (8-6b).

$$F_z = \frac{C_{new}}{C_{old}} = \frac{1\ \mu\text{F}}{3.1831 \times 10^{-4}\text{ F}} = 3.1416 \times 10^{-3}$$

Substituting into Eqs. (8-6a) and (8-7) gives

$$C_{1\,new} = F_z C_{1\,old} = (3.1416 \times 10^{-3})(3.1831 \times 10^{-4}\text{ F}) = 1\ \mu\text{F}$$

$$C_{2\,new} = F_z C_{2\,old} = (3.1416 \times 10^{-3})(3.1831 \times 10^{-5}\text{ F}) = 0.1\ \mu\text{F}$$

$$R_{1\,new} = \frac{R_{1\,old}}{F_z} = \frac{0.8}{3.1416 \times 10^{-3}} = 254.65\ \Omega$$

$$R_{2\,new} = \frac{R_{2\,old}}{F_z} = \frac{1.2}{3.1416 \times 10^{-3}} = 381.97\ \Omega$$

In Example 8-2 we see that we arrive at the same solution when we impedance shift. This is why it is unimportant whether R's or C's are used in frequency shifting. At this point we have all the information to design a practical circuit. The following example demonstrates the design of a low-pass filter from beginning to end by completing an example from Chapter 7.

EXAMPLE 8-3

Complete the design of the third-order Chebyshev filter in Example 7-6 (see Fig. 8-1). Make the 1-F capacitor 1 μF.

SOLUTION First we need to shift the filter's frequency from 1 rad/sec to 1 kHz. Using Eq. (8-3a), we obtain

$$F_f = \frac{f_{\text{new}}}{f_{\text{old}}} = \frac{1 \text{ kHz}}{1/2\pi} = 6283.2$$

Let's change the capacitors using Eq. (8-4b). For the second stage,

$$C_{E\text{ new}} = \frac{C_{E\text{ old}}}{F_f} = \frac{1}{6283.2} = 1.5915 \times 10^{-4} \text{ F}$$

$$C_{D\text{ new}} = \frac{C_{D\text{ old}}}{F_f} = \frac{0.01}{6283.2} = 1.5915 \times 10^{-6} \text{ F}$$

For the first stage

$$C_{B\text{ new}} = \frac{C_{B\text{ old}}}{F_f} = \frac{1}{6283.2} = 1.5915 \times 10^{-4} \text{ F}$$

Next we need to calculate the impedance shifting factor F_z using Eq. (8-6b). Here we want to make the 1-F capacitor 1 μF, but the current value

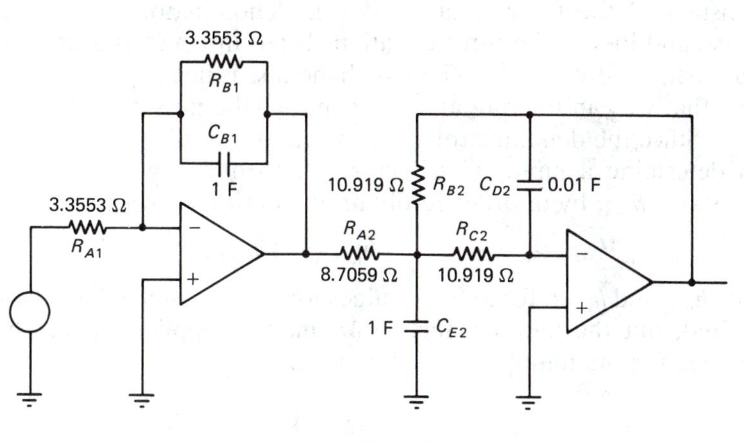

Figure 8-1

of this capacitor is 1.5915×10^{-4}. Since both stages have the same value of capacitors, they will both have the same impedance shifting factor.

$$F_z = \frac{C_{new}}{C_{old}} = \frac{1\ \mu F}{1.5915 \times 10^{-4}\ F} = 6.2832 \times 10^{-3}$$

Changing the components for the second stage using Eqs. (8-6a) and (8-7), we obtain

$$C_{E\ new} = F_z C_{E\ old} = (6.2832 \times 10^{-3})(1.5915 \times 10^{-4}\ F) = 1\ \mu F$$

$$C_{D\ new} = F_z C_{D\ old} = (6.2832 \times 10^{-3})(1.5915 \times 10^{-6}\ F) = 0.01\ \mu F$$

$$R_{new} = \frac{R_{old}}{F_z} = \frac{10.919}{6.2832 \times 10^{-3}} = 1.7378\ k\Omega$$

$$R_B = R_C = 1.7378\ k\Omega$$

$$R_{A\ new} = \frac{R_{A\ old}}{F_z} = \frac{8.7059}{6.2832 \times 10^{-3}} = 1.3856\ k\Omega$$

For the first stage,

$$C_{B\ new} = F_z C_{B\ old} = (6.2832 \times 10^{-3})(1.5915 \times 10^{-4}\ F) = 1\ \mu F$$

$$R_{new} = \frac{R_{old}}{F_z} = \frac{3.3553}{6.2832 \times 10^{-3}} = 534.01\ \Omega$$

$$R_A = R_B = 534.01\ \Omega$$

The finished circuit is shown in Fig. 8-2.

8-3 GAIN SHIFTING

In Chapter 7 we discussed adding an amplifier to obtain the denominator constant of the filter equation. This denominator constant determines the pass-band loss. The filter equations listed in Chapter 7 are normalized for a pass-band loss of 0 dB. The pass-band loss is not L_{max}. L_{max} is the maximum loss that we can tolerate and still consider the pass-band loss to be 0 dB.

Since the denominator constant, h, may not be what we require, we need to determine a multiplying factor, M, to multiply the current denominator constant, h_{old}, by in order to obtain the desired denominator constant, h_{new}.

$$h_{new} = M h_{old} \tag{8-8a}$$

The h_{new} and h_{old} refer to the h values for a function that has not been frequency shifted, but the multiply factor, M, may be applied to the shifted function. Solving for the multiplying factor, we have

$$M = \frac{h_{new}}{h_{old}} \tag{8-8b}$$

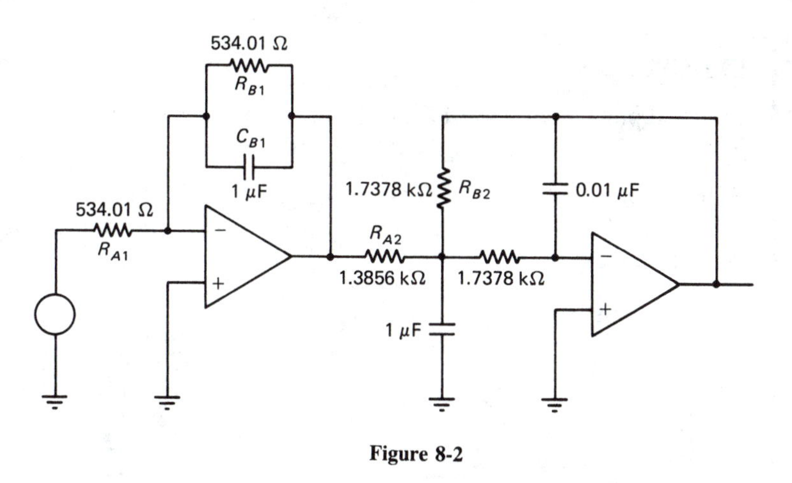

Figure 8-2

This multiplying factor will shift the pass-band loss. The amount of the shift can be expressed as

$$L_{\text{shift}} = -20 \log M \tag{8-9a}$$

If we know the amount we need to shift the loss we may solve this for the multiplying factor

$$M = 10^{-0.05L_{\text{shift}}} \tag{8-9b}$$

We may either multiply this factor times the normalized equation denominator before we start designing, or we may add a stage that has a loss of $-20 \log M$ (or gain of $+20 \log M$).

EXAMPLE 8-4

Modify a third-order elliptic equation with $L_{\max} = 1.5$ dB and $\Omega_s = 2$, so it has a pass-band loss of -4 dB.

SOLUTION From Table 7-8a,

$$\frac{(s + 0.45674)(s^2 + 0.37285s + 0.94651)}{8.3891 \times 10^{-2}(s^2 + 5.1532)}$$

Using Eq. (8-9b) gives

$$M = 10^{-0.05L_{\text{shift}}} = 10^{(-0.05)(-4)} = 1.5849$$

Since the equation from Table 7-8a has a pass-band loss of 0 dB, we simply multiply.

$$\frac{(s + 0.45674)(s^2 + 0.37285s + 0.94651)}{(1.5849)(8.3891 \times 10^{-2})(s^2 + 5.1532)}$$

$$\frac{(s + 0.45674)(s^2 + 0.37285s + 0.94651)}{(0.13296)(s^2 + 5.1532)}$$

EXAMPLE 8-5

Find the pass-band loss of a third-order Chebyshev filter with $L_{\max} = 3.0103$ if the resulting h, before frequency shifting, was 0.105.

SOLUTION From Table 7-4 we see that the denominator constant should have been 0.25. From Eq. (8-8b) we can find the multiplying factor that was effectively used.

$$M = \frac{h_{\text{new}}}{h_{\text{old}}}$$

$$= \frac{0.105}{0.25}$$

$$= 0.42$$

From Eq. (8-9a) we can find the amount the function was shifted.

$$L_{\text{shift}} = -20 \log M$$

$$= -20 \log (0.42)$$

$$= 7.5350 \text{ dB}$$

Since the equation was normalized, the original pass band gain was 0 dB. Therefore, 7.5350 dB more than 0 dB is 7.5350 dB.

An alternative to using a gain shifting stage is to add a resistive voltage divider network on a filter stage. When this is done we can either amplify or attenuate the output, depending on how it is used. We must be careful to use large enough resistors to prevent loading the output. The shifting of the gain by this method is limited to moderate levels of not less than 0.1 or greater than 10.

Figure 8-3 shows how to attenuate the overall gain or to increase the overall loss for inverting and noninverting circuits. We are reducing the output by the voltage divider of two series resistors. The overall loss will be shifted by

$$L_{\text{shift}} = -20 \log \left(\frac{R_\alpha}{R_\alpha + R_\beta} \right) \tag{8-10a}$$

Knowing the shift required, we choose a value for either R_α or R_β and solve for the other resistor. Solving the equation for R_β, we have

$$R_\beta = R_\alpha(10^{0.05L_{\text{shift}}} - 1) \tag{8-10b}$$

Figure 8-4 shows how to amplify the overall gain or decrease the overall loss for inverting and noninverting circuits. In this case we have reduced the feedback by the voltage divider of two series resistors. The op-amp must therefore increase its output in order to balance the reduced feedback voltage.

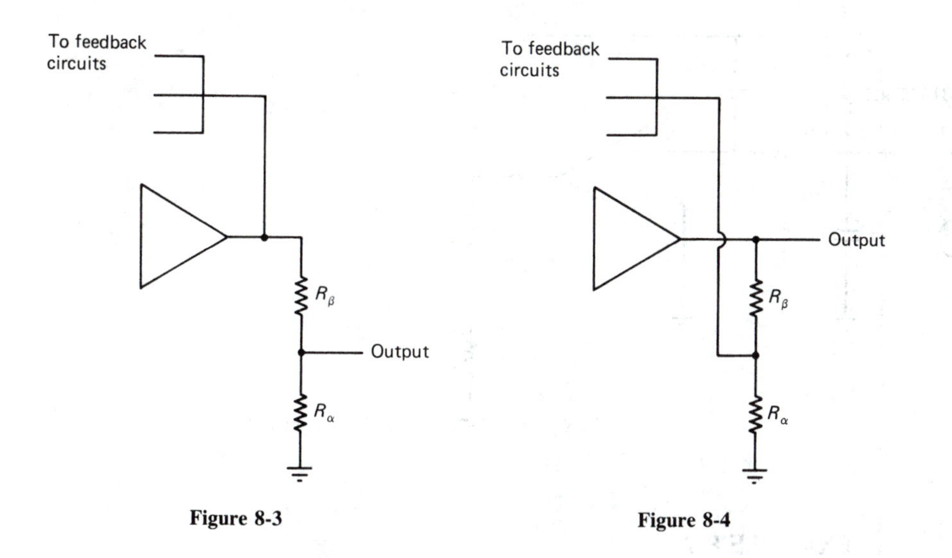

Figure 8-3 **Figure 8-4**

The overall loss will be shifted by

$$L_{\text{shift}} = -20 \log \left(\frac{R_\alpha + R_\beta}{R_\alpha} \right) \quad (8\text{-}11a)$$

Here again if we know the shift required, we assume a value for one of the resistors and solve for the other. Solving the equation for R_β yields

$$R_\beta = R_\alpha(10^{-0.05L_{\text{shift}}} - 1) \quad (8\text{-}11b)$$

This method of gain shifting is very useful in circuit configurations where we cannot match the h in the biquad stage. We design the circuit, determine what multiplying factor to use, and add the voltage divider circuit.

EXAMPLE 8-6

Figure 8-5 is a second-order Butterworth filter with a pass-band gain of 0 dB. Find R_β if we choose R_α to be 100 kΩ, and we want to change the gain to −2 dB.

SOLUTION If we want to shift the gain to be −2 dB, we want to shift the loss by +2 dB. Since we are attenuating the gain or increasing the loss, we use Eq. (8-10b).

$$\begin{aligned} R_\beta &= R_\alpha(10^{0.05L_{\text{shift}}} - 1) \\ &= (100\text{ k})[10^{(0.05)(2)} - 1] \\ &= 25.893\text{ k}\Omega \end{aligned}$$

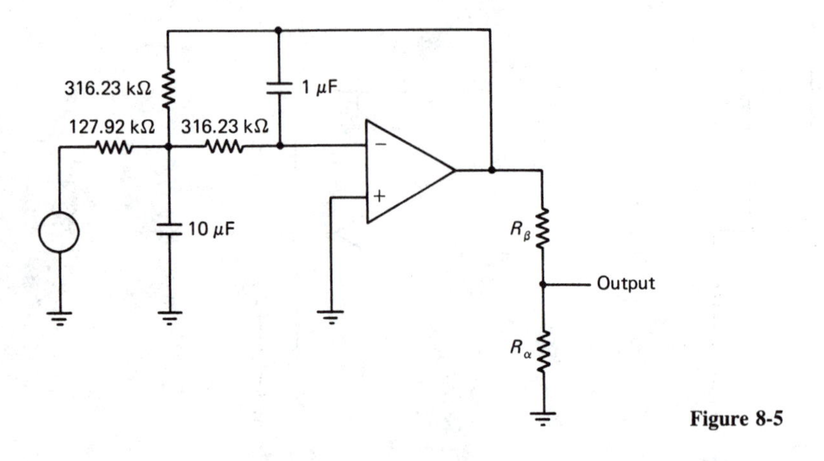

Figure 8-5

EXAMPLE 8-7

Figure 8-6 shows a Chebyshev filter. Design a voltage divider network for the output to increase the gain by 5 dB using $R_\alpha = 10\ \text{k}\Omega$.

SOLUTION Increasing the gain by 5 dB is the same as shifting the loss by -5 dB. Using Eq. (8-11b), we have

$$R_\beta = R_\alpha(10^{-0.05L_{\text{shift}}} - 1)$$

$$= (10\ \text{k})[10^{(-0.05)(-5)} - 1]$$

$$= 7.7828\ \text{k}\Omega$$

Figure 8-7 shows the resulting circuit. Notice that all of the feedback paths are moved to between R_α and R_β.

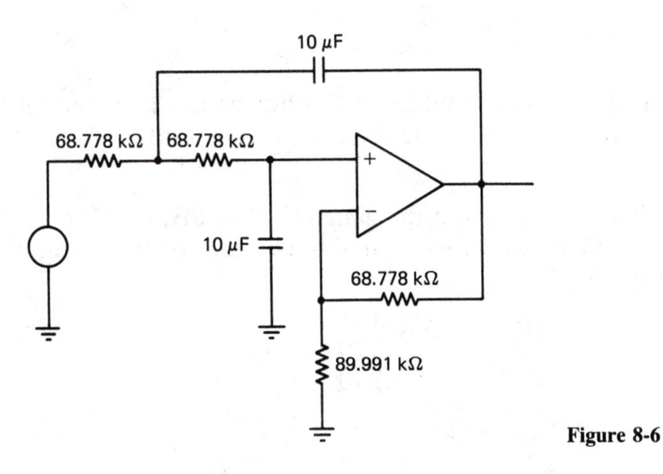

Figure 8-6

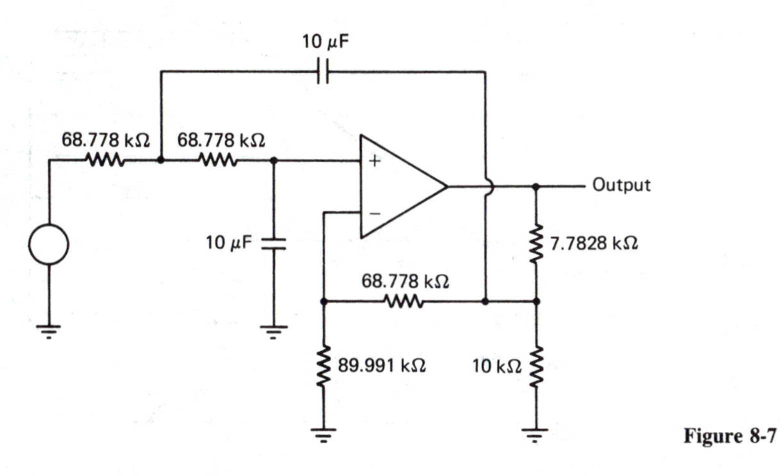

Figure 8-7

8-4 HIGH-PASS FILTER

In Chapter 7 we referred to the low-pass equations for the Butterworth, Chebyshev, and elliptic equations as generic equations. This is because we can transform these equations into high-pass, band-pass, and notch equations. In this section we transform these equations into high-pass equations.

To transform a low-pass equation into a high-pass equation, we make the substitution

$$s = \frac{1}{s} \tag{8-12}$$

This causes the stop-band frequency to occur below 1 rad/sec instead of above 1 rad/sec, while maintaining the pass-band frequency at 1 rad/sec. The low-pass equation is flipped over the 1-rad/sec line. This equation can be applied to the Butterworth, Chebyshev, and elliptic equations equally well.

For example, Fig. 8-8a shows the second-order Chebyshev filter.

$$\frac{s^2 + 0.64359s + 0.70711}{0.5}$$

If we substitute $1/s$ into this, we have the equation transformed into a high-pass equation.

$$\frac{\left(\dfrac{1}{s}\right)^2 + (0.64359)\left(\dfrac{1}{s}\right) + 0.70711}{0.5}$$

$$\frac{s^2 + 0.91017s + 1.4142}{0.70710s^2}$$

which is shown in Fig. 8-8b.

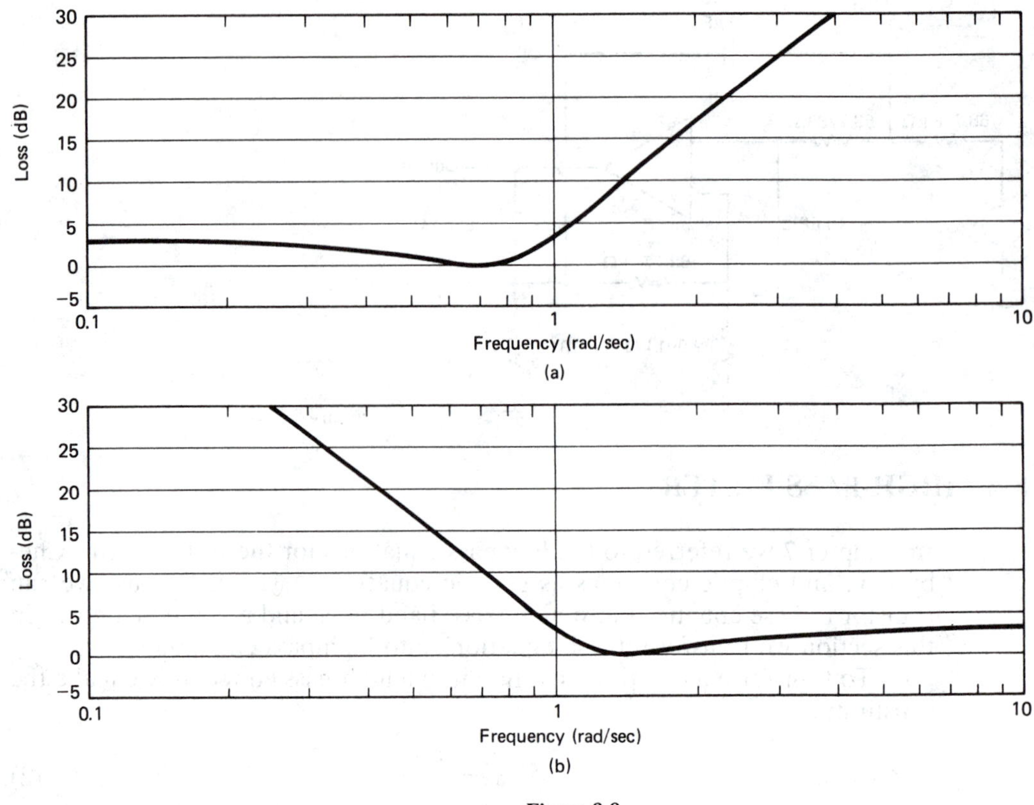

Figure 8-8

We must determine the appropriate low-pass equation to transform into a high-pass equation. The only difference in determining the low-pass requirement for a high-pass equation is in finding Ω_s.

$$\Omega_s = \frac{\omega_p}{\omega_s} \tag{8-13}$$

Notice this is the reciprocal of how Ω_s is found for a low-pass filter. This is not Ω_s for the high-pass filter, but it is the corresponding low-pass Ω_s for determining the low-pass equation to use. Once the Ω_s is calculated, we use the same techniques as presented in Chapter 7.

EXAMPLE 8-8

For the filter specification, design a Butterworth filter using an inverting amplifier with $C = 0.01\ \mu\text{F}$ and the pass-band loss shifted by -10 dB.

$$\omega_p = 8 \text{ krad/sec} \qquad \omega_s = 1 \text{ krad/sec}$$

$$L_{\max} = 3.0103 \text{ dB} \qquad L_{\min} = 15 \text{ dB}$$

SOLUTION From the specifications we see that this is a high-pass filter; therefore, we use Eq. (8-13) to find Ω_s.

$$\Omega_s = \frac{\omega_p}{\omega_s} = \frac{8\text{k}}{1\text{k}} = 8$$

From equations in Chapter 7, we find that

$$\varepsilon = 1$$

and

$$n \geq 0.82276 \qquad \text{use } n = 1$$

From the first entry in Table 7-1, the low-pass loss equation is

$$\frac{s + 1}{1}$$

Substituting $s = 1/s$ gives

$$\frac{s + 1}{s}$$

Going to Section E-2-1, we choose $C_B = 1$ F. Therefore,

$$R_B = \frac{1}{bC_B} = \frac{1}{(1)(1)} = 1\ \Omega$$

$$C_A = hC_B = (1)(1) = 1\text{ F}$$

Next we will frequency shift the filter using Eq. (8-3b).

$$F_f = \frac{\omega_{\text{new}}}{\omega_{\text{old}}}$$

$$= \frac{8\text{k}}{1}$$

$$= 8000$$

Dividing R by this value using Eq. (8-4a), we find that

$$R_{B\text{ new}} = \frac{R_{B\text{ old}}}{F_f} = \frac{1}{8000} = 1.25 \times 10^{-4}\ \Omega$$

We then calculate the frequency shifting factor using Eq. (8-6b).

$$F_z = \frac{C_{\text{new}}}{C_{\text{old}}} = \frac{0.01\ \mu\text{F}}{1\text{ F}} = 1 \times 10^{-8}$$

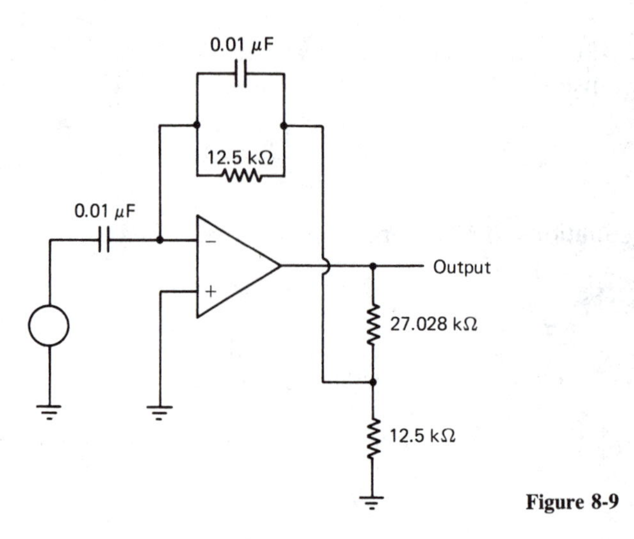

Figure 8-9

Substituting into Eqs. (8-6a) and (8-7) gives

$$C_{A\text{ new}} = C_{B\text{ new}} = F_z C_{1\text{ old}} = (1 \times 10^{-8})(1\text{ F}) = 0.01\ \mu\text{F}$$

$$R_{B\text{ new}} = \frac{R_{B\text{ old}}}{F_z} = \frac{1.25 \times 10^{-4}}{1 \times 10^{-8}} = 12.5\text{ k}\Omega$$

Next we need to determine the voltage divider output network to increase the gain. Using Eq. (8-11b) and choosing $R_\alpha = 12.5\text{ k}\Omega$, we obtain

$$\begin{aligned} R_\beta &= R_\alpha(10^{-0.05L_{\text{shift}}} - 1) \\ &= (12.5\text{k})[10^{(-0.05)(-10)} - 1] \\ &= 27.028\text{ k}\Omega \end{aligned}$$

The final circuit is shown in Fig. 8-9.

8-5 BAND-PASS FILTER

Band-pass filters can be broken into two types: wide band pass and narrow band pass. The wide-band-pass filter can be constructed by cascading low-pass and high-pass filters. The narrow band pass uses biquad band-pass stages that have transfer functions with s appearing in the denominator.

8-5-1 General Band-Pass Equations

Before we begin, we should review some fundamental equations for band-pass filters. Figure 8-10 shows the gain response of a Butterworth band-pass filter. When we use Butterworth, Chebyshev, or elliptic, $L_{\max}$ will define the endpoints of the pass band, f_L or ω_L and f_H or ω_H. In most cases $L_{\max}$ is 3.0103 dB, but we will not limit ourselves to this value.

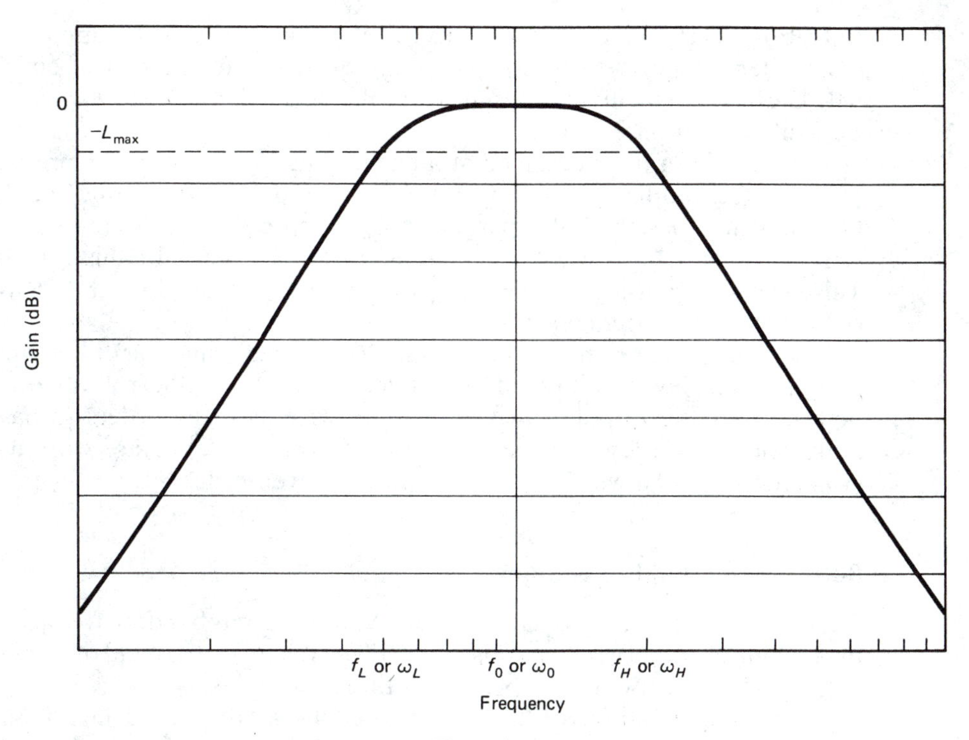

Figure 8-10

The center frequency, f_0 or ω_0, is the geometric mean of the frequencies of the lower point, f_L or ω_L, and the upper point, f_H or ω_H.

$$f_0 = \sqrt{f_L f_H} \tag{8-14a}$$

Expressed in radians,

$$\omega_0 = \sqrt{\omega_L \omega_H} \tag{8-14b}$$

The band width, BW, is

$$\mathrm{BW}_{\mathrm{Hz}} = f_H - f_L \tag{8-15a}$$

Expressed in radians,

$$\mathrm{BW}_{R/S} = \omega_H - \omega_L \tag{8-15b}$$

The quality factor, Q, is

$$Q = \frac{f_0}{\mathrm{BW}_{\mathrm{Hz}}} \tag{8-16a}$$

In radian quantities,

$$Q = \frac{\omega_0}{\mathrm{BW}_{R/S}} \tag{8-16b}$$

Both Eqs. (8-16a) and (8-16b) will yield the same value when applied to the same filter. Since we are not restricting ourselves to the typical endpoints, 3.0103 dB, this Q will be different from the normally defined Q when we use L_{max} different from 3.0103 dB.

The Q for a filter designed at a particular L_{max} cannot be compared to a Q found for another filter designed at another L_{max}. If we find the Q of a filter that was designed at a particular L_{max} and calculate the Q of this filter using a larger L_{max}, we will calculate a smaller Q value for this same filter. This occurs since using a larger L_{max} will increase the band width but will not change the center frequency.

It is important to remember that a second-order band-pass filter will have an upper and lower roll-off of 20 dB/dec each. Since the roll-off occurs at both ends of the response, it will require twice as many orders to have the same roll-off as a low-pass or a high-pass filter. Since we are working with symmetrical filters, we will always require an even-order filter.

8-5-2 Wide-Band-Pass Filters

There are a number of situations that require a large band of frequencies to pass. One example would be voice filtering. When a large band of frequencies are required, we use a wide-band-pass filter.

The wide-band-pass filter is constructed by cascading a high-pass filter designed at f_L or ω_L and a low-pass filter designed at f_H or ω_H. These stages are designed separately without consideration to the other filter. When cascading, we must be sure that f_H or ω_H and f_L or ω_L are separated sufficiently far apart. When they are not sufficiently separated, the two stages will interfere with each other, causing a lower center frequency gain and an affective shift in the break frequencies. Since we are designing the two filters separately, we must make sure that they have the same pass-band loss or gain.

To determine whether to use a wide-band filter, we will look at the Q. If Q is less than or equal to 0.5 ($Q \leq 0.5$), we will use a wide-band-pass filter. This will keep the two filters' pass-band frequencies sufficiently separated.

Since a band-pass filter does not pass dc signals, we must put the high-pass filter first. A high-pass filter has a capacitor input and a low-pass will amplify dc values. With the high-pass filter first, we will be able to block any dc voltages that could saturate a low-pass filter.

EXAMPLE 8-9

Design a Butterworth filter with the following specifications.

$$\omega_{p1} = 3 \text{ rad/sec} \qquad \omega_{p2} = 48 \text{ rad/sec}$$

$$\omega_{s1} = 0.25 \text{ rad/sec} \qquad \omega_{s2} = 576 \text{ rad/sec}$$

$$L_{max} = 0.25 \text{ dB} \qquad L_{min} = 30 \text{ dB}$$

SOLUTION We must first determine whether this is a wide-band or a narrow-band filter from Eqs. (8-14b), (8-15b), and (8-16b).

$$\begin{aligned}\omega_0 &= \sqrt{\omega_L \omega_H}\\ &= \sqrt{(3)(48)}\\ &= 12 \text{ rad/sec}\end{aligned}$$

$$\begin{aligned}BW_{R/S} &= \omega_H - \omega_L\\ &= 48 - 3\\ &= 45 \text{ rad/sec}\end{aligned}$$

$$\begin{aligned}Q &= \frac{\omega_0}{BW_{R/S}}\\ &= \frac{12}{45}\\ &= 0.26667\end{aligned}$$

This is ≤ 0.5, and is, therefore, a wide-band-pass filter.

Next we split this specification into a low-pass filter specification and a high-pass filter specification. The low-pass specification is

$$\omega_p = 48 \text{ rad/sec} \qquad \omega_s = 576 \text{ rad/sec}$$

$$L_{\max} = 0.25 \text{ dB} \qquad L_{\min} = 30 \text{ dB}$$

The high-pass specification is therefore

$$\omega_p = 3 \text{ rad/sec} \qquad \omega_s = 0.25 \text{ rad/sec}$$

$$L_{\max} = 0.25 \text{ dB} \qquad L_{\min} = 30 \text{ dB}$$

Since the specification is symmetrical, we will require the same-order filter. This is apparent when we find the stop-band frequency, Ω_s, for both filters. For the low-pass filter,

$$\Omega_s = \frac{\omega_s}{\omega_p} = \frac{576}{48} = 12$$

For the high-pass filter, the corresponding low-pass stop-band frequency using Eq. (8-13) is

$$\Omega_s = \frac{\omega_p}{\omega_s} = \frac{3}{0.25} = 12$$

For both stages we calculate the order required using equations from Chapter 7:

$$\varepsilon = 0.24342$$

$$n \geq 1.9584 \qquad \text{use } n = 2$$

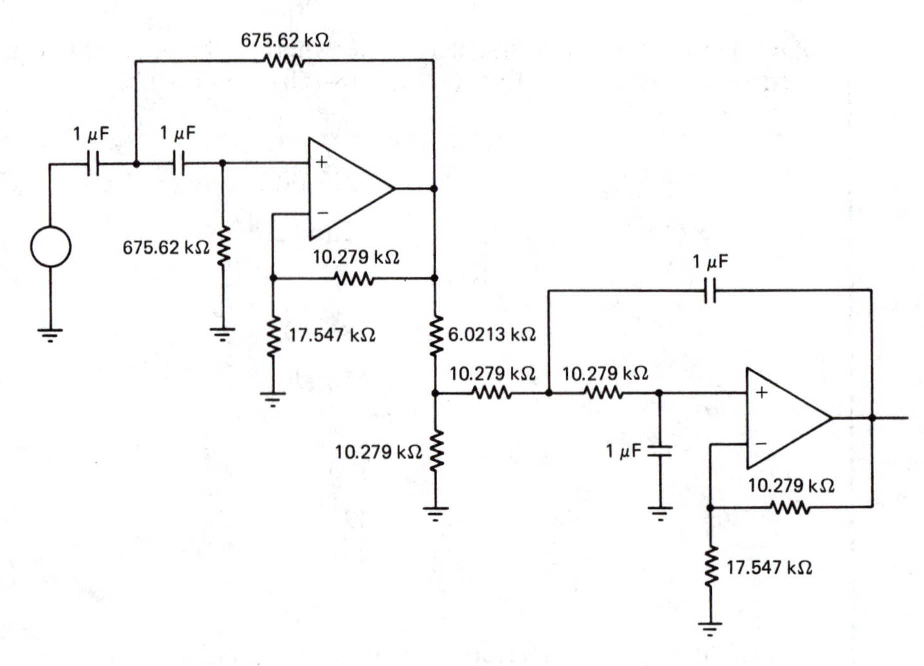

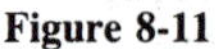

Figure 8-11

Therefore, we use a second-order Butterworth. The equation from Table 7-3 is

$$\frac{s^2 + 2.8664s + 4.1081}{4.1081}$$

This equation will apply to the low-pass filter and the high-pass filter. For the high-pass filter, we must transform this equation using Eq. (8-12). The result is

$$\frac{s^2 + 0.69774s + 0.24342}{s^2}$$

At this point the design follows the procedure outlined in previous sections. We must remember that the low-pass filter must be frequency shifted to 48 rad/sec, and the high-pass filter must be frequency shifted to 3 rad/sec. Figure 8-11 shows the wide-band-pass filter. Notice gain shifting was used to match the passband loss of the two stages.

8-5-3 Narrow-Band-Pass Filter

Narrow-band-pass filters pass a smaller band width of frequencies, and will have a $Q > 0.5$. This type of filter may be very useful when a small range of frequencies must be detected in a group of frequencies.

Narrow-band-pass filters may be constructed by using a biquad band-pass filter. These may, also, be cascaded to form higher-order band-pass filters.

To transform a low-pass equation into a high-pass equation with a center frequency of 1 rad/sec, we make the substitution

$$s = \frac{s^2 + 1}{\left(\dfrac{\omega_H - \omega_L}{\omega_0}\right)s} \tag{8-17}$$

where $\dfrac{\omega_H - \omega_L}{\omega_0}$ is the normalized radian bandwidth.

This will cause the low-pass function to shift above 1 rad/sec and cause a second set of poles and zeros in a high-pass form to be generated. The high-pass poles and zeros will be shifted below 1 rad/sec and the low-pass poles and zero will be shifted above 1 rad/sec such that the center frequency will be 1 rad/sec. A Chebyshev band-pass filter response is shown in Fig. 8-12. Once this substitution is made, we design the filter from the equations in Appendix E like any other filter.

These band-pass filters are assumed to be symmetrical, but if they are not, we should apply the techniques shown in Chapter 6 to make the specifications symmetrical. Since the filters will be symmetrical, we may use either the high-pass side or the low-pass side to determine the filter order required (just as we did for the wide-band-pass filter).

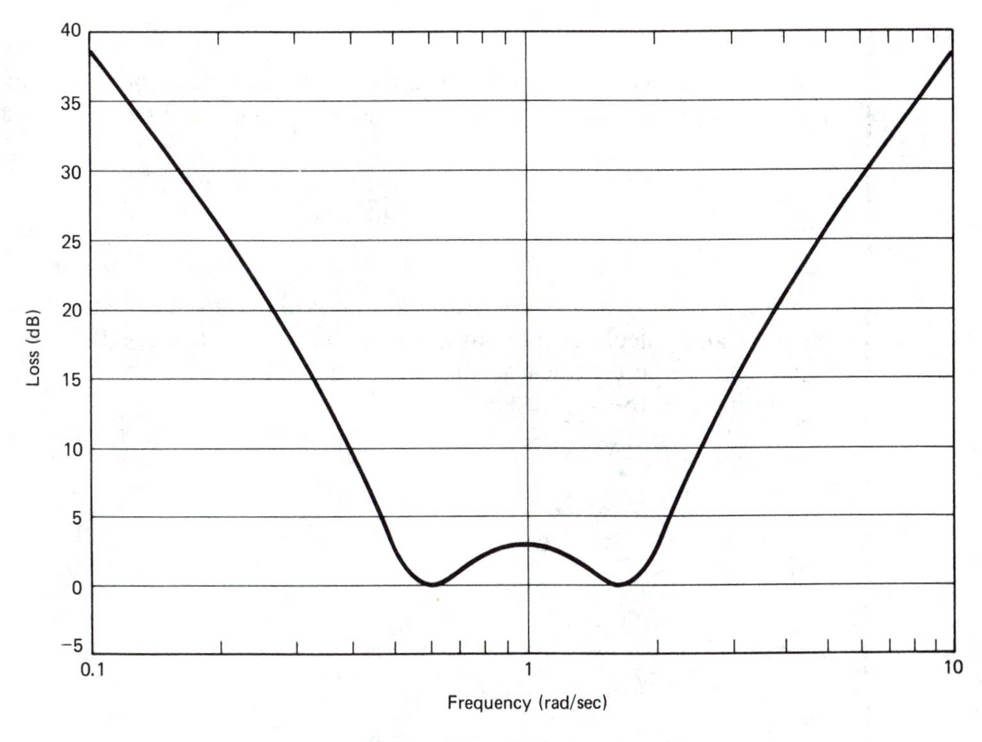

Figure 8-12

EXAMPLE 8-10

Design a Chebyshev filter with the following specifications.

$$\omega_{p1} = 400 \text{ rad/sec} \qquad \omega_{p2} = 10 \text{ krad/sec}$$

$$\omega_{s1} = 200 \text{ rad/sec} \qquad \omega_{s2} = 20 \text{ krad/sec}$$

$$L_{max} = 3.0103 \text{ dB} \qquad L_{min} = 5 \text{ dB}$$

Use $C = 0.22\ \mu$F. Calculate the resulting loss at the center frequency.

SOLUTION First we find the order of filter required using the specifications on the low-pass side (the upper break-frequency side). Using the equations in Chapter 7, we find that

$$\Omega_s = \frac{\omega_{s2}}{\omega_{p2}} = \frac{20\text{k}}{10\text{k}} = 2$$

$$\varepsilon = 1$$

and

$$n \geq 0.71037 \qquad \text{use } n = 1$$

The equation for a first-order Chebyshev from Table 7-4 is

$$\frac{s + 1}{1}$$

Next we calculate ω_0. We use the pass-band frequencies ω_{p1} and ω_{p2} since they are the upper and lower break frequencies. From Eq. (8-14b),

$$\begin{aligned}\omega_0 &= \sqrt{\omega_L \omega_H} \\ &= \sqrt{(400)(10\text{k})} \\ &= 2 \text{ krad/sec}\end{aligned}$$

Since the filter is symmetrical, we could have used the stop-band frequencies and calculated the same value, but we should use the pass-band frequencies to be technically correct.

Using Eq. (8-17), we let

$$\begin{aligned}s &= \frac{s^2 + 1}{\left(\dfrac{\omega_H - \omega_L}{\omega_0}\right)s} \\ &= \frac{s^2 + 1}{\left(\dfrac{10\text{k} - 400}{2\text{k}}\right)s} \\ &= \frac{s^2 + 1}{4.8s}\end{aligned}$$

Substituting this into the equation form of Table 7-4 gives us

$$\frac{\dfrac{s^2 + 1}{4.8s} + 1}{1}$$

which becomes

$$\frac{s^2 + 4.8s + 1}{4.8s}$$

Notice that this has changed the equation order to second order.

Using an inverting dual-feedback topology from Appendix E, we may select

$$C_C = C_B = 1\ \text{F}$$

and we find that

$$R_A = R_E = 4.8\ \Omega$$

$$R_D = 0.41667\ \Omega$$

Currently, the filter is designed with a center frequency of $\omega_0 = 1$ rad/sec. The desired center frequency is $\omega_0 = 2$ krad/sec. We calculate the frequency shifting factor F_f using Eq. (8-3b).

$$F_f = \frac{\omega_{\text{new}}}{\omega_{\text{old}}}$$

$$= \frac{2\text{k}}{1}$$

$$= 2 \times 10^3$$

Shifting the capacitors using Eq. (8-4b) yields

$$C_{\text{new}} = \frac{C_{\text{old}}}{F_f} = \frac{1}{2\text{k}} = 5 \times 10^{-4}\ \text{F}$$

Finally, we impedance shift the circuit. Using Eq. (8-6b), we find the shifting factor:

$$F_z = \frac{C_{\text{new}}}{C_{\text{old}}} = \frac{0.22\ \mu\text{F}}{5 \times 10^{-4}\ \text{F}} = 4.4 \times 10^{-4}$$

Substituting into Eqs. (8-6a) and (8-7) gives

$$C_{\text{new}} = F_z C_{\text{old}} = (4.4 \times 10^{-4})(5 \times 10^{-4}\ \text{F}) = 0.22\ \mu\text{F}$$

$$R_A = R_E = R_{\text{new}} = \frac{R_{\text{old}}}{F_z} = \frac{4.8}{4.4 \times 10^{-4}} = 10.909\ \text{k}\Omega$$

$$R_{D\ \text{new}} = \frac{R_{D\ \text{old}}}{F_z} = \frac{0.41667}{4.4 \times 10^{-4}} = 946.97\ \Omega$$

The final circuit is shown in Fig. 8-13.

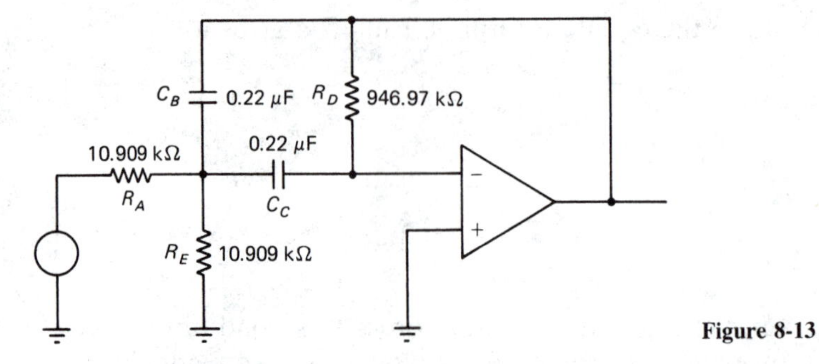

Figure 8-13

We did not take into account the overall gain, but we were not given a gain requirement. The resulting h, before frequency shifting, for this design was 0.20833. Using Eq. (8-8b), we can determine the effective multiplying factor.

$$M = \frac{h_{\text{new}}}{h_{\text{old}}}$$

$$= \frac{0.20833}{4.8}$$

$$= 0.043403$$

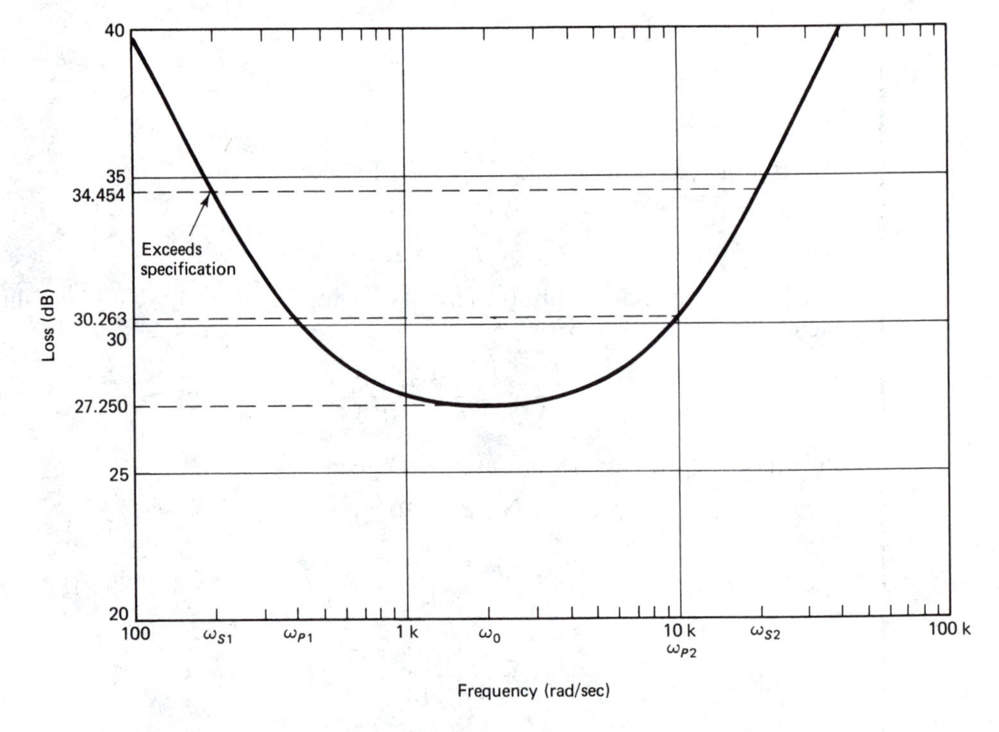

Figure 8-14

Using Eq. (8-9a), we can determine the dB shift from 0 dB or the overall loss.

$$L_{\text{shift}} = -20 \log (M)$$
$$= -20 \log (0.043403)$$
$$= 27.250 \text{ dB}$$

Figure 8-14 shows the loss response of this filter.

8-6 NOTCH FILTER

In this section we examine two ways to design a notch filter. The first way is to use a band-pass filter and a summing amplifier. The other way is to cascade single biquad notch filters.

Figure 8-15 shows the response of a Butterworth notch filter. The center frequency, bandwidth, and Q are calculated with the same equation for a band-pass filter: Eqs. (8-14), (8-15), and (8-16). The only difference is that the stop-band frequencies are used for f_L or ω_L and f_H or ω_H. The same conditions for Q in relation to $L_{\max}$ apply for the notch.

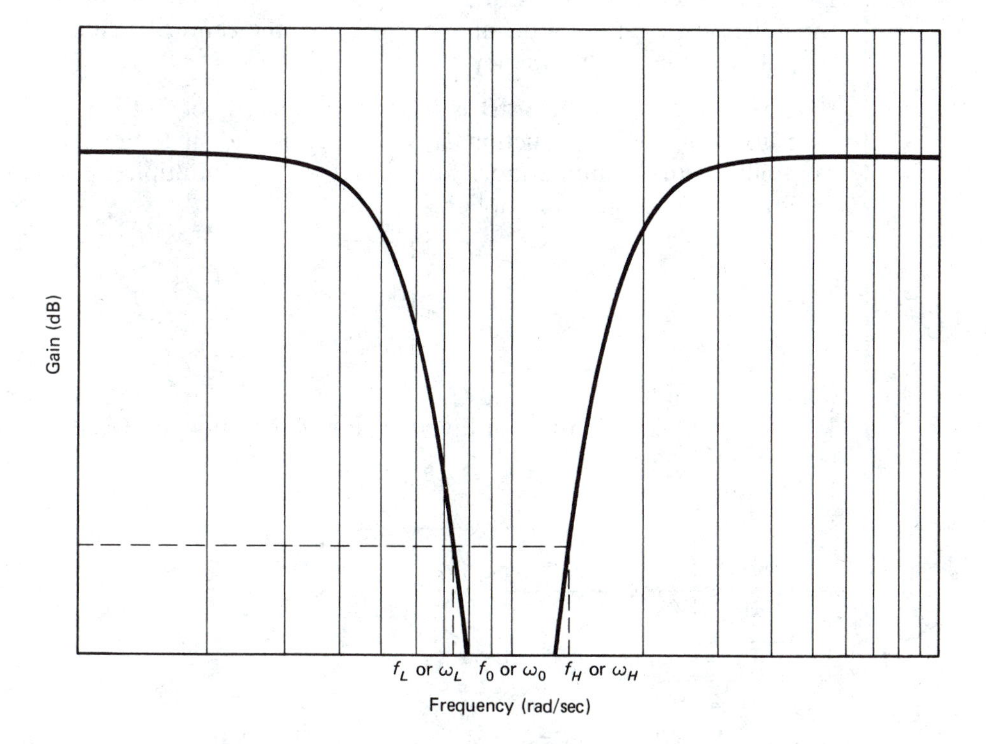

Figure 8-15

8-6-1 Parallel Design

Figure 8-16 shows the basic parallel design. An inverting band-pass filter is designed using the center frequency and bandwidth requirements for the notch. With the band-pass filter as one path, two parallel paths for the signal are made using a summing amplifier.

An inverting version of a band-pass filter must be used. Since the path around the band-pass filter is not inverted, the two paths will subtract. Figure 8-17 shows the result of this action. The output gain due to the inverting band-pass filter is shown in Fig. 8-17a. Figure 8-17b shows the output gain due to the path through R_2. When these two are combined, the result is as shown in Fig. 8-17c. When the peak gain of Fig. 8-17a is exactly equal to the gain level in Fig. 8-17b, the notch will have no output, as shown in Fig. 8-17c.

We may determine the transfer function for this very easily by subtracting 1 from the band-pass equation. Since the loss is the reciprocal of the gain, we must find a common numerator instead of a common denominator to perform this subtraction. If we assume that the loss is a numerator polynomial divided by a denominator polynomial [$L_{\text{band pass}} = N(s)/D(s)$], the resulting transfer function will be $L_{\text{notch}} = N(s)/(D(s) - N(s))$.

EXAMPLE 8-11

Design a notch filter using the band-pass filter in Example 8-10. Design the notch for a gain of 4 dB above and below the center frequency, and no output at the center frequency.

SOLUTION A gain of 4 dB is the same as a loss of -4 dB. Since the gain function and loss functions are reciprocals and since the gain function multiplier is in the numerator, we can use the loss multiplier equation Eq. (8-9b) to find the gain multiplier.

$$M = 10^{-0.05L_{\text{shift}}}$$

$$= 10^{(-0.05)(-4)}$$

$$= 1.5849$$

[This could be found by using $G_{\text{dB}} = 20 \log (G)$.] From Fig. 8-16 and our knowledge of the gain for an inverting amplifier, the gain (R_F/R_2) must be

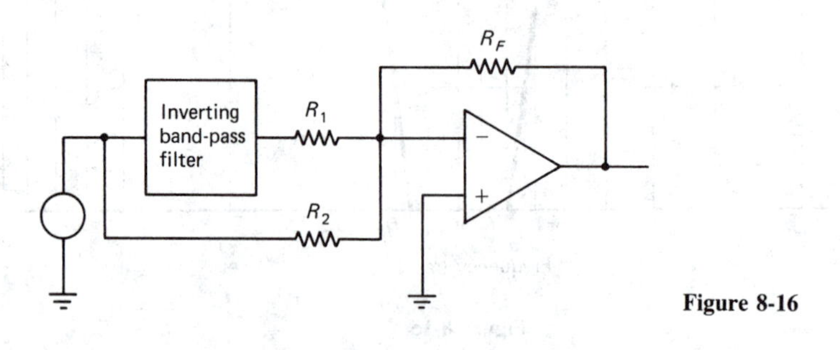

Figure 8-16

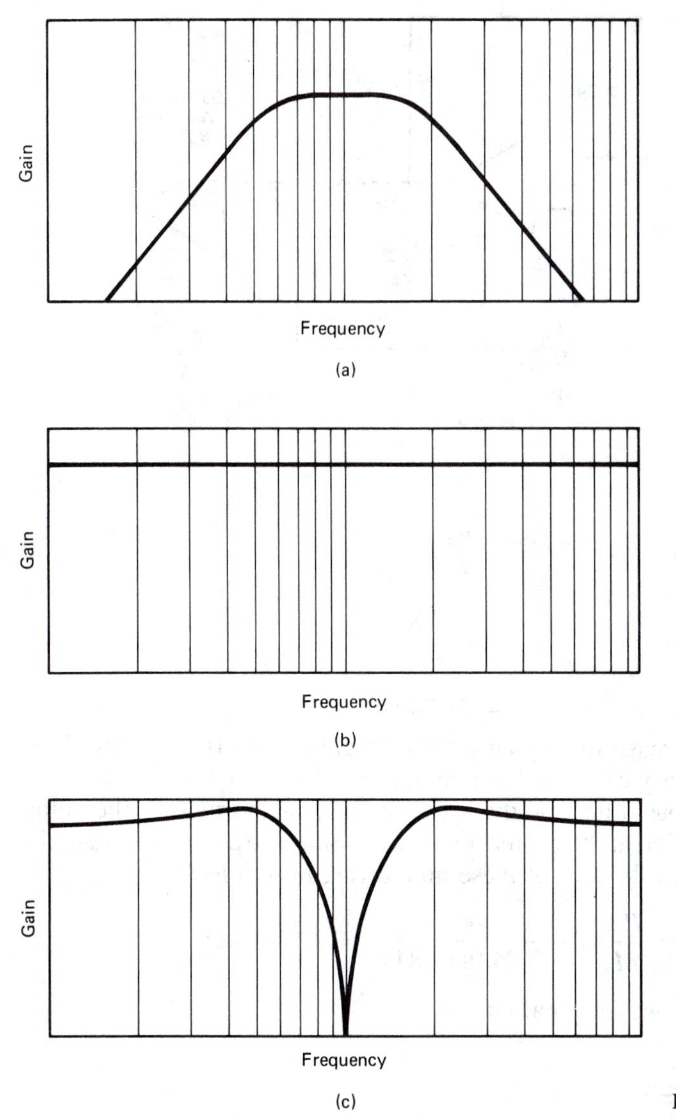

Figure 8-17

1.5849. If we let R_F be equal to 100 kΩ, then

$$R_2 = \frac{R_F}{M} = \frac{100\text{ k}}{1.5849} = 63.096\text{ k}\Omega$$

From Example 8-10 the denominator factor h, before frequency shifting, was 0.20833 for the band-pass filter. In order for the band-pass filter to have a pass-band gain of 0 dB, the denominator factor, before frequency shifting, should have been 4.8. Using Eq. (8-8b), we therefore need a

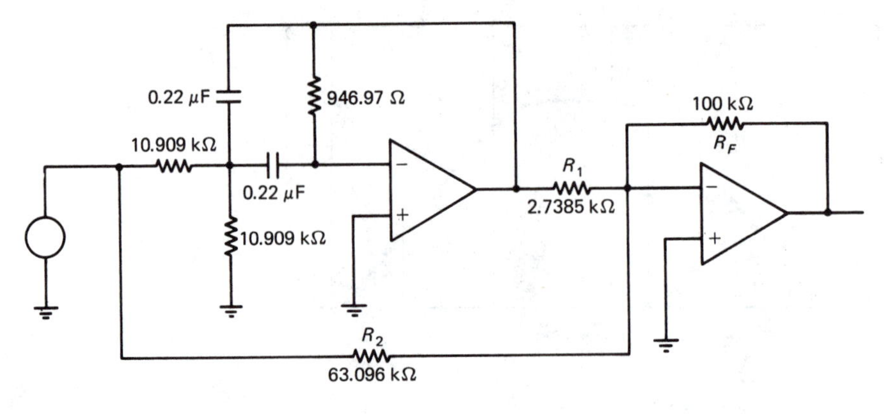

Figure 8-18

multiplying factor of

$$M = \frac{h_{\text{new}}}{h_{\text{old}}}$$

$$= \frac{4.8}{0.20833}$$

$$= 23.040$$

We also need to increase this by an additional gain of 4 dB or a loss of -4 dB which we previously calculated as a multiplying factor of 1.5849. We must have both the band-pass filter and the bypass path equal in order for the output of the notch to be nulled at the center frequency. Since we are using a summing amplifier, we can include both of these multiplying factors in R_1.

$$R_1 = \frac{R_F}{M} = \frac{100\text{k}}{(23.040)(1.5849)} = 2.7385\ \text{k}\Omega$$

Figure 8-18 shows the final circuit.

8-6-2 Cascaded Design

Using biquads, we can cascade notch filter stages to obtain a filter with the desired characteristics. In this design we will find, using the topologies in Appendix E, that the twin-T is the only possibility.

To convert a low-pass equation from Chapter 7 to a notch equation with a center frequency of 1 rad/sec, we make the substitution

$$s = \frac{\left(\dfrac{\omega_H - \omega_L}{\omega_0}\right)s}{s^2 + 1} \tag{8-18}$$

where $\dfrac{\omega_H - \omega_L}{\omega_0}$ is the normalized radian bandwidth.

We notice that this is the reciprocal of the band-pass filter transformation equation (8-17). This substitution will cause a second set of poles and zeros to appear in the low-pass equation similar to the band-pass filter transformation, but this substitution will also flip the function response upside down.

EXAMPLE 8-12

Design a fourth-order Chebyshev notch filter with $Q = 5$, $\omega_0 = 4$ krad/sec, and $L_{max} = 1.5$ dB.

SOLUTION Since we require a fourth order, we will use a second-order low-pass equation to transform into a notch equation. From Table 7-5,

$$\frac{s^2 + 0.92218s + 0.92521}{0.77846}$$

To transform this equation, we need to know ω_L and ω_H. From Eq. (8-16b),

$$Q = \frac{\omega_0}{\text{BW}_{R/S}}$$

Solving for the bandwidth gives

$$\text{BW}_{R/S} = \frac{\omega_0}{Q} = \frac{4\text{k}}{5} = 800 \text{ rad/sec}$$

and we know from Eq. (8-15b) that

$$\text{BW}_{R/S} = \omega_H - \omega_L = 800$$

Therefore, substituting into Eq. (8-18), we obtain

$$s = \frac{\left(\dfrac{\omega_H - \omega_L}{\omega_0}\right)s}{s^2 + 1}$$

$$= \frac{\left(\dfrac{800}{4\text{k}}\right)s}{s^2 + 1}$$

$$= \frac{0.2s}{s^2 + 1}$$

Substituting this into the low-pass equation, we have

$$\frac{\left(\dfrac{0.2s}{s^2 + 1}\right)^2 + 0.92218\left(\dfrac{0.2s}{s^2 + 1}\right) + 0.92521}{0.77846}$$

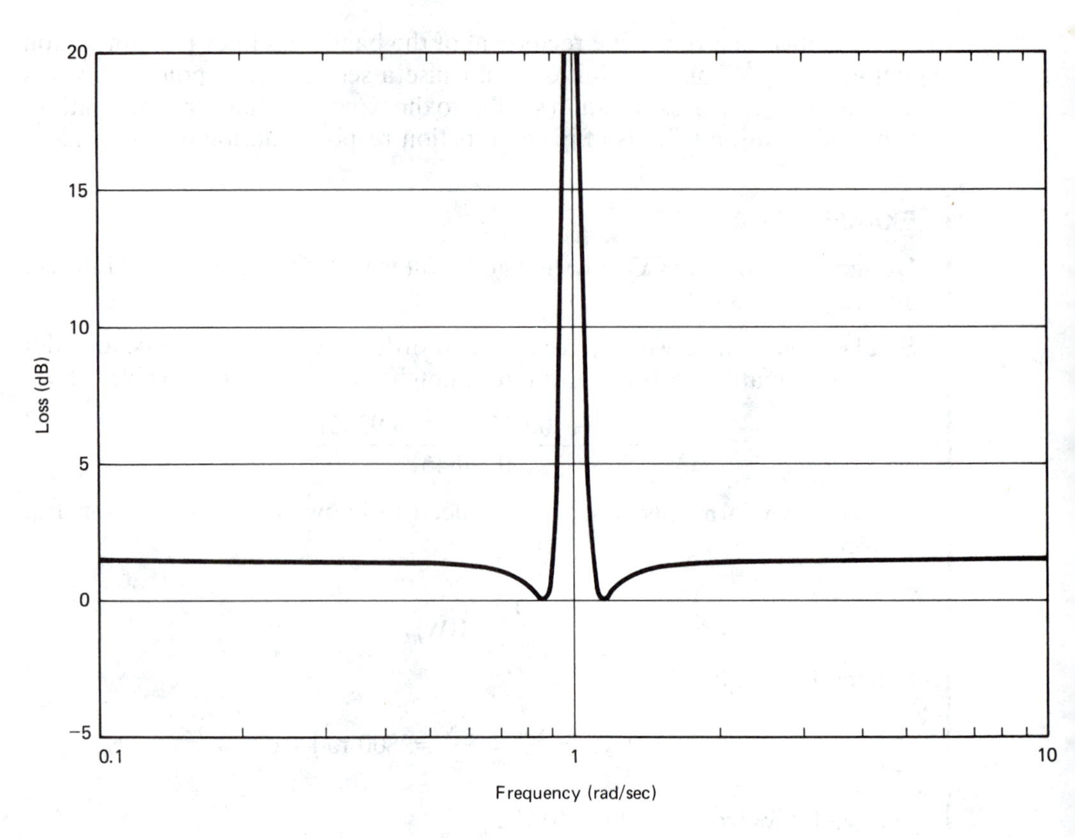

Figure 8-19

Simplifying, we have

$$\frac{s^4 + 0.19935s^3 + 2.0432s^2 + 0.19935s + 1}{0.84139(s^2 + 1)^2}$$

Using Bairstow's method shown in Appendix F, we can factor this into biquads.

$$\frac{1}{0.84139}\left(\frac{s^2 + 0.090605s + 0.83323}{s^2 + 1} + \frac{s^2 + 0.10874s + 1.2002}{s^2 + 1}\right)$$

Figure 8-19 shows the response of this equation.

Following the design equations in Section E-5, frequency shifting, and impedance shifting, we will find the values shown in Fig. 8-20. The resulting h, before frequency shifting, for this circuit was 1.6613.

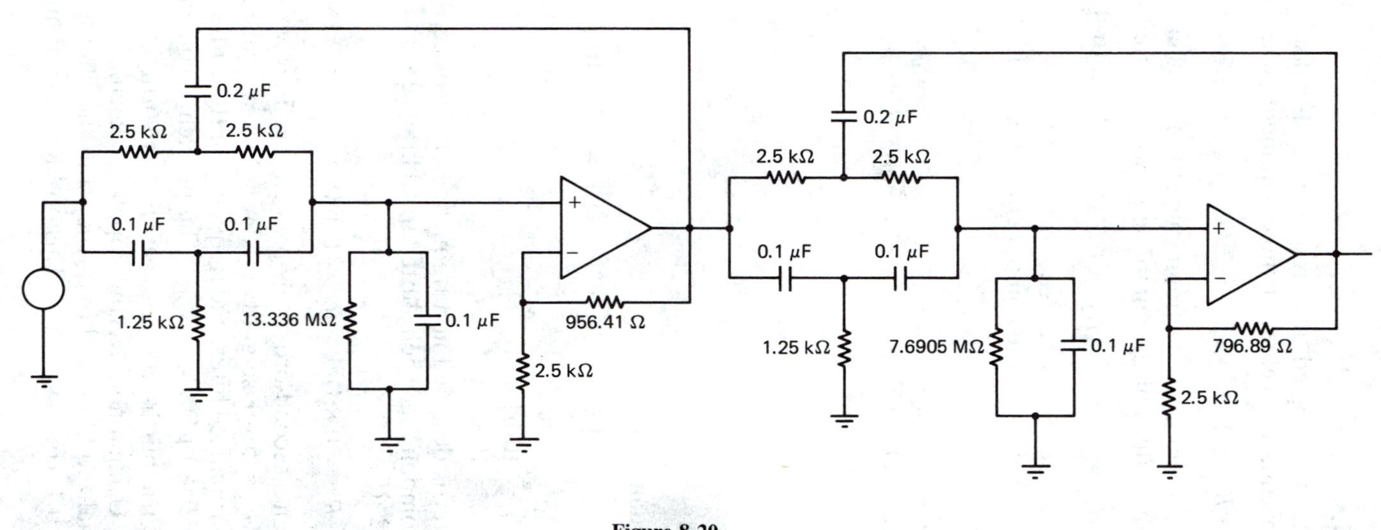
0.2 μF
2.5 kΩ
2.5 kΩ
0.1 μF
0.1 μF
1.25 kΩ
13.336 MΩ
0.1 μF
956.41 Ω
2.5 kΩ
0.2 μF
2.5 kΩ
2.5 kΩ
0.1 μF
0.1 μF
1.25 kΩ
7.6905 MΩ
0.1 μF
796.89 Ω
2.5 kΩ

Figure 8-20

PROBLEMS

Section 8-1

1. A filter is designed at a frequency of 1 rad/sec. Changing the resistors, find the values of the components when shifted to 2 kHz.

$$R_1 = 1\ \Omega \qquad R_2 = 2\ \Omega \qquad C_1 = 2\text{ F} \qquad C_2 = 0.2\text{ F}$$

2. A filter is designed at a frequency of 10 Hz. Changing the capacitors, find the values of the components when shifted to 1 rad/sec.

$$R_1 = 1.8\ \Omega \qquad R_2 = 0.2\ \Omega \qquad C_1 = 10\text{ F} \qquad C_2 = 1\text{ F}$$

3. A filter is designed at a frequency of 1 kHz. Changing the capacitors, find the values of the components when shifted to 5 kHz.

$$R_1 = 3\ \Omega \qquad R_2 = 2.2\ \Omega \qquad C_1 = 0.01\text{ F} \qquad C_2 = 0.1\text{ F}$$

4. A filter is designed at a frequency of 1 rad/sec. Changing the resistors, find the values of the components when shifted to 2.5 krad/sec.

$$R_1 = 2\text{ k}\Omega \qquad R_2 = 2.2\text{ k}\Omega \qquad C_1 = 2\ \mu\text{F} \qquad C_2 = 0.1\ \mu\text{F}$$

Section 8-2

5. Impedance shift

$$R_1 = 4\ \Omega \qquad R_2 = 0.2\ \Omega \qquad C_1 = 10\text{ F} \qquad C_2 = 1\text{ F}$$

so that:
(a) $C_1 = 10\ \mu\text{F}$
(b) $C_2 = 10\ \mu\text{F}$

6. A filter is designed at a frequency of 1 rad/sec. Find the values of the components when shifted to 10 krad/sec, and impedance shift C_1 to 2.2 μF.

$$R_1 = 0.8\ \Omega \qquad R_2 = 0.1\ \Omega \qquad C_1 = 2\text{ F} \qquad C_2 = 1\text{ F}$$

7. Design a second-order low-pass elliptic filter with $L_{max} = 3.0103$ dB and $\Omega_s = 5$. Using a twin-T topology, calculate the resulting h of the normalized filter, and frequency shift the filter to 4.2 kHz. Make $C_A = 0.001\ \mu\text{F}$, $R_2 = R_D$. (This problem is continued in Problem 13.)

8. Design a fourth-order low-pass Butterworth filter with $L_{max} = 3.0103$ dB. Using a noninverting topology, calculate the resulting h of the normalized filter, and frequency shift the filter to 2 krad/sec. Make $C = 0.1\ \mu\text{F}$, $R_2 = R_A$. (This problem is continued in Problem 14.)

Section 8-3

9. Modify a first-order Butterworth equation with $L_{max} = 1.5$ dB so that it has a pass-band loss of:
(a) -5 dB
(b) $+5$ dB

10. Modify a fourth-order elliptic equation with $L_{max} = 0.25$ dB and $\Omega_s = 5$ so that it has a pass-band loss of:
(a) -8 dB
(b) $+10$ dB

11. Showing the points to connect the op-amp output and the feedback circuit, design a resistive voltage divider network (with $R_\alpha = 22$ kΩ) to shift the filter's loss by:
(a) $+3.0103$ dB
(b) -2.5 dB

12. Showing the points to connect the op-amp output and the feedback circuit, design a resistive voltage divider network (with $R_\beta = 22$ kΩ) to shift the filter's gain by:
(a) $+3.0103$ dB
(b) -2.5 dB

13. Using a resistive voltage divider network ($R_\alpha = R_B$), shift the pass-band gain of the elliptic filter in Problem 7 to a pass-band gain of:
(a) 20 dB
(b) 45 dB

14. Using a resistive voltage divider network ($R_\alpha = R$), shift the pass-band gain of the Butterworth filter in Problem 8 to a pass-band gain of:
(a) -2 dB
(b) 10 dB

Section 8-4

15. Transform a third-order Chebyshev equation with $L_{max} = 1.5$ dB into a high-pass Chebyshev equation.

16. Transform a third-order elliptic equation with $L_{max} = 0.25$ dB and $\Omega_s = 2$ into a high-pass elliptic equation.

17. For the filter specification, design a Butterworth filter using a non-inverting amplifier with $C = 0.22$ μF and the pass-band loss shifted to -6 dB.

$$\omega_p = 440 \text{ rad/sec} \qquad \omega_s = 110 \text{ rad/sec}$$

$$L_{max} = 0.25 \text{ dB} \qquad L_{min} = 12 \text{ dB}$$

18. For the filter specification, design an elliptic filter. Use a twin-T topology. Make $C_A = 0.1\ \mu F$, and $R_2 = R_B = R_\alpha$.

$$\omega_p = 9 \text{ krad/sec} \qquad \omega_s = 4.5 \text{ krad/sec}$$

$$L_{max} = 3.0103 \text{ dB} \qquad L_{min} = 20 \text{ dB}$$

Section 8-5

Section 8-5-1

19. Find the Q, $BW_{R/S}$, BW_{Hz}, f_0, and ω_0 for the specifications shown.

$$\omega_{p1} = 2.4 \text{ krad/sec} \qquad \omega_{p2} = 3.75 \text{ krad/sec}$$

$$\omega_{s1} = 1.2 \text{ krad/sec} \qquad \omega_{s2} = 7.5 \text{ krad/sec}$$

$$L_{max} = 0.25 \text{ dB} \qquad L_{min} = 40 \text{ dB}$$

20. Find ω_L and ω_H for a $BW_{R/S}$ of 500 rad/sec at a center frequency of 2 krad/sec.

Section 8-5-2

21. For the filter specification, design a Chebyshev filter. Use inverting topologies. Make $C_A = C_E = 0.022\ \mu F$, and $R_\alpha = 10\ k\Omega$.

$$f_{p1} = 720 \text{ Hz} \qquad f_{p2} = 4.5 \text{ kHz}$$

$$f_{s1} = 160 \text{ Hz} \qquad f_{s2} = 20.25 \text{ kHz}$$

$$L_{max} = 1.5 \text{ dB} \qquad L_{min} = 12 \text{ dB}$$

22. For the filter specification, design an elliptic filter using a twin-T topology with a pass-band loss of -10 dB. Make $C_A = 0.01\ \mu F$, and $R_2 = R_\alpha = 10\ k\Omega$.

$$\omega_{p1} = 1.35 \text{ krad/sec} \qquad \omega_{p2} = 9.6 \text{ krad/sec}$$

$$\omega_{s1} = 675 \text{ rad/sec} \qquad \omega_{s2} = 19.2 \text{ krad/sec}$$

$$L_{max} = 0.25 \text{ dB} \qquad L_{min} = 8 \text{ dB}$$

Section 8-5-3

23. Transform a third-order Butterworth equation with $L_{max} = 1.5$ dB into a band-pass Butterworth equation with $\omega_0 = 5$ krad/sec having a Q of 10.

24. Transform a fourth-order elliptic equation with $L_{max} = 3.0103$ dB and $\Omega_s = 2$ into a band-pass elliptic equation with $\omega_0 = 50$ rad/sec having a Q of 2.

25. For the filter specification, write the normalized Chebyshev equation.

$$f_{p1} = 2 \text{ kHz} \qquad f_{p2} = 4.5 \text{ kHz}$$

$$f_{s1} = 400 \text{ Hz} \qquad f_{s2} = 22.5 \text{ kHz}$$

$$L_{max} = 3.0103 \text{ dB} \qquad L_{min} = 30 \text{ dB}$$

26. For the filter specification, design a Butterworth filter using a non-inverting topology. Make $C = 2.2\ \mu\text{F}$ and $R_2 = R$. Calculate the resulting pass-band gain.

$$\omega_{p1} = 37.5 \text{ rad/sec} \qquad \omega_{p2} = 54 \text{ rad/sec}$$

$$\omega_{s1} = 1 \text{ rad/sec} \qquad \omega_{s2} = 2.025 \text{ krad/sec}$$

$$L_{max} = 0.25 \text{ dB} \qquad L_{min} = 10 \text{ dB}$$

Section 8-6

Section 8-6-1

27. Using parallel design, design a notch filter using a second-order Butterworth band-pass filter having $L_{max} = 0.25$ dB, $\omega_0 = 200$ rad/sec, and $BW_{R/S} = 200$ rad/sec. Design the gain outside the notch to be 5 dB and design the notch center to have no output. Make $C_1 = 1\ \mu\text{F}$, $R_2 = R$, and the summing amplifier have $R_F = 100\ \text{k}\Omega$.

28. Using parallel design, design a notch filter using a second-order Chebyshev band-pass filter having $L_{max} = 0.25$ dB, $\omega_0 = 200$ rad/sec, and $BW_{R/S} = 200$ rad/sec. Design the gain outside the notch to be 5 dB and design the notch center to have no output. Make $C_1 = 1\ \mu\text{F}$, $R_2 = R$, and the summing amplifier have $R_F = 100\ \text{k}\Omega$. (This is the same as Problem 27 except using a Chebyshev circuit.)

Section 8-6-2

29. Transform a third-order Chebyshev equation with $L_{max} = 3.0103$ dB into a normalized notch Chebyshev equation with $\omega_0 = 200$ rad/sec having a Q of 5.

30. Design a second-order Butterworth notch having $L_{max} = 3.0103$ dB, $Q = 2$, and $BW_{R/S} = 500$ rad/sec. Make $C_A = 0.1\ \mu\text{F}$ and $R_2 = R_B$.

31. For the filter specification, design an elliptic filter using any reasonable topology with any reasonable capacitor values.

$$\omega_{p1} = 1.96 \text{ krad/sec} \qquad \omega_{p2} = 9 \text{ krad/sec}$$

$$\omega_{s1} = 3.92 \text{ krad/sec} \qquad \omega_{s2} = 4.5 \text{ krad/sec}$$

$$L_{max} = 3.0103 \text{ dB} \qquad L_{min} = 20 \text{ dB}$$

APPENDIX A

Transform Tables

A-1 TRANSFORM PAIRS

$f(t)$	$F(s)$	
$\delta(t)$	1	(P-1)
$u(t)$	$\frac{1}{s}$	(P-2)
$tu(t)$	$\frac{1}{s^2}$	(P-3)
$e^{-bt}u(t)$	$\frac{1}{s+b}$	(P-4)
$\sin(\omega t)\, u(t)$	$\frac{\omega}{s^2+\omega^2}$	(P-5)
$\cos(\omega t)\, u(t)$	$\frac{s}{s^2+\omega^2}$	(P-6)

A-2 TRANSFORM OPERATIONS

$f(t)$	$F(s)$	
$h(t-a)u(t-a)$	$e^{-as}H(s)$	(O-1)
$e^{-bt}h(t)u(t)$	$H(s+b)$	(O-2)
$th(t)u(t)$	$-\frac{d}{ds}[H(s)]$	(O-3)
$\left\{\frac{d}{dt}[h(t)]\right\}u(t)$	$sH(s)-h(0)$	(O-4a)
$\frac{d}{dt}[h(t)u(t)]$	$sH(s)$	(O-4b)
$\int_0^t h(t)u(t)\,dt$	$\frac{H(s)}{s}$	(O-5)

A-3 TRANSFORM IDENTITIES

$$\mathcal{L}[f(t)] = F(s) \tag{I-1}$$

$$\mathcal{L}^{-1}[F(s)] = f(t)$$

$$f(t) \Leftrightarrow F(s) \tag{I-2}$$

$$\mathcal{L}[Kf(t)] = K\mathcal{L}[f(t)] \tag{I-3}$$

$$\mathcal{L}^{-1}[KF(s)] = K\mathcal{L}^{-1}[F(s)]$$

$$\mathcal{L}[f_1(t) + f_2(t) + \cdots] = \mathcal{L}[f_1(t)] + \mathcal{L}[f_2(t)] + \cdots \tag{I-4}$$

$$\mathcal{L}^{-1}[F_1(s) + F_2(s) + \cdots] = \mathcal{L}^{-1}[F_1(s)] + \mathcal{L}^{-1}[F_2(s)] + \cdots$$

The following are often mistaken as identities but they are not.

$$\mathcal{L}[f_1(t)f_2(t)] \quad \textit{is not} \quad \mathcal{L}[f_1(t)]\mathcal{L}[f_2(t)]$$

$$f(t) \quad \textit{is not} \quad F(s)$$

APPENDIX B

Laplace Derivations

This appendix derives the Laplace transform pairs, the Laplace operations, and the inverse formula for complex poles.

B-1 DERIVING LAPLACE TRANSFORM PAIRS

B-1-1 $\delta(t)$ Function (P-1 Pair)

$$\mathcal{L}[\delta(t)] = \int_0^\infty \delta(t)e^{-st}\,dt$$

For an impulse, the value is zero except at $t = 0$. Therefore, the integral will obtain a value only at $t = 0$. Since e^{-st} has a value of 1 at $t = 0$ and the impulse has an area of 1 at $t = 0$, then

$$\int_0^\infty \delta(t)e^{-st}\,dt = 1$$

B-1-2 $u(t)$ Function (P-2 Pair)

$$\mathcal{L}[u(t)] = \int_0^\infty u(t)e^{-st}\,dt$$

After $t = 0$, $u(t)$ is equal to a constant value of 1.

$$\int_0^\infty e^{-st}\,dt$$

If we let

$$u = -st$$

then

$$du = -s\,dt$$

Rearranging the function yields

$$\frac{1}{-s}\int_0^\infty e^{-st}(-s)\,dt = \left.\frac{e^{-st}}{-s}\right|_0^\infty = \frac{1}{s}$$

B-1-3 t Function (P-3 Pair)

$$\mathscr{L}[tu(t)] = \int_0^\infty tu(t)e^{-st}\,dt$$

After $t = 0$, $u(t)$ is equal to 1.

$$\int_0^\infty te^{-st}\,dt$$

Using

$$\int u\,dv = uv - \int v\,du$$

we let

$$u = t \qquad \text{and} \qquad dv = e^{-st}\,dt$$

Therefore,

$$du = dt \qquad \text{and} \qquad v = \frac{-e^{-st}}{s}$$

Substituting into the new form gives us

$$\left.\frac{-te^{-st}}{s}\right|_0^\infty - \int_0^\infty \frac{-e^{-st}}{s}\,dt$$

$$0 - \frac{1}{s^2}\int_0^\infty e^{-st}(-s)\,dt$$

$$\left.\frac{-e^{-st}}{s^2}\right|_0^\infty = \frac{1}{s^2}$$

B-1-4 e^{-bt} Function (P-4 Pair)

$$\mathscr{L}[e^{-bt}u(t)] = \int_0^\infty e^{-bt}u(t)e^{-st}\,dt$$

After $t = 0$, $u(t)$ is equal to 1, and we may combine the two exponential functions

$$\int_0^\infty e^{-t(s+b)}\,dt$$

If we let

$$u = -t(s + b)$$

then

$$du = -(s + b)\,dt$$

Rearranging the function yields

$$\frac{1}{-(s + b)} \int_0^\infty e^{-t(s+b)}[-(s + b)]\,dt$$

$$\left.\frac{e^{-t(s+b)}}{-(s + b)}\right|_0^\infty = \frac{1}{s + b}$$

B-1-5 sin (ωt) Function (P-5 Pair)

$$\mathcal{L}[\sin(\omega t)u(t)] = \int_0^\infty \sin(\omega t)u(t)e^{-st}\,dt$$

After $t = 0$, $u(t)$ is equal to 1, and using Euler's identities, we obtain

$$\int_0^\infty \frac{e^{j\omega t} - e^{-j\omega t}}{2j} e^{-st}\,dt$$

Combining exponentials and splitting up the integral, we have

$$\frac{1}{2j}\left[\int_0^\infty e^{-t(s-j\omega)}\,dt - \int_0^\infty e^{-t(s+j\omega)}\,dt\right]$$

Both integrals are integrated using the same process

$$u = -t(s - j\omega) \qquad \text{and} \qquad u = -t(s + j\omega)$$

Therefore,

$$du = -(s - j\omega)\,dt \qquad \text{and} \qquad du = -(s + j\omega)\,dt$$

This results in

$$\frac{1}{2j}\left[\frac{e^{-t(s-j\omega)}}{-(s - j\omega)} - \frac{e^{-t(s+j\omega)}}{-(s + j\omega)}\right]_0^\infty$$

$$\frac{1}{2j}\left(\frac{1}{s - j\omega} - \frac{1}{s + j\omega}\right)$$

$$\frac{\omega}{s^2 + \omega^2}$$

B-1-6 cos (ωt) Function (P-6 Pair)

$$\mathcal{L}[\cos(\omega t)u(t)] = \int_0^\infty \cos(\omega t)u(t)e^{-st}\,dt$$

After $t = 0$, $u(t)$ is equal to 1, and using Euler's identities, we obtain

$$\int_0^\infty \frac{e^{j\omega t} + e^{-j\omega t}}{2} e^{-st}\,dt$$

Combining the exponentials, we have

$$\frac{1}{2}\left[\int_0^\infty e^{-t(s-j\omega)}\,dt + \int_0^\infty e^{-t(s+j\omega)}\,dt\right]$$

Both integrals are integrated using the same process

$$u = -t(s - j\omega) \qquad \text{and} \qquad u = -t(s + j\omega)$$

Therefore,

$$du = -(s - j\omega)\,dt \qquad \text{and} \qquad du = -(s + j\omega)\,dt$$

This results in

$$\frac{1}{2}\left[\frac{e^{-t(s-j\omega)}}{-(s - j\omega)} + \frac{e^{-t(s+j\omega)}}{-(s + j\omega)}\right]_0^\infty$$

$$\frac{1}{2}\left(\frac{1}{s - j\omega} + \frac{1}{s + j\omega}\right)$$

$$\frac{s}{s^2 + \omega^2}$$

B-2 DERIVING LAPLACE TRANSFORM OPERATIONS

B-2-1 $h(t - a)u(t - a)$ (O-1 Operation)

$$\mathcal{L}[h(t - a)u(t - a)] = \int_0^\infty h(t - a)u(t - a)e^{-st}\,dt$$

After $t = a$, $u(t - a)$ is equal to 1. Since $u(t - a)$ is zero before $t = a$, the integral is zero between $t = 0$ and $t = a$. We may therefore change the lower limit of the integral so that it starts at $t = a$, and drop the $u(t - a)$.

$$\int_a^\infty h(t - a)e^{-st}\,dt$$

We may change the variable to x by letting

$$x = t - a$$

Then

$$dx = dt$$

The lower limit of the integral then becomes zero.

Rewriting the function yields

$$\int_0^\infty h(x)e^{-s(x+a)}\,dx$$

When we split up the exponential, the e^{-sa} part is a constant and may be factored out of the integral. The remaining part, by definition, is $H(s)$.

$$e^{-sa} \int_0^{\infty} h(x)e^{-sx}\, dx = e^{-sa}H(s)$$

B-2-2 $e^{-bt}h(t)u(t)$ (O-2 Operation)

$$\mathcal{L}[e^{-bt}h(t)u(t)] = \int_0^{\infty} e^{-bt}h(t)u(t)e^{-st}\, dt$$

Removing the $u(t)$, we may rewrite this as

$$\int_0^{\infty} h(t)e^{-(s+b)t}\, dt$$

By definition this is

$$H(s + b)$$

B-2-3 $th(t)u(t)$ (O-3 Operation)

$$\mathcal{L}[th(t)u(t)] = \int_0^{\infty} th(t)u(t)e^{-st}\, dt$$

Since $u(t)$ is 1 after $t = 0$, we may rewrite this as

$$\int_0^{\infty} th(t)e^{-st}\, dt$$

If we take the derivative of this function with respect to s, the time variable is considered to be a constant. When the derivative is performed on the $H(s)$ integral and multiplied by -1, we will see that the result is equal to the integral we require.

$$\begin{aligned}
\int_0^{\infty} th(t)e^{-st}\, dt &\stackrel{?}{=} \frac{-d}{ds} F(s) \\
&\stackrel{?}{=} \frac{-d}{ds} \int_0^{\infty} h(t)e^{-st}\, dt \\
&\stackrel{?}{=} \int_0^{\infty} \frac{-d}{ds} h(t)e^{-st}\, dt \\
&= \int_0^{\infty} th(t)e^{-st}\, dt
\end{aligned}$$

B-2-4 $\left\{\dfrac{d}{dt}[h(t)]\right\}u(t)$ **(O-4a Operation)**

$$\mathcal{L}\left[\left\{\frac{d}{dt}[h(t)]\right\}u(t)\right] = \int_0^\infty \left\{\frac{d}{dt}[h(t)]\right\}u(t)e^{-st}\,dt$$

After $t = 0$, $u(t)$ is equal to 1.

$$\int_0^\infty \frac{d}{dt}[h(t)]e^{-st}\,dt$$

Using

$$\int u\,dv = uv - \int v\,du$$

we let

$$u = e^{-st} \qquad \text{and} \qquad dv = \frac{d}{dt}h(t)\,dt$$

Therefore,

$$du = -se^{-st}\,dt \qquad \text{and} \qquad v = h(t)$$

Substituting into the new form gives us

$$h(t)e^{-st}\Big|_0^\infty - \int_0^\infty -sh(t)e^{-st}\,dt$$

$$sH(s) - h(0)$$

B-2-5 $\dfrac{\mathbf{d}}{\mathbf{dt}}$ **[$h(t)u(t)$] (O-4b Operation)**

$$\mathcal{L}\left[\frac{d}{dt}[h(t)u(t)]\right] = \int_0^\infty \frac{d}{dt}[h(t)u(t)]e^{-st}\,dt$$

In this case we must take the derivative before integrating.

$$\int_0^\infty \left\{\left[\frac{d}{dt}h(t)\right]u(t) + h(0)\delta(t)\right\}e^{-st}\,dt$$

Separating the into two integrals, we have

$$\int_0^\infty \left[\frac{d}{dt}h(t)\right]u(t)e^{-st} + \int_0^\infty h(0)\delta(t)e^{-st}\,dt$$

The first term, as seen in Section B-2-4 is $sH(s) - h(0)$, and the second term, as seen in Section B-1-1, is $h(0)$ since $h(0)$ is a constant. Adding these two terms together, we have

$$[sH(s) - h(0)] + h(0)$$

Therefore,

$$sH(s)$$

B-2-6 $\int_0^t h(t)u(t)\,dt$ (O-5 Operation)

$$\mathcal{L}\left[\int_0^t h(t)u(t)\,dt\right] = \int_0^\infty \int_0^t h(t)u(t)\,dt\,e^{-st}\,dt$$

After $t = 0$, $u(t)$ is equal to 1.

$$\int_0^\infty \int_0^t h(t)\,dt\,e^{-st}\,dt$$

Using

$$\int u\,dv = uv - \int v\,du$$

we let

$$u = \int_0^t h(t)\,dt \qquad \text{and} \qquad dv = e^{-st}\,dt$$

Therefore,

$$du = h(t)\,dt \qquad \text{and} \qquad v = \frac{e^{-st}}{-s}$$

Substituting into the new form gives us

$$\left.\frac{\left[\int_0^t h(t)\,dt\right] e^{-st}}{-s}\right|_0^\infty - \int_0^\infty \frac{e^{-st}h(t)}{-s}\,dt$$

$$0 - \frac{1}{-s}\int_0^\infty e^{-st}h(t)\,dt$$

This becomes

$$\frac{H(s)}{s}$$

B-3 DERIVING COMPLEX POLES FORMULA

$R(s)$ for a complex pole of any order is

$$R(s)_{-\propto+j\omega} = F(s)[(s+\propto)^2+\omega^2]^r$$

In this section we drop the $R(s)$ subscript $-\propto + j\omega$ to simplify the equations.

$$R(s) = R(s)_{-\propto+j\omega} \qquad \text{(for this section)}$$

The function $F(s)$ can be represented as

$$F(s) = \frac{k_1}{(s+\propto-j\omega)^r} + \frac{k_2}{(s+\propto-j\omega)^{r-1}} + \cdots + \frac{k_r}{(s+\propto-j\omega)} + \cdots$$

$$+ \frac{k_1^*}{(s+\propto+j\omega)^r} + \frac{k_2^*}{(s+\propto+j\omega)^{r-1}} + \cdots + \frac{k_r^*}{(s+\propto+j\omega)} + \cdots$$

$$+ \text{ rest of the function}$$

The "*" indicates the constant associated with the conjugate root. From this point on we will drop the notation "+ rest of the function" and just remember that it is always present.

We may write these factors in a summation form as

$$F(s) = \sum_{n=1}^{r} \frac{k_n}{(s+\propto-j\omega)^{r-n+1}} + \sum_{n=1}^{r} \frac{k_n^*}{(s+\propto+j\omega)^{r-n+1}}$$

Using Eq. (3-7a), we can solve for k_n. Since $R(s)_{-\propto+j\omega}$ contains both roots, we must supply the conjugant root.

$$k_n = \frac{d^{n-1}}{ds^{n-1}} \frac{R(s)}{(s+\propto+j\omega)^r}\Bigg|_{s=-\propto+j\omega}$$

Applying this equation, we will see a pattern developing as n goes from 1 to r.

$$k_1 = \frac{R(s)}{(s+\propto+j\omega)^r}\Bigg|_{s=-\propto+j\omega}$$

$$= \frac{R(-\propto+j\omega)}{(2j\omega)^r}$$

$$k_2 = \frac{d^1}{ds^1} \frac{R(s)}{(s+\propto+j\omega)^r}\Bigg|_{s=-\propto+j\omega}$$

$$= \frac{R'(s)(s+\propto+j\omega)^r - R(s)r(s+\propto+j\omega)^{r-1}}{(s+\propto+j\omega)^{2r}}\Bigg|_{s=-\propto+j\omega}$$

$$= \frac{1}{(2j\omega)^r}\left[R'(-\propto+j\omega) - \frac{rR(-\propto+j\omega)}{2j\omega}\right]$$

Continuing the procedure, k_3 and k_4 are found to be

$$k_3 = \frac{1}{(2j\omega)^r}\left[R''(-\alpha + j\omega) - \frac{2rR'(-\alpha + j\omega)}{2j\omega} + \frac{r(r+1)R(-\alpha + j\omega)}{(2j\omega)^2}\right]$$

$$k_4 = \frac{1}{(2j\omega)^r}\left[R'''(-\alpha + j\omega) - \frac{3rR''(-\alpha + j\omega)}{2j\omega} + \frac{3r(r+1)R'(-\alpha + j\omega)}{(2j\omega)^2} - \frac{r(r+1)(r+2)R(-\alpha + j\omega)}{(2j\omega)^3}\right]$$

By a similar process we can find the conjugate k values.

$$k_1^* = \frac{R(-\alpha - j\omega)}{(-2j\omega)^r}$$

$$k_2^* = \frac{1}{(-2j\omega)^r}\left[R'(-\alpha - j\omega) - \frac{rR(-\alpha - j\omega)}{-2j\omega}\right]$$

$$k_3^* = \frac{1}{(-2j\omega)^r}\left[R''(-\alpha - j\omega) - \frac{2rR'(-\alpha - j\omega)}{-2j\omega} + \frac{r(r+1)R(-\alpha - j\omega)}{(-2j\omega)^2}\right]$$

$$k_4^* = \frac{1}{(-2j\omega)^r}\left[R'''(-\alpha - j\omega) - \frac{3rR''(-\alpha - j\omega)}{-2j\omega} + \frac{3r(r+1)R'(-\alpha - j\omega)}{(-2j\omega)^2} - \frac{r(r+1)(r+2)R(-\alpha - j\omega)}{(-2j\omega)^3}\right]$$

Here we see a similar pattern being established.

In both the k and k^*, the part of the expression inside the brackets will be a magnitude, M, with a phase angle. Since k and k^* are complex conjugates, the magnitudes inside the brackets will be the same with the phase angles equal and opposite. We may therefore express the k's as

$$k_n = \frac{1}{(2j\omega)^r} M_n \angle\theta_n$$

$$k_n^* = \frac{1}{(-2j\omega)^r} M_n \angle -\theta_n$$

where

$$M_1\angle\theta_1 = R(-\alpha + j\omega)$$

$$M_2\angle\theta_2 = R'(-\alpha + j\omega) - \frac{rR(-\alpha + j\omega)}{2j\omega}$$

and so on.

We may write the magnitude and phase angle in exponential form as

$$M_n e^{\pm j\theta_n}$$

using Eq. (3-8b) and combining the k_n and k_n^*, we may write

$$F_n(s) = \frac{M_n e^{j\theta_n}}{(n-1)!(2j\omega)^r(s+\propto-j\omega)^{r-n+1}} + \frac{(-1)^r M_n e^{-j\theta_n}}{(n-1)!(2j\omega)^r(s+\propto+j\omega)^{r-n+1}}$$

Note that $1/(-1)^r = (-1)^r$. Rearranging this, we have

$$F_n(s) = \frac{M_n}{(n-1)!(2j\omega)^r}\left[\frac{e^{j\theta_n}}{(s+\propto-j\omega)^{r-n+1}} + \frac{(-1)^r e^{-j\theta_n}}{(s+\propto+j\omega)^{r-n+1}}\right]$$

Using Eq. (3-9b), we can find the inverse of this.

$$f_n(t) = \frac{M_n}{(n-1)!(r-n)!(2j\omega)^r}\left[e^{j\theta_n}e^{-(\propto-j\omega)t}t^{r-n} + (-1)^r e^{-j\theta_n}e^{-(\propto+j\omega)t}t^{r-n}\right]$$

Rearranging the exponentials yields

$$f_n(t) = \frac{M_n e^{-\propto t}t^{r-n}}{(n-1)!(r-n)!(2j\omega)^r}\left[e^{j(\omega t+\theta_n)} + (-1)^r e^{-j(\omega t+\theta_n)}\right]$$

Using Euler's identities gives us

$$= \frac{M_n e^{-\propto t}t^{r-n}}{(n-1)!(r-n)!(2j\omega)^r}\{j[1-(-1)^r]\sin(\omega t+\theta_n)$$
$$+ [1+(-1)^r]\sin(\omega t+\theta_n+90°)\}$$

We notice that for odd values of r the second term inside the braces, {}, goes to zero, and for even values of r the first term inside the braces goes to zero. The difference between these two terms is a "j" and a "90°." We may develop a term that will cause the j and 90° to appear with odd or even r. To make the j we may use

$$(-1)^{\text{int}(r/2)}$$

where "int" means the integer portion. This takes into account the j inside the braces in conjunction with the j in the "$(2j\omega)^r$." To make the 90° term, we may use

$$\frac{1+(-1)^r}{2}(90°)$$

Combining these into our equation, we have

$$f_n(t) = \frac{(2)(-1)^{\text{int}(r/2)}M_n}{(2\omega)^r(n-1)!(r-n)!}t^{r-n}e^{-\propto t}$$
$$\times \sin\left[\omega t+\theta_n+\frac{1+(-1)^r}{2}(90°)\right]u(t)$$

where $n = 1, 2, 3, \cdots, r$

int(r/2) means the integer portion

APPENDIX C

Basic DC Circuit Equations

This appendix is a review of basic dc circuit equations. Only dc sources and resistors are used, so the review can be done before or during the introduction of Laplace transforms.

C-1 IDENTIFYING SERIES CIRCUITS

An element is in series with another element if and only if both elements have the same current. The current must not simply be the same value of current, but it must be the same flow of current.

In Fig. C-1a the 10-Ω resistors have the same current value flowing through them, but not the same flow of current. In Fig. C-1b the two resistors have the same current and are therefore in series.

C-2 IDENTIFYING PARALLEL CIRCUITS

An element is in parallel with another element if and only if both elements have the same potential across them. The potential must not simply be the same value of potential, but it must be the same potential.

In Fig. C-1b the 10-Ω resistors have the same value of potential across them, but not the same potential across them. In Fig. C-1a the two resistors have the same potential across them, and are therefore in parallel.

C-3 SERIES VOLTAGE SOURCES

Voltage sources in series may be replaced by a single voltage source having the magnitude of the algebraic sum of the individual sources. The algebraic sum is the addition of potential rises in a loop and the subtraction of potential drops in that loop.

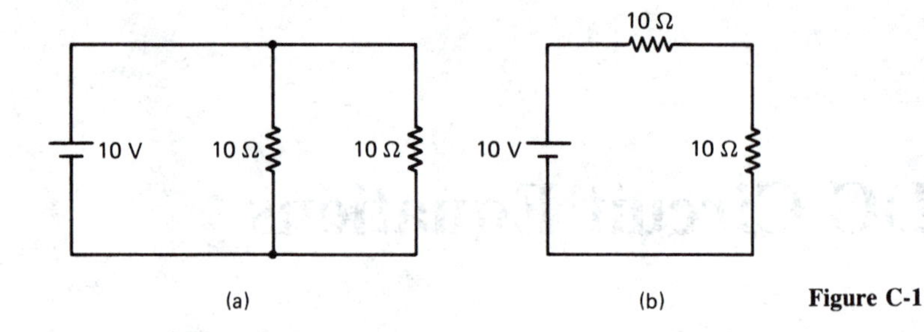

Figure C-1

C-4 PARALLEL CURRENT SOURCES

Current sources in parallel may be replaced by a single current source having the magnitude of the algebraic sum of the individual sources. The algebraic sum is the addition of currents entering a node and the subtraction of currents leaving that node.

C-5 OHM'S LAW

The voltage across a resistor is equal to the current through it times the resistor's value.

$$V = IR \tag{C-1}$$

EXAMPLE C-1

What is the voltage across R_2 in Fig. C-2?

SOLUTION Since R_1 is in series with R_2, the same 5 A flows in both elements.

$$V = IR_2 = (5)(10) = 50 \text{ V}$$

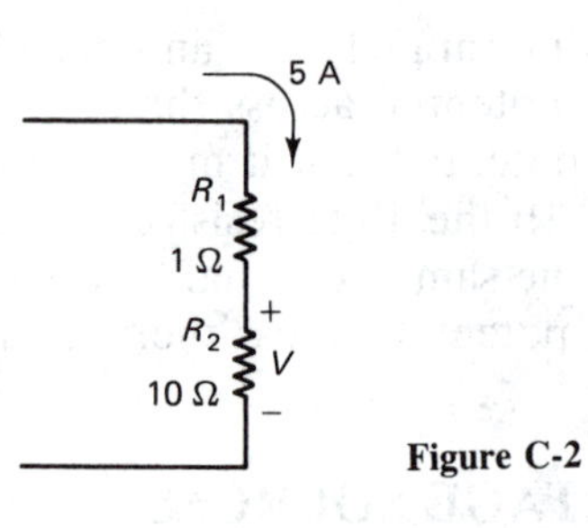

Figure C-2

C-6 VOLTAGE AND CURRENT MEASUREMENTS

The "+"/"–" signs in this book do *not* refer to the polarity of the voltage. They refer to the polarity of a voltmeter used to test the point. The "+" indicates the test probe (or red test lead), and the "–" indicates the reference (or black test lead). This means that the + is measured with respect to the −. If the + side is more positive than the reference (−), the value is positive. If the + side is more negative than the reference, the value is negative.

EXAMPLE C-2

Find the voltage across the resistors in Fig. C-3.

SOLUTION Taking a clockwise loop using Kirchhoff's voltage law in Fig. C-3a, we have

$$E - V = 0$$
$$V = E = 10\text{ V}$$

Taking a clockwise loop using Kirchhoff's voltage law in Fig. C-3b yields

$$E + V = 0$$
$$V = -E = -10\text{ V}$$

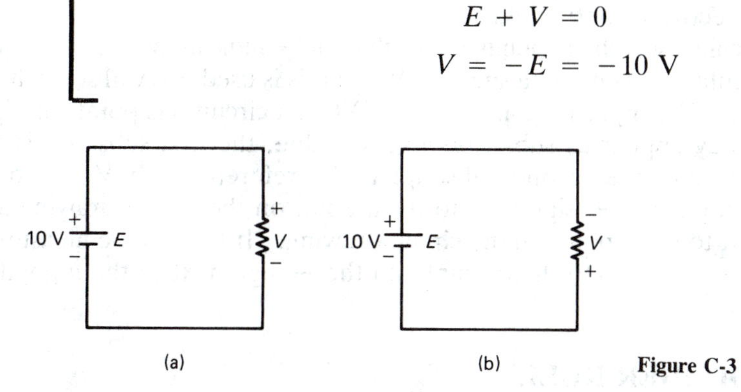

Figure C-3

A similar situation exists with current. The current arrow does *not* refer to current direction but rather to the direction to measure. If the current flows in the direction of the arrow, the value is positive. If the current flows opposite the direction of the arrow, the value is negative.

Using the +/− and arrows this way, we can determine if a component is taking energy from the circuit or supplying energy to the circuit. When we draw the current arrow between the voltage +/− signs so that the current value and voltage value have the same sign (both positive or both negative), the following observations can be made.

1. If the arrowhead is closest to the − sign, energy is being removed from the circuit.
2. If the arrowhead is closest to the + sign, energy is being supplied to the circuit.

If we measure the current and voltage as indicated in Fig. C-4, the values of the current and voltage will both be positive or both negative since resistors always remove energy from the circuit.

A voltage source, current source, capacitor, or inductor can take energy from or supply energy to the circuit. In Fig. C-5, if the voltage and current

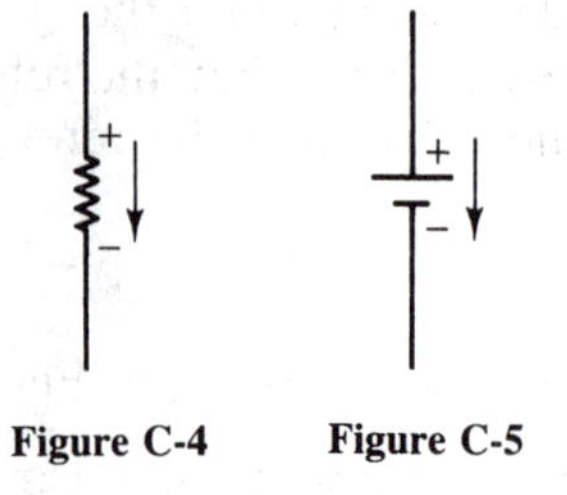

Figure C-4 Figure C-5

are measured and both values have the same sign, the source is being charged by removing energy from the circuit. If the signs are opposite, the voltage source is supplying energy to the circuit.

The capacitor and inductor work the same way. The only difference is that the capacitor stores energy in an electrostatic field and the inductor stores energy in an electromagnetic field.

Some circuits use voltage points rather than +/− indications. This method is used by manufacturers on data sheets. When this is used, we will see voltage indicated as V_{ab}, V_{ba}, V_{BE}, V_{GS}, and so on. When a circuit has points marked a and b and they appear as subscripts in a variable, the first subscript is the point measured and the second subscript is the reference. If V_{ab} is to be measured, then put the + sign next to the a point on the circuit drawing and the − sign next to the b point on the circuit drawing. If V_{ba} is to be measured, then put the + sign next to the b point and the − sign next to the a point.

C-7 VOLTAGE DIVIDER RULE

The voltage across any element in a series network is equal to the total voltage across the series network, times the element you want to know the voltage across, divided by the total resistance of the series network.

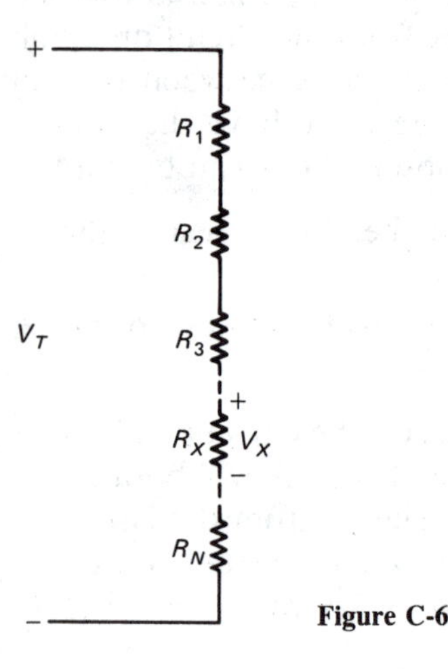

Figure C-6

For the circuit in Fig. C-6, the voltage V_x across R_x is

$$V_x = \frac{V_T R_x}{R_1 + R_2 + \cdots + R_N} \tag{C-2}$$

For two resistors, R_1 and R_2, in series, Eq. (C-2) becomes

$$V_1 = \frac{V_T R_1}{R_1 + R_2} \tag{C-3a}$$

$$V_2 = \frac{V_T R_2}{R_1 + R_2} \tag{C-3b}$$

C-8 CURRENT DIVIDER RULE

The current through any element in a parallel network is equal to the total current entering the parallel network, times the total resistance of the parallel network, divided by the element you want to know the current through.

For the circuit in Fig. C-7 the current I_x through R_x is

$$I_x = \frac{I_T R_T}{R_x} \tag{C-4}$$

where R_T is the resistance the current I_T sees.

For two resistors, R_1 and R_2, in parallel, the total resistance becomes $R_T = (R_1R_2)/(R_1 + R_2)$, and substituting into Eq. (C-4) gives

$$I_1 = \frac{I_T R_2}{R_1 + R_2} \tag{C-5a}$$

$$I_2 = \frac{I_T R_1}{R_1 + R_2} \tag{C-5b}$$

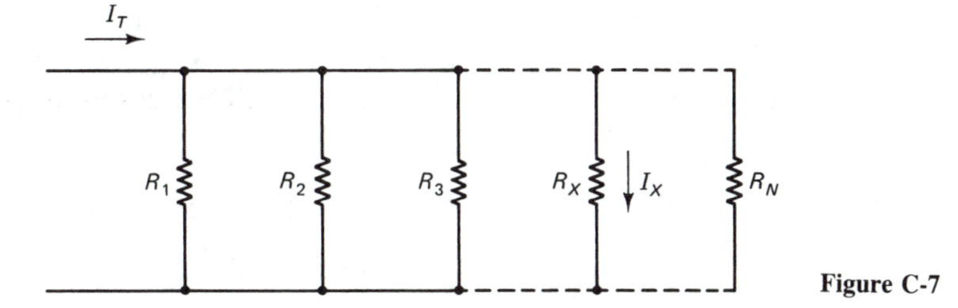

Figure C-7

C-9 KIRCHHOFF'S VOLTAGE LAW

The sum of the voltages around a complete path is zero. The path must start at a point and return to that same point without going through a component or section of wire more than once. We determine whether to add or subtract, not by the physical polarity of the voltages, but rather by the way the voltages were measured.

EXAMPLE C-3

Find V_1 in Fig. C-8.

SOLUTION The path is shown in Fig. C-8 by the dashed line. We start at the × and return to the × in a clockwise direction. This could also be done in a counterclockwise direction with the same result.

$$E + V_1 - V_2 = 0$$

$$10 + V_1 - 6 = 0$$

$$V_1 = -10 + 6$$

$$V_1 = -4 \text{ V}$$

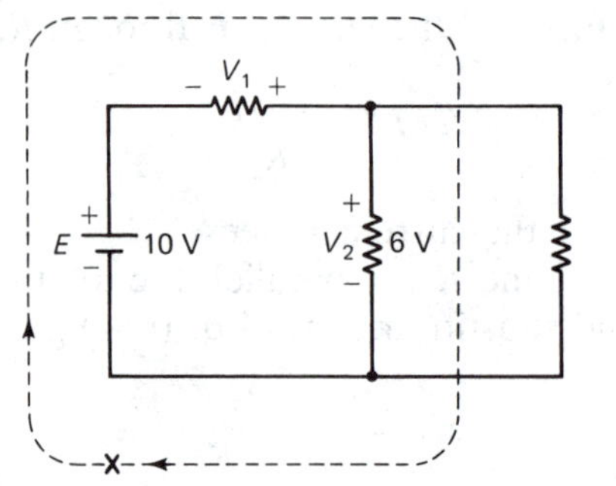

Figure C-8

EXAMPLE C-4

Find the voltage V_{ab} in Fig. C-9.

SOLUTION The a will be assigned "+" and the b will be assigned "−" due to the order of the subscript in V_{ab}. This time we will use a counterclockwise direction.

$$-V_2 - V_{ab} + V_1 - E = 0$$

$$-2 - V_{ab} + 5 - 15 = 0$$

$$V_{ab} = -2 + 5 - 15$$

$$V_{ab} = -12 \text{ V}$$

If we had been looking for V_{ba}, we would assign the +/− the opposite way and the result would be $V_{ba} = +12$ V.

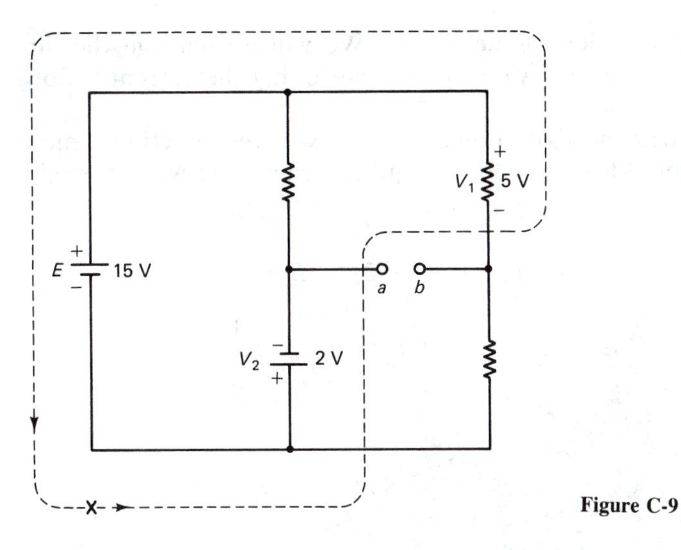

Figure C-9

C-10 KIRCHHOFF'S CURRENT LAW

The sum of the currents at a node is equal to zero. We determine whether to add or subtract by the direction the current was measured. A current measured as entering a node is added, and a current measured as leaving a node is subtracted.

EXAMPLE C-5

Find I_3 and I_5 in Fig. C-10.

SOLUTION At node N_1,

$$I_1 - I_2 + I_3 = 0$$

$$4 - 3 + I_3 = 0$$

$$I_3 = -4 + 3$$

$$I_3 = -1 \text{ A}$$

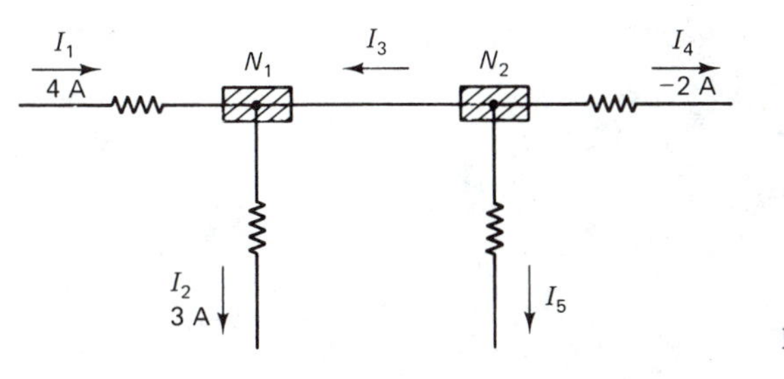

Figure C-10

Notice that the value is negative. We will not change the direction we measured I_3. Instead, we will just realize that the current is flowing in the opposite direction.

Since Kirchhoff's current law is done with the direction of measurement, I_3 is considered leaving node N_2 and entering node N_1. At node N_2,

$$-I_3 - I_4 - I_5 = 0$$

$$-(-1) - (-2) - I_5 = 0$$

$$I_5 = 1 + 2$$

$$I_5 = 3 \text{ A}$$

APPENDIX D

Semilog Graphs

When we graph a function's response to different frequencies, we use semilog or log-log graph paper. Since log-log graphs are rarely seen, we will concentrate on semilog graphs. Semilog graphs have the horizontal axis based on a logarithmic ($\log_{10}$) scale and the vertical axis based on a linear scale. Frequently, the only information we may have about a circuit is a graph. We must therefore know how to work with semilog graphs.

D-1 HOW TO READ A LOG SCALE

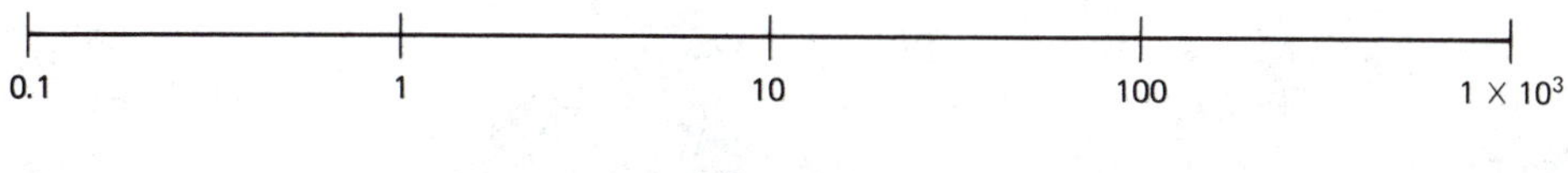

Figure D-1

Figure D-1 shows only the major divisions of a log scale. Each major increment represents a power of 10. Since it is based on powers of 10, the log scale will never start at 0. The scale will start at 1×10 to some positive or negative integer power. Mathematically, we could start at any value, except zero, with each increment 10 times the previous major division, but we exclusively use 1×10 to some integer power.

Each major division is called a cycle even though the divisions have nothing to do with a periodic waveform. Graph paper with four major divisions marked off is called four-cycle paper, which is used in Fig. D-1. It is called five-cycle paper when five areas are marked off. Linear values are marked off between major divisions, which will, therefore, have logarithmic spacing.

Each cycle is divided into 10 secondary steps as show in Fig. D-2. These steps are the linear increments from one major division to the next major division. The increments always get smaller as we increase the frequency to approach the next major division. The secondary marks are often numbered, and are multiplied times the major division value to the left. For example, if

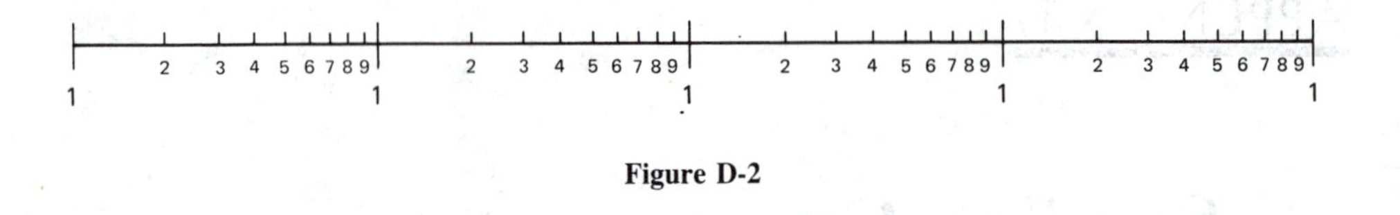

Figure D-2

the major division to the left of the secondary mark 6 is 1×10^{-3}, the value is $6 \times (1 \times 10^{-3}) = 6 \times 10^{-3}$.

The marks between the secondary division are marked in different ways depending on the manufacture and the number of cycles. These marks between the secondary marks are also linear increments. To find the linear value increment of each steps, we count the number of spaces between the secondary marks and use the reciprocal. If there are 10 spaces, the linear increment value is 0.1, and if there are 5 spaces, the increment value is 0.2. These are multiplied by the value of the major division to the left and added to the secondary value.

EXAMPLE D-1

Figure D-3 shows one cycle. Find the value of the point marked.

SOLUTION We are in the secondary division 2, and there are 10 spaces between the secondary division 2 and the secondary division 3. The value is

$$(2 \times 100) + \left[\left(4 \times \frac{1}{10}\right) \times 100\right] = 240$$

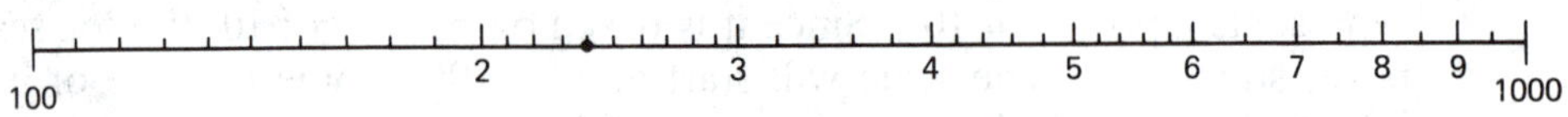

100 2 3 4 5 6 7 8 9 1000

Figure D-3

D-2 CALCULATING DISTANCES ON A LOG SCALE

Measurements on a log axis are identical to measurements on a linear axis except that we use the logarithm of the value instead of the value. This means that what we do on a linear scale we do on the log scale using the logarithm of the number. Since this is the case, we should review how we measure distance on a linear scale first.

Figure D-4 shows a linear scale. To find the distance between two points X_1 and X_2 we find the number of units between the two points, $X_2 - X_1$. For

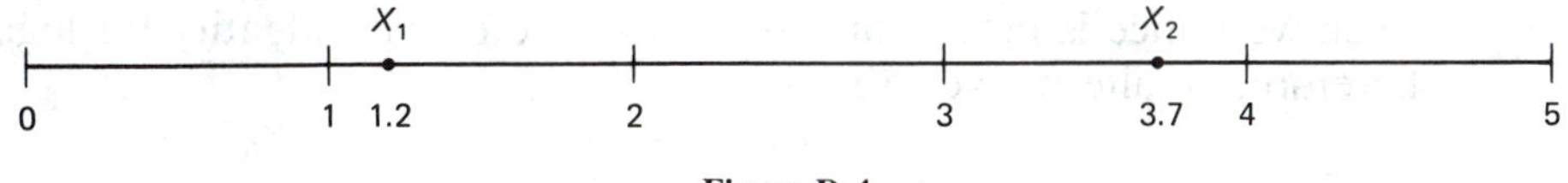

Figure D-4

example, if $X_1 = 1.2$, $X_2 = 3.7$, the number of units between the points is

$$\begin{aligned} \text{NU} &= X_2 - X_1 \\ &= 3.7 - 1.2 \qquad \text{(D-1)} \\ &= 2.5 \text{ units} \end{aligned}$$

On a log scale, the unit is a decade since it is a change in frequency by a factor of 10. On a linear scale to find the number of decades, ND, we solve for ND in the equation

$$10^{\text{ND}} = \frac{X_1}{X_2}$$

By using the log of the values, we can find the same quantities on a log scale. Since we are using frequency we will use f_1 and f_2 instead of X_1 and X_2, Where f_1 will be less than f_2 and both must be in the same units.

The number of decades between two frequencies will be

$$\log 10^{\text{ND}} = \log(f_2) - \log(f_1)$$

$$\text{ND} = \log(f_2) - \log(f_1) \qquad \text{[compare with Eq. (D-1)]}$$

or

$$\text{ND} = \log\left(\frac{f_2}{f_1}\right) \qquad f_1 < f_2 \qquad \text{(D-2)}$$

Sometimes units of octaves are used. This term comes from the octave in music, and is usually applied to electronic circuits that reproduce music. In electronic terms this means "to double the frequency." On a linear scale, to find the number of octaves we would solve for the number of octaves, NO, in the equation

$$2^{\text{NO}} = \frac{X_2}{X_1}$$

On a log scale we do the same thing except that we use the log of the values.

$$\text{NO} \log(2) = \log(f_2) - \log(f_2)$$

$$\text{NO} = \frac{\log(f_2) - \log(f_1)}{\log(2)} \qquad \text{(D-3a)}$$

$$= \frac{\log\left(\dfrac{f_2}{f_1}\right)}{\log(2)} \qquad f_1 < f_2$$

which we notice is in the form of the change of base identity for logarithms; therefore, an alternative form is

$$\text{NO} = \log_2\left(\frac{f_2}{f_1}\right) \qquad f_1 < f_2 \tag{D-3b}$$

Each equation has been restricted to $f_1 < f_2$, but this is for conveyance in working on the semilog graphs of frequency responses. Without this restriction we would calculate a negative value when f_1 is greater than f_2. This would indicate a movement to the left of f_1 just as a negative value on a linear scale would indicate. Also, each equation is a ratio of frequencies, which means that f could be in hertz or rad/sec.

EXAMPLE D-2

Find the number of decades and the number of octaves between:

a. $f_1 = 2$ rad/sec and $f_2 = 8$ rad/sec
b. $f_1 = 200$ rad/sec and $f_2 = 800$ rad/sec

SOLUTION

a. Using Eq. (D-2) gives

$$\begin{aligned}\text{ND} &= \log\left(\frac{f_2}{f_1}\right)\\ &= \log\left(\frac{8}{2}\right)\\ &= 0.60206\end{aligned}$$

The number of octaves using Eq. (D-3a) are

$$\begin{aligned}\text{NO} &= \frac{\log\left(\frac{f_2}{f_1}\right)}{\log(2)}\\ &= \frac{0.60206}{\log(2)}\\ &= 2\end{aligned}$$

This stands to reason since two octaves means to multiply by 2, two times.

b. These two frequencies are different by a factor of 100 from the frequencies in part a. It is important to notice that these frequencies are separated by the same number of decades or octaves as the frequencies in part a. Using Eq. (D-2), we have

$$\begin{aligned}
\text{ND} &= \log\left(\frac{f_2}{f_1}\right) \\
&= \log\left(\frac{800}{200}\right) \\
&= 0.60206
\end{aligned}$$

The number of octaves using Eq. (D-3a) is

$$\begin{aligned}
\text{NO} &= \frac{\log\left(\frac{f_2}{f_1}\right)}{\log(2)} \\
&= \frac{0.60206}{\log(2)} \\
&= 2
\end{aligned}$$

D-3 CALCULATING ROLL-OFF RATES ON SEMILOG GRAPHS

The roll-off rate, RR, is measured in dB/decade or dB/octave. This is the slope of a straight line on semilog graphs. A straight line on semilog graph paper is not a straight line when graphed on linear paper. Since the x-axis is linear, we first find the change in value of the x-axis the same as we do on any linear axis.

$$\Delta A_{\text{dB}} = A_{2\text{dB}} - A_{1\text{dB}} \tag{D-4}$$

where $A_{2\text{dB}}$ corresponds to the dB value at f_2
$A_{1\text{dB}}$ corresponds to the dB value at f_1
and $f_1 < f_2$

We then find the number of decades or number of octaves between f_1 and f_2 using Eq. (D-2) or (D-3). This value is divided into the change in db, ΔA, giving us the roll-off rate

$$\text{RR} = \frac{\Delta A_{\text{dB}}}{\text{ND}} \tag{D-5a}$$

$$\text{RR} = \frac{\Delta A_{\text{dB}}}{\text{NO}} \tag{D-5b}$$

Equation (D-5a) has the units of dB/dec, and Eq. (D-5b) has the units of dB/oct.

EXAMPLE D-3

For Fig. D-5, find the roll-off rate in dB/dec, and find the dB value at f_3.

SOLUTION There is enough information using f_1 and f_2 to find the slope of the line. To find the slope we first find the change in dB using Eq. (D-4).

$$\begin{aligned}\Delta A_{dB} &= A_{2dB} - A_{1dB} \\ &= (5.5630) - (-4.8945) \\ &= 10.4575 \text{ dB}\end{aligned}$$

Then using Eq. (D-2), we find the number of decades between these frequencies.

$$\begin{aligned}ND &= \log\left(\frac{f_2}{f_1}\right) \\ &= \log\left(\frac{3000}{900}\right) \\ &= 0.52288 \text{ decade}\end{aligned}$$

Substituting into Eq. (D-5a), we will find the roll-off rate.

$$\begin{aligned}RR &= \frac{\Delta A_{dB}}{ND} \\ &= \frac{10.4575 \text{ dB}}{0.52288 \text{ dec}} \\ &= 20 \text{ dB/dec}\end{aligned}$$

Since roll-off rates will be in multiples of 20, we usually round off to the nearest multiple of 20.

Now that we know the roll-off rate we can apply Eqs. (D-2), (D-3), (D-4), and/or Eq. (D-5) to find any point on the slope, but we will have to modify our equations to accommodate f_2 and f_3.

First we find the number of decades between the two points using Eq. (D-2).

$$\begin{aligned}ND &= \log\left(\frac{f_3}{f_2}\right) \qquad f_2 < f_3 \\ &= \log\left(\frac{5000}{3000}\right) \\ &= 0.22185 \text{ decade}\end{aligned}$$

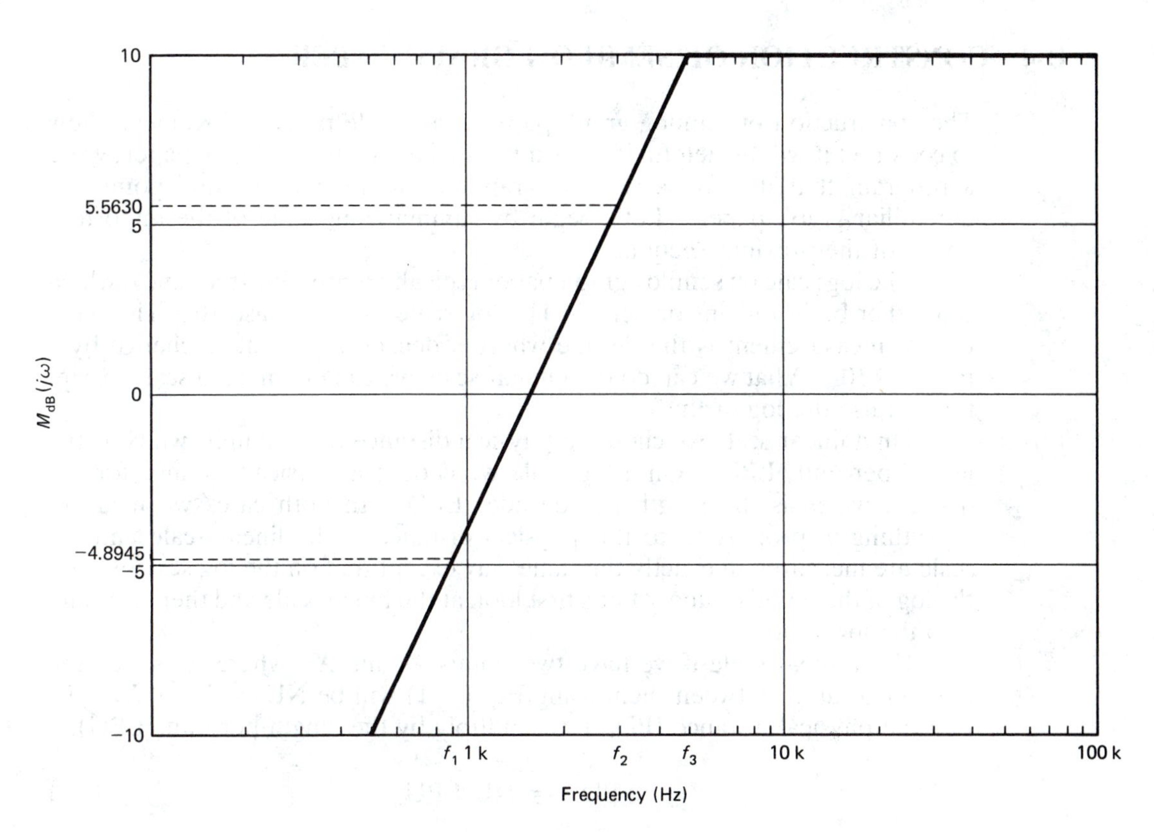

Figure D-5

Using Eq. (D-5a), we can find the change in dB between the two points.

$$\mathrm{RR} = \frac{\Delta A_{\mathrm{dB}}}{\mathrm{ND}}$$

$$\Delta A_{\mathrm{dB}} = (\mathrm{RR})(\mathrm{ND})$$

$$= (20\ \mathrm{dB/dec})(0.22185\ \mathrm{dec})$$

$$= 4.4370\ \mathrm{dB}$$

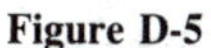

This tells us the dB difference between the two points. To find the value of the point, we use a modified Eq. (D-4)

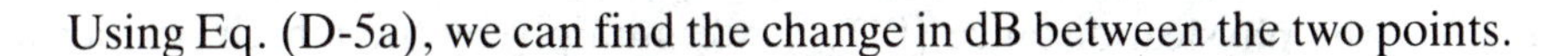

$$\Delta A_{\mathrm{dB}} = A_{3\mathrm{dB}} - A_{2\mathrm{dB}}$$

and solve for $A_{3\mathrm{dB}}$.

$$A_{3\mathrm{dB}} = \Delta A_{\mathrm{dB}} + A_{2\mathrm{dB}}$$

$$= 4.4370\ \mathrm{dB} + 5.5630\ \mathrm{dB}$$

$$= 10\ \mathrm{dB}$$

D-4 CONSTRUCTION OF SEMILOG GRAPH PAPER

The construction of semilog graph paper is a simple process. Knowing how to construct it will be helpful if we want to make a sketch on blank paper, write a program that displays a semilog graph, or if we need to find points on unfamiliar graph paper. Let's begin by summarizing some of the important points of the previous sections.

The log scale on semilog graph paper typically represents frequency, which can either be in radians or hertz. The log scale uses log base 10. The basic unit of measurement is the decade where a decade represents a change by a factor of 10. What we can do on a linear scale we can do on a log scale except that we use the log of the value.

On a linear scale we choose a physical distance for one unit, which is the length per unit, LPU. On a log scale we choose a physical distance for the decade, which is the length per decade, LPD. In both cases we measure everything in proportion to this physical distance. The linear scale and log scale are measured in exactly the same way except that on the log scale we use the log of the point's value. Let's first look at the linear scale and then translate it to the log scale.

On a linear scale if we have two points X_1 and X_2, where $X_1 < X_2$, the number of units between them using Eq. (D-1) will be $\text{NU} = X_2 - X_1$. To find the physical distance (PD_{lin}) we multiply by the length per unit (LPU).

$$\text{PD}_{\text{lin}} = \text{NU LPU} \tag{D-6}$$

On a log scale if we have two points f_1 and f_2 where $f_1 < f_2$, the number of decades between them using Eq. (D-2) will be $\text{ND} = \log (f_2/f_1)$. To find the physical distance (PD_{log}), we multiply by the length per decade (LPD).

$$\text{PD}_{\text{log}} = \text{ND LPD} \tag{D-7}$$

From Eqs. (D-6) and (D-7) we can find the physical distance between any two points. Let's first examine the physical distance of secondary divisions between major divisions.

On a linear scale if we want to mark off the values in tenths, we can find the physical distance of each tenth by finding the physical distance between 0 and 0.1, 0 and 0.2, 0 and 0.3, and each other interval. We then plug into Eq. (D-6) to find where to mark it off. Since the number of units between any one-tenth interval using Eq. (D-1) will give us the same value, the spacing will be equal in each one-tenth interval.

On a log scale each interval using Eq. (D-2) will give us a different value, and we must measure each one individually. We must therefore find the physical distance from the left major division to each point we want. We could calculate the distance from the last interval to the next interval, but more accuracy is obtained if we always start at the left major division.

EXAMPLE D-4

Where are the secondary divisions between major divisions of a log scale when the physical distance is 2 inches for each decade?

SOLUTION Using Eq. (D-2) we can find the number of decades from the major division to the increment we want. We can assume any starting value for the major division as seen in Example D-2. For conveyance we will assume that we are starting at 1. Then using Eq. (D-7), we multiply the number of decades by 2 inches per decade.

$$\text{ND} = \log\left(\frac{f_2}{f_1}\right) \qquad \text{PD}_{\log} = \text{ND LPD}$$

$$\text{ND} = \log\left(\frac{2}{1}\right) = 0.30103 \qquad \text{PD} = (0.30103)(2) = 0.60206 \text{ in.}$$

$$\text{ND} = \log\left(\frac{3}{1}\right) = 0.47712 \qquad \text{PD} = (0.47712)(2) = 0.95424 \text{ in.}$$

$$\text{ND} = \log\left(\frac{4}{1}\right) = 0.60206 \qquad \text{PD} = (0.60206)(2) = 1.2041 \text{ in.}$$

$$\text{ND} = \log\left(\frac{5}{1}\right) = 0.69897 \qquad \text{PD} = (0.69897)(2) = 1.3979 \text{ in.}$$

$$\text{ND} = \log\left(\frac{6}{1}\right) = 0.77815 \qquad \text{PD} = (0.77815)(2) = 1.5563 \text{ in.}$$

$$\text{ND} = \log\left(\frac{7}{1}\right) = 0.84510 \qquad \text{PD} = (0.84510)(2) = 1.6902 \text{ in.}$$

$$\text{ND} = \log\left(\frac{8}{1}\right) = 0.90309 \qquad \text{PD} = (0.90309)(2) = 1.8062 \text{ in.}$$

$$\text{ND} = \log\left(\frac{9}{1}\right) = 0.95424 \qquad \text{PD} = (0.95424)(2) = 1.9085 \text{ in.}$$

From this last example we can see an easy technique for approximating a log scale without measuring. If we take the log of the increment we want and multiply it by 100%, we have the percent of the total distance to the next major division. For example, if we are looking for 31.2×10^3, the increment is 3.12, and $\log(3.12) = 0.49415$. So we go about 50% of the distance to the next major division.

APPENDIX E

OP-AMP Topologies

This appendix shows the topologies used in this book. Each section starts with the general topology and the transfer function in terms of admittances. Each subsection shows specific cases that implement different types of filters (low pass, high pass, and band pass). The subsections use the following format:

1. The admittances are assigned as R and/or C components.
2. The resulting transfer function is written.
3. Some components are assigned to be equal in value.
4. The equations to solve for comonent values are listed. The equations refer to the general biquad,

$$\frac{ms^2 + as + b}{\pm h(ns^2 + ds + e)} \tag{E-1}$$

E-1 NONINVERTING SINGLE-FEEDBACK TOPOLOGY

See Fig. E-1.

$$\frac{V_{\text{in}}}{V_o} = \frac{Y_A + Y_B}{KY_A} \tag{E-2}$$

where

$$K = 1 + \frac{R_1}{R_2}$$

E-1-1 Low Pass

1. $$Y_A = \frac{1}{R_A} \qquad Y_B = sC_B \tag{E-3}$$

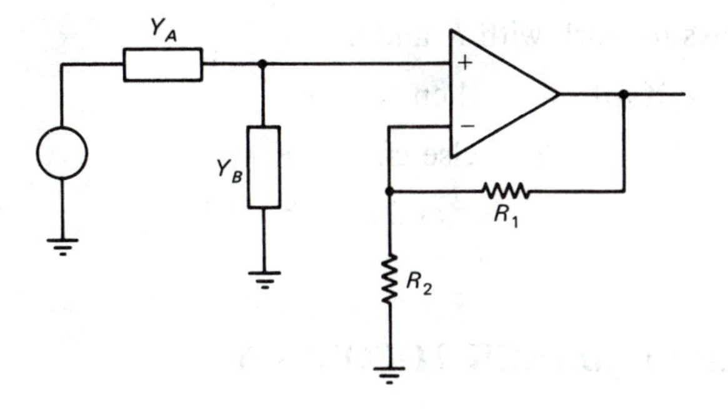

Figure E-1

2.
$$\frac{V_{\text{in}}}{V_o} = \frac{S + \dfrac{1}{C_B R_A}}{\dfrac{K}{C_B R_A}} \tag{E-4}$$

3. There are no components to set equal to each other.

4. Note that in Eq. (E-1), $m = 0$, $a = 1$, $n = 0$, $d = 0$, and $e = 1$.

(**a**) Choose C_B.

(**b**) Solve for R_A

$$R_A = \frac{1}{bC_B} \tag{E-5a}$$

(**c**) There are two ways to work with K and h.

$$\text{If } h \geq \frac{1}{R_A C_B} \qquad \text{then } K = hR_A C_B$$

$$\text{else choose } K \geq 1 \tag{E-5b}$$

$$\text{and } h = \frac{K}{R_A C_B}$$

E-1-2 High Pass

1.
$$Y_A = sC_A \qquad Y_B = \frac{1}{R_B} \tag{E-6}$$

2.
$$\frac{V_{\text{in}}}{V_o} = \frac{S + \dfrac{1}{C_A R_B}}{Ks} \tag{E-7}$$

3. There are no components to set equal to each other.

4. Note that in Eq. (E-1), $m = 0$, $a = 1$, $n = 0$, $d = 1$, and $e = 0$.

(**a**) Choose C_A.

(**b**) Solve for R_B.

$$R_B = \frac{1}{bC_A} \tag{E-8a}$$

(c) There are two ways to work with K and h.

$$\text{If } h \geq 1 \quad \text{then } K = h$$
$$\text{else choose } K \geq 1 \tag{E-8b}$$
$$\text{and } h = k$$

E-2 INVERTING SINGLE-FEEDBACK TOPOLOGY

See Fig. E-2.

$$\frac{V_{\text{in}}}{V_o} = \frac{Y_B}{-Y_A} \tag{E-9}$$

E-2-1 Low Pass

1.
$$Y_A = \frac{1}{R_A} \qquad Y_B = sC_B + \frac{1}{R_B} \tag{E-10}$$

Notice that Y_B is a capacitor in parallel with a resistor.

2.
$$\frac{V_{\text{in}}}{V_o} = \frac{S + \dfrac{1}{C_B R_B}}{\dfrac{-1}{C_B R_A}} \tag{E-11}$$

3. There are no components to set equal to each other.
4. Note that in Eq. (E-1), $m = 0$, $a = 1$, $n = 0$, $d = 0$, and $e = 1$.
(a) Choose C_B.
(b) Solve for R_B.

$$R_B = \frac{1}{bC_B} \tag{E-12a}$$

(c) There are two ways to work with R_A and h.

Either calculate R_A:

$$R_A = \frac{1}{hC_B} \tag{E-12b}$$

or choose R_A and calculate the resulting h:

$$h = \frac{1}{R_A C_B}$$

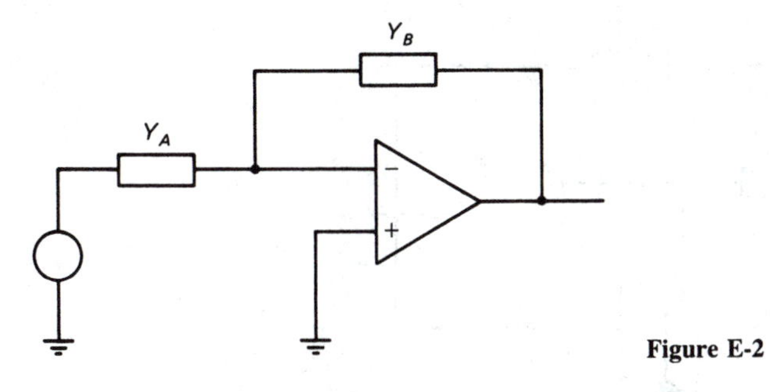

Figure E-2

E-2-2 High Pass

1.

$$Y_A = sC_A \qquad Y_B = sC_B + \frac{1}{R_B} \tag{E-13}$$

Notice that Y_B is a capacitor in parallel with a resistor.

2.

$$\frac{V_{\text{in}}}{V_o} = \frac{s + \dfrac{1}{C_B R_B}}{\dfrac{-C_A}{C_B}s} \tag{E-14}$$

3. There are no components to set equal to each other.

4. Note that in Eq. (E-1), $m = 0$, $a = 1$, $n = 0$, $d = 1$, and $e = 0$.

(a) Choose C_B.

(b) Solve for R_B.

$$R_B = \frac{1}{bC_B} \tag{E-15a}$$

(c) There are two ways to work with C_A and h.

Either calculate C_A:

$$C_A = hC_B \tag{E-15b}$$

or choose C_A and calculate the resulting h:

$$h = \frac{C_A}{C_B}$$

E-3 NONINVERTING DUAL-FEEDBACK TOPOLOGY

See Fig. E-3.

$$\frac{V_{\text{in}}}{V_o} = \frac{(Y_A + Y_B + Y_E)(Y_C + Y_D) + Y_C(Y_D - Y_B K)}{Y_A Y_C K} \tag{E-16}$$

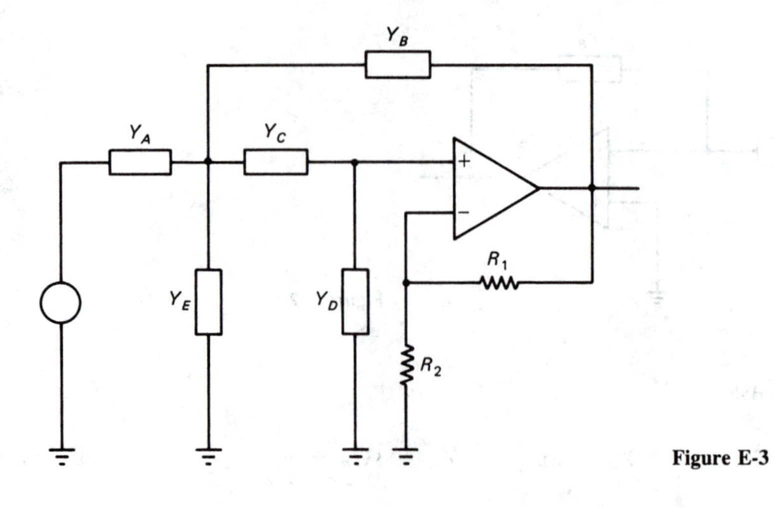

Figure E-3

where

$$K = 1 + \frac{R_1}{R_2}$$

E-3-1 Low Pass

1.

$$Y_A = \frac{1}{R_A} \qquad Y_B = sC_B$$

$$Y_C = \frac{1}{R_C} \qquad Y_D = sC_D \tag{E-17}$$

$$Y_E = 0 \qquad \text{(no component)}$$

2.

$$\frac{V_{\text{in}}}{V_o} = \frac{s^2 + \left(\dfrac{1}{C_BR_A} + \dfrac{1}{C_DR_C} + \dfrac{1 - K}{C_BR_C}\right)s + \dfrac{1}{C_BC_DR_AR_C}}{\dfrac{K}{C_BC_DR_AR_C}} \tag{E-18}$$

3.

$$C_B = C_D = C \qquad \text{and} \qquad R_A = R_C = R \tag{E-19}$$

4. Note that in Eq. (E-1), $m = 1$, $n = 0$, $d = 0$, and $e = 1$.

(a) To use this circuit, the following must be true.

$$0 \le 2 - \frac{a}{\sqrt{b}} \tag{E-20a}$$

(b) Choose C.

(c) Calculate R.

$$R = \frac{1}{C\sqrt{b}} \tag{E-20b}$$

(d) Calculate K.

$$K = 3 - \frac{a}{\sqrt{b}} \tag{E-20c}$$

(e) Calculate the resulting h.

$$h = \frac{K}{C_B C_D R_A R_C} \tag{E-20d}$$

E-3-2 High Pass

1.
$$Y_A = sC_A \qquad Y_B = \frac{1}{R_B}$$

$$Y_C = sC_C \qquad Y_D = \frac{1}{R_D} \tag{E-21}$$

$$Y_E = 0 \qquad \text{(no component)}$$

2.
$$\frac{V_{\text{in}}}{V_o} = \frac{s^2 + \left(\dfrac{1}{C_C R_D} + \dfrac{1}{C_A R_D} + \dfrac{1 - K}{C_A R_B}\right)s + \dfrac{1}{C_A C_C R_B R_D}}{Ks^2} \tag{E-22}$$

3.
$$C_A = C_C = C \qquad \text{and} \qquad R_B = R_D = R \tag{E-23}$$

4. Note that in Eq. (E-1), $m = 1$, $n = 1$, $d = 0$, and $e = 0$.
   (a) To use this circuit, the following must be true.

$$0 \leq 2 - \frac{a}{\sqrt{b}} \tag{E-24a}$$

   (b) Choose C.
   (c) Calculate R.

$$R = \frac{1}{C\sqrt{b}} \tag{E-24b}$$

   (d) Calculate K.

$$K = 3 - \frac{a}{\sqrt{b}} \tag{E-24c}$$

   (e) Calculate the resulting h.

$$h = K \tag{E-24d}$$

E-3-3 Band Pass

1.

$$Y_A = \frac{1}{R_A} \qquad Y_B = \frac{1}{R_B}$$

$$Y_C = sC_C \qquad Y_D = \frac{1}{R_D} \tag{E-25}$$

$$Y_E = sC_E$$

2.

$$\frac{V_{\text{in}}}{V_o} = \frac{s^2 + \left(\dfrac{1}{C_E R_A} + \dfrac{1}{C_C R_D} + \dfrac{1}{C_E R_D} + \dfrac{1-K}{C_E R_B}\right)s + \dfrac{1}{C_C C_E}\left(\dfrac{1}{R_A R_D} + \dfrac{1}{R_B R_D}\right)}{\left(\dfrac{K}{C_E R_A}\right)s} \tag{E-26}$$

3.

$$C_C = C_E = C \quad \text{and} \quad R_A = R_B = R_D = R \tag{E-27}$$

4. Note that in Eq. (E-1), $m = 1$, $n = 0$, $d = 1$, and $e = 0$.

(a) To use this circuit, the following must be true.

$$0 \le 3 - \frac{a}{\sqrt{b}} \tag{E-28a}$$

(b) Choose C.

(c) Calculate R.

$$R = \frac{1}{C\sqrt{b}} \tag{E-28b}$$

(d) Calculate K.

$$K = 4 - \frac{a}{\sqrt{b}} \tag{E-28c}$$

(e) Calculate the resulting h.

$$h = \frac{K}{C_E R_A} \tag{E-28d}$$

E-4 INVERTING DUAL-FEEDBACK TOPOLOGY

See Fig. E-4.

$$\frac{V_{\text{in}}}{V_o} = \frac{Y_D(Y_A + Y_B + Y_C + Y_E) + Y_B Y_C}{-Y_A Y_C} \tag{E-29}$$

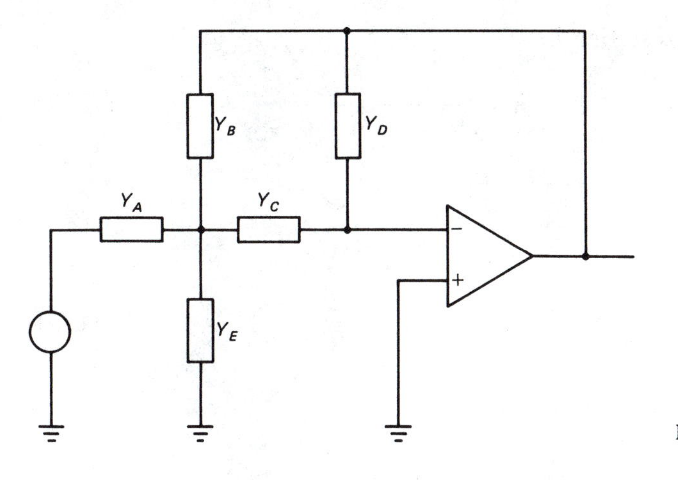

Figure E-4

E-4-1 Low Pass

1.

$$Y_A = \frac{1}{R_A} \qquad Y_B = \frac{1}{R_B}$$

$$Y_C = \frac{1}{R_C} \qquad Y_D = sC_D \tag{E-30}$$

$$Y_E = sC_E$$

2.

$$\frac{V_{\text{in}}}{V_o} = \frac{s^2 + \dfrac{1}{C_E}\left(\dfrac{1}{R_A} + \dfrac{1}{R_B} + \dfrac{1}{R_C}\right)s + \dfrac{1}{C_D C_E R_B R_C}}{\dfrac{-1}{C_D C_E R_B R_C}} \tag{E-31}$$

3.

$$R_B = R_C = R \tag{E-32}$$

4. Note that in Eq. (E-1), $m = 1$, $n = 0$, $d = 0$, and $e = 1$.

(a) Choose C_E.

(b) Choose C_D such that

$$C_D \leq \frac{C_E a^2}{4b} \tag{E-33a}$$

(c) Calculate R.

$$R = \frac{1}{\sqrt{bC_D C_E}} \tag{E-33b}$$

(d) Calculate R_A.

$$R_A = \frac{R}{RC_E a - 2} \tag{E-33c}$$

(e) Calculate the resulting h.

$$h = \frac{1}{C_D C_E R_B R_C} \tag{E-33d}$$

E-4-2 High Pass

1.

$$Y_A = sC_A \qquad Y_B = sC_B$$

$$Y_C = sC_C \qquad Y_D = \frac{1}{R_D} \tag{E-34}$$

$$Y_E = \frac{1}{R_E}$$

2.

$$\frac{V_{\text{in}}}{V_o} = \frac{s^2 + \left(\frac{C_A}{C_B C_C R_D} + \frac{1}{C_C R_D} + \frac{1}{C_B R_D}\right)s + \frac{1}{C_B C_C R_D R_E}}{\left(\frac{-C_A}{C_B}\right)s^2} \tag{E-35}$$

3.

$$C_A = C_B = C_C = C \tag{E-36}$$

4. Note that in Eq. (E-1), $m = 1$, $n = 1$, $d = 0$, and $e = 0$.
   (a) Choose C.
   (b) Calculate R_D.

$$R_D = \frac{3}{Ca} \tag{E-37a}$$

   (c) Calculate R_E.

$$R_E = \frac{a}{3Cb} \tag{E-37b}$$

   (d) Calculate the resulting h.

$$h = \frac{C_A}{C_B} = 1 \tag{E-37c}$$

E-4-3 Band Pass

1.

$$Y_A = \frac{1}{R_A} \qquad Y_B = sC_B$$

$$Y_C = sC_C \qquad Y_D = \frac{1}{R_D} \tag{E-38}$$

$$Y_E = \frac{1}{R_E}$$

2.
$$\frac{V_{\text{in}}}{V_o} = \frac{s^2 + \left(\dfrac{1}{C_C R_D} + \dfrac{1}{C_B R_D}\right)s + \dfrac{1}{C_B C_C}\left(\dfrac{R_A + R_E}{R_A R_D R_E}\right)}{\left(\dfrac{-1}{C_B R_A}\right)s} \tag{E-39}$$

3.
$$C_C = C_B = C \quad \text{and} \quad R_A = R_E = R \tag{E-40}$$

4. Note that in Eq. (E-1), $m = 1$, $n = 0$, $d = 1$, and $e = 0$.

(a) Choose C.

(b) Calculate R.

$$R = \frac{a}{Cb} \tag{E-41a}$$

(c) Calculate R_D.

$$R_D = \frac{2}{Ca} \tag{E-41b}$$

(d) Calculate the resulting h.

$$h = \frac{1}{C_B R_A} \tag{E-41c}$$

E-5 TWIN-T TOPOLOGY

See Fig. E-5.

$$\frac{V_{\text{in}}}{V_o} = \frac{Y_G Y_1 Y_2 + Y_F Y_1 (Y_B + Y_C - KY_C) + Y_E Y_2 (Y_A + Y_D)}{K(Y_A Y_E Y_2 + Y_B Y_F Y_1)} \tag{E-42}$$

where

$$Y_1 = Y_A + Y_D + Y_E$$

$$Y_2 = Y_B + Y_C + Y_F$$

$$K = 1 + \frac{R_1}{R_2}$$

E-5-1 Low Pass, High Pass, and Band Pass

1.
$$\begin{aligned} Y_A &= sC_A & Y_B &= \frac{1}{R_B} \\ Y_C &= sC_C & Y_D &= \frac{1}{R_D} \\ Y_E &= sC_E & Y_F &= \frac{1}{R_F} \\ Y_G &= sC_G + \frac{1}{R_G} \end{aligned} \tag{E-43}$$

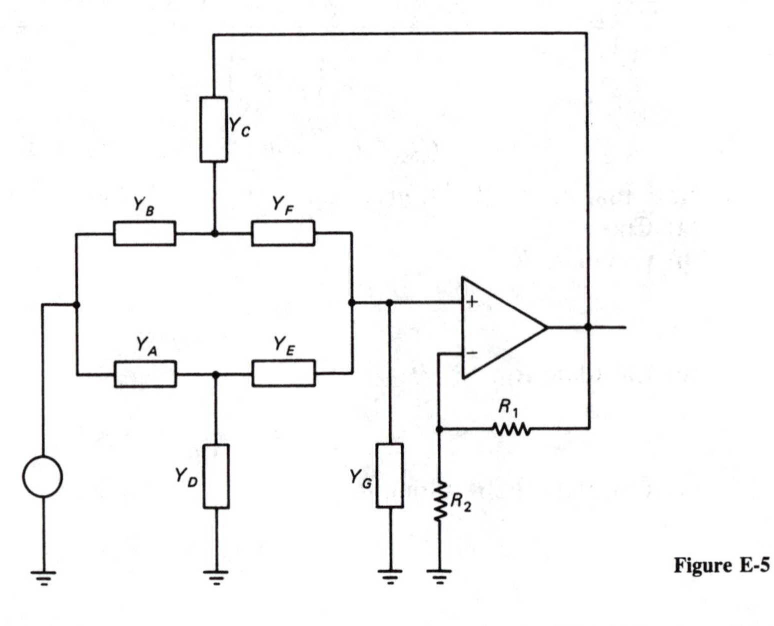

Figure E-5

3. Due to the complexity of this circuit's transfer function, we will do step 3 first, and then step 2 will be the transfer function with these assumptions.

$$C_A = C_E = C \qquad R_B = R_F = R$$
$$C_C = 2C \qquad R_D = \frac{R}{2} \tag{E-44}$$

2.
$$\frac{V_{\text{in}}}{V_o} = \frac{s^2 + C_X\left(\dfrac{2}{CR_G} + \dfrac{4 - 2K}{CR} + \dfrac{2C_G}{C^2R}\right)s + C_X\left(\dfrac{2}{C^2R_GR} + \dfrac{1}{C^2R^2}\right)}{KC_X\left(s^2 + \dfrac{1}{C^2R^2}\right)} \tag{E-45}$$

where

$$C_X = \frac{C}{C + 2C_G}$$

4. Note that in Eq. (E-1), $m = 1$, $n = 1$, and $d = 0$.
(a) To use this circuit, the following must be true.

$$0 \leq e + b - a\sqrt{e} \tag{E-46a}$$

(b) Choose C.

(c) Calculate R.

$$R = \frac{1}{C\sqrt{e}} \tag{E-46b}$$

(d) Choose C_G such that

$$C_G \geq \frac{C(e - b)}{2b} \tag{E-46c}$$

(e) Calculate R_G.

$$R_G = \frac{2\sqrt{e}}{C(b - e) + 2C_Gb} \tag{E-46d}$$

(f) Calculate K.

$$K = \frac{\dfrac{2}{R_G} + C(4\sqrt{e} - a) + 2C_G(\sqrt{e} - a)}{2C\sqrt{e}} \tag{E-46e}$$

(g) Calculate the resulting h.

$$h = KC_X \tag{E-46f}$$

APPENDIX F

Filter Table Calculations

This appendix shows how to make tables of the Butterworth, Chebyshev, and elliptic equations, and how to derive the Butterworth and Chebyshev equations. From this information a computer program could be written to make tables for all three filter equations. First we will look at the general derivation for these filter, and then we will look at each one specifically.

F-1 GENERAL PROCEDURE

The Butterworth, Chebyshev, and elliptic filter equations are defined by the magnitude of the loss function squared.

$$ML^2(j\Omega) = 1 + \varepsilon^2 F^2(j\Omega) \tag{F-1}$$

where ε = a numerical constant
$F(j\Omega)$ = the particular function that determines whether the loss function will be Butterworth, Chebyshev, or elliptic

In all of these filters, ε will be determined in the same way. When $F(j\Omega)$ is equal to 1, we are at a maximum in the pass band. Therefore, 20 log $(ML(j\Omega))$ will be equal to L_{max}. Since we know what L_{max} is required and we know that $F(j\Omega)$ is 1 at a maximum, we can solve Eq. (F-1) for ε.

$$L_{max} = 20 \log \sqrt{1 + \varepsilon^2}$$

$$\varepsilon = \sqrt{10^{0.1L_{max}} - 1} \tag{F-2}$$

When the function $F(j\Omega)$ is defined in Eq. (F-1), we replace $j\Omega$ with s, making the magnitude function squared into a loss function squared. When we take the square root of any function, we find two possible solutions. When we find the square root of the loss function, we will select only the stable roots (the roots to the left of the $j\omega$-axis). Let's examine how this is done.

We know that when we replace s with $j\Omega$ in the loss function, we will have a real part, R, and an imaginary part, I.

$$L(j\Omega) = R + jI \tag{F-3}$$

Using polar-to-rectangular conversion, we have

$$L(j\Omega) = \sqrt{R^2 + I^2} \angle \tan^{-1}(I/R) \qquad \text{(F-4)}$$

We are concerned only with the magnitude part and will therefore not put any conditions on the phase angle.

$$ML(j\Omega) = \sqrt{R^2 + I^2} \qquad \text{(F-5a)}$$

Since Eq. (F-1) is based on the magnitude squared, we square Eq.(F-5a).

$$ML^2(j\Omega) = R^2 + I^2 \qquad \text{(F-5b)}$$

We can factor this into complex conjugates.

$$ML^2(j\Omega) = (R + jI)(R - jI) \qquad \text{(F-5c)}$$

The first factor is $L(j\Omega)$, as seen from Eq. (F-3). The second term is equal to $L(-j\Omega)$. We may then write Eq. (F-5c) as

$$ML^2(j\Omega) = L(j\Omega)L(-j\Omega) \qquad \text{(F-6)}$$

We can show that $L(-j\Omega)$ is equal to $(R - jI)$ by example.

EXAMPLE F-1

Show that $L(-j\Omega)$ is the complex conjugate of $L(j\Omega)$.

SOLUTION We start with $L(s)$, which will be a numerator polynomial divided by a denominator polynomial.

$$L(s) = \frac{A_0 + A_1 s + A_2 s^2 + A_3 s^3 + \cdots}{B_0 + B_1 s + B_2 s^2 + B_3 s^3 + \cdots}$$

Then substituting $j\Omega$ for s,

$$L(j\Omega) = \frac{A_0 + jA_1\Omega - A_2\Omega^2 - jA_3\Omega^3 + \cdots}{B_0 + jB_1\Omega - B_2\Omega^2 - jB_3\Omega^3 + \cdots}$$

$$L(j\Omega) = \frac{(A_0 - A_2\Omega^2 + \cdots) + j(A_1\Omega - A_3\Omega^3 + \cdots)}{(B_0 - B_2\Omega^2 + \cdots) + j(B_1\Omega - B_3\Omega^3 + \cdots)}$$

We notice that all the odd powers of s will contain a j factor and the even powers of s do not. When we use $s = -j\Omega$, the even powers of s (the real part) will still have the same sign since a negative value to an even power is positive, but the odd powers of s will change sign.

$$L(-j\Omega) = \frac{A_0 - jA_1\Omega - A_2\Omega^2 + jA_3\Omega^3 + \cdots}{B_0 - jB_1\Omega - B_2\Omega^2 + jB_3\Omega^3 + \cdots}$$

$$L(-j\Omega) = \frac{(A_0 - A_2\Omega^2 + \cdots) - j(A_1\Omega - A_3\Omega^3 + \cdots)}{(B_0 - B_2\Omega^2 + \cdots) - j(B_1\Omega - B_3\Omega^3 + \cdots)}$$

Therefore, $L(j\Omega)$ and $L(-j\Omega)$ are complex conjugates.

Going back to Eq. (F-6), we can substitute s in for $j\Omega$.

$$ML^2(j\Omega) \mid_{j\Omega=s} = L(s)L(-s) \tag{F-7}$$

When $j\Omega$ is replaced by s in the squared magnitude equation, $ML^2(j\Omega)$, we can solve for the roots. These roots will either be part of $L(s)$ or $L(-s)$. We use the roots to the left of the $j\omega$-axis in the s-plane that are the roots of $L(s)$. We then have the loss function.

The loss function is normally factored with complex conjugate roots paired and can be written as

$$L(s) = \frac{(s + a_0)^J \prod_{i=1}^{\text{int}(n/2)} (s^2 + a_i s + b_i)}{C_D \prod_{i=1}^{\text{int}(n/2)} (s^2 + c_i)} \tag{F-8}$$

where n is the order of the filter

int(n/2) is the integer portion of $n/2$

$J = 0$ for even n, and $J = 1$ for odd n

For the Butterworth and Chebyshev filter, there are no roots in the denominator.

We can find the denominator constant, C_D, if we know the value of $L(s)$ at a given frequency. For the Butterworth, Chebyshev, and elliptic filter equations, the magnitude of the loss function, Eq. (F-8), will be equal to $L_{\max}$ at $s = j$.

$$L_{\max} = 20 \log \left| \frac{(j + a_0)^J \prod_{i=1}^{\text{int}(n/2)} (-1 + a_i j + b_i)}{C_D \prod_{i=1}^{\text{int}(n/2)} (-1 + c_i)} \right|_{\text{mag}}$$

Solving for C_D gives

$$C_D = \left| \frac{(j + a_0)^J \prod_{i=1}^{\text{int}(n/2)} (-1 + a_i j + b_i)}{\prod_{i=1}^{\text{int}(n/2)} (-1 + c_i)} \right|_{\text{mag}} \left(\frac{1}{10^{L_{\max}/2}} \right) \tag{F-9}$$

Notice that we need only the magnitude.

Following is a summary of the procedure for finding the filter equation.

1. Determine the value of ε using Eq. (F-2)
2. Select $F^2(j\Omega)$ in Eq. (F-1). This function will determine whether the filter is a Butterworth, Chebyshev, or elliptic.
3. Replace $j\Omega$ with s to form Eq. (F-7).
4. We then find the roots of Eq. (F-7). Different methods are used for this,

depending on what is convenient. The rest of this appendix shows this step for each filter equation.

5. From the roots we choose the roots that are stable (to the left of the $j\omega$-axis).
6. Calculate the denominator constant using Eq. (F-9).

F-2 BUTTERWORTH FILTER EQUATION

For the Butterworth filter equation we start with an even-power polynomial for $F^2(j\Omega)$ in Eq.(F-1) and force it to have as few maximums and minimums as possible. (This filter is sometimes called the maximally flat filter because of this.) We therefore make, at $\Omega = 0$, as many derivatives equal to zero as possible. This polynomial is then substituted into Eq. (F-1) for $F^2(j\Omega)$. The magnitude squared for this function must contain only even powers.

$$F^2(j\Omega) = 1 + A_1\Omega^2 + A_2\Omega^4 + A_3\Omega^6 + \cdots + A_n\Omega^{2n} \tag{F-10}$$

We can now evaluate the function by making, at $\Omega = 0$, as many derivatives equal to zero as possible.

$$\frac{d}{d\Omega} ML^2(j\Omega) = 2A_1\Omega + 4A_2\Omega^3 + 6A_3\Omega^5 + \cdots$$

Therefore,

$$\frac{d}{d\Omega} ML^2(j\Omega) \mid_{\Omega=0} = 0$$

In this case the function is already zero for any value of Ω.

Finding the second derivative, we have

$$\frac{d^2}{d\Omega^2} ML^2(j\Omega) = 2A_1 + 12A_2\Omega^2 + 30A_3\Omega^4 + \cdots$$

Therefore,

$$\frac{d^2}{d\Omega^2} ML^2(j\Omega) \mid_{\Omega=0} = 2A_1$$

For this to be zero, A_1 must equal zero.

As each derivative is set equal to zero, we will notice that odd derivatives are automatically equal to zero, and even derivative will require another A coefficient to be zero.

We cannot let all the derivatives go to zero or our function will not be frequency dependent. The last derivative possible will have to be a constant times Ω^{2n}. We will choose A_n such that this constant will be equal to 1. Substituting this into Eq. (F-1) yields

$$ML^2(j\Omega) = 1 + \varepsilon^2\Omega^{2n} \tag{F-11}$$

We replace $j\Omega$ with s as shown in Eq. (F-7), but this equation does not have a $j\Omega$ term. We must rearrange the equation to have a $j\Omega$ term to substitute into.

$$1 + \varepsilon^2[(-1)(j)^2\Omega^2]^n$$
$$1 + \varepsilon^2[(-1)(j\Omega)^2]^n$$

Equation (F-7) therefore becomes

$$ML^2(j\Omega) \mid_{j\Omega=s} = L(s)L(-s) = 1 + \varepsilon^2[-(s)^2]^n \qquad \text{(F-12)}$$

Since Eq. (F-2) can be used to find ε, then ε is a known value.

We then must factor this equation and solve for the stable roots. To solve the roots, we set the right side of the equation equal to zero.

$$1 + \varepsilon^2(-s^2)^n = 0 \qquad \text{(F-13)}$$

We now need to focus our attention on solving the equation to find the $L(s)L(-s)$. First we need to list a few trigonometric identities that are required to solve Eq. (F-13).

$$-1 = \cos(-\pi) + j\sin(-\pi) \qquad \text{(F-14a)}$$
$$= \cos(\pi) + j\sin(\pi)$$
$$e^{jx} = \cos(x) + j\sin(x) \qquad \text{(F-14b)}$$

In Eq. (F-13) we move the 1 to the right side of the equation, which will make it -1. Using Eq. (F-14a), we can express this -1 as

$$-1 = \cos(-\pi) + j\sin(-\pi)$$
$$= \cos(-\pi + 2\pi k) + j\sin(-\pi + 2\pi k) \qquad \text{where } k = 0, 1, 2, 3, \cdots$$

Using Eq. (F-14b), we can express this as an exponential function.

$$-1 = e^{j(2\pi k - \pi)}$$

Therefore, Eq. (F-13) becomes

$$\varepsilon^2(-s^2)^n = e^{j(2\pi k - \pi)}$$

or

$$\varepsilon^2(-1)^n(s^2)^n = e^{j(2\pi k - \pi)}$$

Since -1 and its reciprocal are the same value, we can move $(-1)^n$ to the right side of the equation.

$$\varepsilon^2(s^2)^n = e^{j(2\pi k - \pi)}(-1)^n$$

Using Eqs. (F-14a) and (F-14b), we can express $(-1)^n$ in an exponential form.

$$\varepsilon^2(s^2)^n = e^{j(2\pi k - \pi)}e^{j\pi n}$$
$$= e^{j(2\pi k - \pi + \pi n)}$$

Taking the $(2n)^{\text{th}}$ root of both sides gives

$$\varepsilon^{1/n} s = e^{j\pi(2k-1+n)/2n}$$

Solving for s and using Eq. (F-14b), we obtain

$$s = \varepsilon^{-1/n}\left\{\cos\left[\frac{\pi}{2}\left(\frac{2k-1+n}{n}\right)\right] + j\sin\left[\frac{\pi}{2}\left(\frac{2k-1+n}{n}\right)\right]\right\} \tag{F-15}$$

Since s was to the $2n$ power, there are $2n$ roots. Therefore, $k = 0, 1, 2, 3, \ldots, (2n - 1)$, but since the roots repeat after $2n - 1$, we may use $k = 1, 2, 3, \ldots, 2n$ for convenience.

Equation (F-15) will be the roots for both $L(s)$ and $L(-s)$ in Eq. (F-12). When we evaluate the roots, we must choose only the roots to the left of the $j\omega$-axis.

Summarizing how to find the roots for the Butterworth filter equations:

1. Select an $L_{\max}$ and filter order, n.
2. Calculate ε from Eq. (F-2).

$$\varepsilon = \sqrt{10^{0.1L_{\max}} - 1}$$

3. Calculate the numerator roots from Eq. (F-15) using only the roots to the left of the $j\omega$-axis.

$$s = \varepsilon^{-1/n}\left\{\cos\left[\frac{\pi}{2}\left(\frac{2k-1+n}{n}\right)\right] + j\sin\left[\frac{\pi}{2}\left(\frac{2k-1+n}{n}\right)\right]\right\} \qquad \text{where } k = 1, 2, 3, \ldots, 2n$$

4. Pair the complex conjugates roots to form Eq. (F-8). Notice that the Butterworth filter equation does not have denominator factors.

$$L(s) = \frac{(s + a_0)^J \prod_{i=1}^{\text{int}(n/2)} (s^2 + a_i s + b_i)}{C_D}$$

where n is the order of the filter
int $(n/2)$ is the integer portion of $n/2$
$J = 0$ for even n, and $J = 1$ for odd n

5. Determine the denominator constant using Eq. (F-9).

$$C_D = \left|\frac{(j + a_0)^J \prod_{i=1}^{\text{int}(n/2)} (-1 + a_i j + b_i)}{1}\right|_{\text{mag}} \left(\frac{1}{10^{L_{\max}/2}}\right)$$

where n is the order of the filter
int $(n/2)$ is the integer portion of $n/2$
$J = 0$ for even n, and $J = 1$ for odd n

6. Complete the loss function by substituting the value of C_D into Eq. (F-8).

$$L(s) = \frac{(s + a_0)^J \prod_{i=1}^{\text{int}\,(n/2)} (s^2 + a_i s + b_i)}{C_D}$$

F-3 CHEBYSHEV FILTER EQUATION

The Chebyshev filter equation is made by using the Chebyshev function substituted in for $F(j\Omega)$ in Eq. (F-1). The Chebyshev function can be found in almost any filter textbook. The nth-order Chebyshev function is

$$C_n(\Omega) = \begin{cases} \cos\,(n \cos^{-1} \Omega) & 0 \le |\Omega| \le 1 \quad \text{(F-16a)} \\ \cosh\,(n \cosh^{-1} \Omega) & |\Omega| > 1 \quad \text{(F-16b)} \end{cases}$$

The Chebyshev function can also be expressed in a polynomial form. Since the polynomial will not be used to find the Chebyshev filter equation, it will not be presented here.

The Chebyshev function is substituted into Eq. (F-1) for $F(j\Omega)$.

$$ML^2(j\Omega) = 1 + \varepsilon^2 C_n^2(\Omega) \tag{F-17}$$

We will set the right side of Eq. (F-17) equal to zero, substitute s in for $j\Omega$, and solve for the roots of s.

$$1 + \varepsilon^2 C_n^2(\Omega) = 0$$

$$C_n^2\left(\frac{j\Omega}{j}\right) = \frac{-1}{\varepsilon^2}$$

$$C_n\left(\frac{s}{j}\right) = \frac{\pm j}{\varepsilon} \tag{F-18}$$

Since there are two possibilities for $C_n(s/j)$ as shown in Eqs. (F-16a) and (F-16b), we need to solve Eq. (F-18) for both cases. However, we will only solve the case of $0 \le |\Omega| \le 1$ since the solution using the other case will be similar. Therefore, we have

$$C_n\left(\frac{s}{j}\right) = \cos\left[n \cos^{-1}\left(\frac{s}{j}\right)\right] \tag{F-19}$$

The inverse cosine will have a real part, R_1, and an imaginary part, I_1.

$$R_1 + jI_1 = \cos^{-1}\left(\frac{s}{j}\right) \tag{F-20}$$

Substituting into Eq. (F-19) yields

$$C_n\left(\frac{s}{j}\right) = \cos\,[n(R_1 + jI_1)]$$

Substituting this into Eq. (F-18) gives

$$\cos\,[n(R_1 + jI_1)] = \frac{\pm j}{\varepsilon} \tag{F-21}$$

Using the following trigonometric identities.

$$\cos\,(x + y) = \cos\,(x)\cos\,(y) - \sin\,(x)\sin\,(y) \tag{F-22a}$$

$$\cos\,(jx) = \cosh\,(x) \tag{F-22b}$$

$$\sin\,(jx) = j\sinh\,(x) \tag{F-22c}$$

Equation (F-21) becomes

$$\cos\,(nR_1)\cosh\,(nI_1) - j\sin\,(nR_1)\sinh\,(nI_1) = \frac{\pm j}{\varepsilon}$$

We can associate the real part and imaginary part on the left side of the equation to the real part and imaginary part on the right side of the equation.

$$\cos\,(nR_1)\cosh\,(nI_1) = 0 \tag{F-23a}$$

$$-\sin\,(nR_1)\sinh\,(nI_1) = \frac{\pm 1}{\varepsilon} \tag{F-23b}$$

Since, in Eq. (F-23a), $\cosh\,(nI_1) \geq 1$ for all values of nI_1, then $\cos\,(nR_1)$ must equal zero. Therefore, we want nR_1 to be odd multiples of $\pi/2$. There are many ways to express this, but a convenient way for our purpose is

$$nR_1 = \left(k - \frac{1}{2}\right)\pi \qquad k = 1, 2, \ldots, 2n$$

$$R_1 = \left(\frac{2k - 1}{2n}\right)\pi \qquad k = 1, 2, \ldots, 2n \tag{F-24}$$

Looking at the imaginary part, we can substitute Eq. (F-24) in Eq. (F-23b).

$$-\sin\left[\left(\frac{2k - 1}{2}\right)\pi\right]\sinh\,(nI_1) = \frac{\pm 1}{\varepsilon} \qquad k = 1, 2, \ldots, 2n$$

For the values of k, the sine part on the left will be either ± 1. Since we will only use positive ε, we can simplify our equation to

$$\sinh\,(nI_1) = \frac{1}{\varepsilon}$$

$$I_1 = \frac{1}{n}\sinh^{-1}\left(\frac{1}{\varepsilon}\right) \tag{F-25}$$

where

$$\sinh^{-1}(x) = \ln\left(x + \sqrt{x^2 + 1}\right)$$

We can calculate the real part, R_1, and imaginary part, I_1, using Eqs. (F-24) and (F-25), and substitute back into Eq. (F-20).

$$\cos^{-1}\left(\frac{s}{j}\right) = R_1 + jI_1$$

Solving for s and applying Eqs. (F-22a), (F-22b), and (F-22c) gives

$$s = j\cos(R_1 + jI_1)$$

$$s = \sin(R_1)\sinh(I_1) + j\cos(R_1)\cosh(I_1) \qquad \text{(F-26)}$$

where R_1 and I_1 are defined by Eqs. (F-24) and (F-25)

With a similar procedure we could derive Eq. (F-17) using the Chebyshev function of Eq. (F-16b). For $|\Omega| > 1$, the resulting equation would be identical.

Following is a summary of how to find the roots for the Chebyshev filter equations:

1. Select an L_{max} and filter order, n.
2. Calculate ε from Eq. (F-2).

$$\varepsilon = \sqrt{10^{0.1L_{max}} - 1}$$

3. Calculate the numerator roots from Eqs. (F-26), (F-24), and (F-25) using only the roots to the left of the $j\omega$-axis.

$$s = \sin(R_1)\sinh(I_1) + j\cos(R_1)\cosh(I_1)$$

$$R_1 = \left(\frac{2k - 1}{2n}\right)\pi \qquad k = 1, 2, \ldots, 2n$$

$$I_1 = \frac{1}{n}\sinh^{-1}\left(\frac{1}{\varepsilon}\right)$$

where

$$\sinh^{-1}(x) = \ln\left(x + \sqrt{x^2 + 1}\right)$$

4. Pair the complex-conjugate roots to form Eq. (F-8). Notice that the Chebyshev filter equation does not have denominator factors.

$$L(s) = \frac{(s + a_0)^J \prod_{i=1}^{\text{int}(n/2)} (s^2 + a_i s + b_i)}{C_D}$$

where n is the order of the filter
int $(n/2)$ is the integer portion of $n/2$
$J = 0$ for even n, and $J = 1$ for odd n

5. Determine the denominator constant using Eq. (F-9).

$$C_D = \left| \frac{(j + a_0)^J \prod_{i=1}^{\text{int }(n/2)} (-1 + a_i j + b_i)}{1} \right|_{\text{mag}} \left(\frac{1}{10^{L_{\max}/2}} \right)$$

where n is the order of the filter
int $(n/2)$ is the integer portion of $n/2$
$J = 0$ for even n, and $J = 1$ for odd n

6. Complete the loss function by substituting the value of C_D into Eq. (F-8).

$$L(s) = \frac{(s + a_0)^J \prod_{i=1}^{\text{int }(n/2)} (s^2 + a_i s + b_i)}{C_D}$$

F-4 ELLIPTIC FILTER EQUATION

The elliptic filter equations uses the Chebyshev rational function for $F(j\Omega)$ in Eq. (F-1). Since this function is complex, we will concentrate on how to find the elliptic function equations. A complete discussion of the elliptic function equations can be found in Daniels.[1] We will only summarize and make note of what is required to solve for the roots of the elliptic function equation.

The nth-order Chebyshev rational function can be written with $j\Omega$ replaced by s as

$$R_n(s) = C_R s^J \prod_{i=1}^{\text{int }(n/2)} \frac{s^2 - r_{zi}^2}{s^2 - r_{pi}^2} \tag{F-27}$$

where $J = 0$ for even n, and $J = 1$ for odd n

$$C_R = \prod_{i=1}^{\text{int }(n/2)} \frac{1 - r_{pi}^2}{1 - r_{zi}^2}$$

int $(n/2)$ is the integer portion of $n/2$

This is substituted back into Eq. (F-1) to form the loss function.

$$F^2(s) = 1 + \varepsilon^2 R_n^2(s) \tag{F-28}$$

It is important to notice that the denominator of the Chebyshev rational function will also be the denominator of the elliptic filter equation.

[1] R. W. Daniels, *Approximation Methods for Electronic Filter Design*, McGraw-Hill Book Company, New York, 1974.

The roots of the Chebyshev rational function are found by

$$r_{zi}z = \text{sn}\,(y_i;\, \Omega_s^{-1}) \tag{F-29a}$$

$$r_{pi} = \frac{1}{\Omega_s^{-1} r_{zi}} \tag{F-29b}$$

where *sn* is the elliptic sine function
Ω_s is the normalized stop-band edge frequency
$i = 1, 2, 3, \ldots, \text{int}(n/2)$

The calculation of the elliptic sine function of Eq. (F-29a) is quite involved, and the procedure is as follows:

1. Evaluate the complete Jacobian elliptic function of the first kind, and then the complementary form.

The complete Jacobian elliptic function is defined as

$$K(\Omega_s^{-1}) = \int_0^{\pi/2} \frac{d\theta}{[1 - \Omega_s^{-1}\sin^2(\theta)]^{1/2}} \tag{F-30}$$

Using a binomial series expansion, we can evaluate this integral by

$$K(\Omega_s^{-1}) = \frac{\pi}{2}\left[1 + \left(\frac{1}{2}\right)^2 (\Omega_s^{-1})^2 + \left(\frac{1\cdot 3}{2\cdot 4}\right)^2 (\Omega_s^{-1})^4 + \left(\frac{1\cdot 3\cdot 5}{2\cdot 4\cdot 6}\right)^2 (\Omega_s^{-1})^6 + \cdots\right] \tag{F-31}$$

The complementary form, $K'(\Omega_s^{-1})$, uses $\Omega_s^{-1'}$ substituted into Eqs. (F-30) and (F-31) for Ω_s^{-1}, where

$$\Omega_s^{-1'} = \sqrt{1 - (\Omega_s^{-1})^2} \tag{F-32}$$

This changes Eq. (F-31) into

$$K'(\Omega_s^{-1}) = \frac{\pi}{2}\left[1 + \left(\frac{1}{2}\right)^2 (\Omega_s^{-1'})^2 + \left(\frac{1\cdot 3}{2\cdot 4}\right)^2 (\Omega_s^{-1'})^4 + \left(\frac{1\cdot 3\cdot 5}{2\cdot 4\cdot 6}\right)^2 (\Omega_s^{-1'})^6 + \cdots\right] \tag{F-33}$$

2. The values of $K(\Omega_s^{-1})$ and $K'(\Omega_s^{-1})$ are used to find the value of y in the Jacobian elliptic integral.

$$y_i = \int_0^{u_i} \frac{d\theta}{[1 - \Omega_s^{-1}\sin^2(\theta)]^{1/2}} \tag{F-34a}$$

where for odd n

$$y_i = i\,\frac{2K(\Omega_s^{-1})}{n} \tag{F-34b}$$

and for even n,

$$y_i = (i - 0.5)\,\frac{2K(\Omega_s^{-1})}{n} \tag{F-34c}$$

$$i = 1, 2, 3, \ldots, \mathrm{int}(n/2)$$

3. Knowing the specific value the Jacobian elliptic integral must be, we solve for the upper limit u_i required to obtain this specific value. Using a numerical process from the *Handbook of Mathematical Functions*,[2] we obtain

$$u_i = v + \sum_{m=1}^{\infty} \frac{2q^m \sin(2mv)}{m(1 + q^{2m})} \tag{F-35}$$

where

$$v = \frac{\pi y_i}{2K(\Omega_s^{-1})}$$

$$q = e^{-\pi K'(\Omega_s^{-1})/K(\Omega_s^{-1})}$$

4. Last, we find the sine of the upper limit for each y_i. Therefore,

$$\mathrm{sn}\,(y_i; \Omega_s^{-1}) = \sin(u_i) \tag{F-36}$$

Once this is found, we can solve Eqs. (F-29a) and (F-29b), which are the roots of Eq. (F-27). At this point we have found the roots of the Chebyshev rational function, $R_n(s)$. We still must find the roots of the loss function $L(s)$. Solving the roots of the loss function is not a trivial problem.

Daniels[3] points out that since the roots in the rational function are close together, they should be spread apart to increase the accuracy of finding the roots of the loss function. This transformation we will call Z-transformation. This is not the same as Z-transforms and should not be thought of as being even remotely related.

We see in Eq. (F-28) that the rational function is squared. We will therefore define the Z-transformation of the Chebyshev rational function squared as

$$R_n^2(Z) = \frac{C_{ZN}^2}{C_{ZD}^2(Z^2 - 1)^J} \prod_{i=1}^{\mathrm{int}(n/2)} \frac{(Z^2 - r_{zzi}^2)^2}{(Z^2 - r_{zpi}^2)^2} \tag{F-37}$$

[2]L. M. Milne-Thomson, *Handbook of Mathematical Functions with Formulas, Graphs, and Mathematical Tables*, National Bureau of Standards Applied Mathematics Series 55, Section 17, p. 591, issued June 1964, third printing March 1965.

[3]R. W. Daniels, *Approximation Methods for Electronic Filter Design*, McGraw-Hill Book Company, New York, 1974.

where

$$C_{ZN}^2 = C_R^2 \prod_{i=1}^{\text{int }(n/2)} (r_{zi}^2)^2$$

C_R from Eq. (F-27)

$$C_{ZD}^2 = (-1)^J \prod_{i=1}^{\text{int }(n/2)} (r_{pi}^2)^2$$

$$r_{zzi}^2 = 1 - \frac{1}{r_{zi}^2}$$

$$r_{zpi}^2 = 1 - \frac{1}{r_{pi}^2}$$

$J = 0$ for even n, and $J = 1$ for odd n

The factors of Eq. (F-37) can be multiplied out to form a numerator polynomial, $N(Z)$, divided by a denominator polynomial, $D(Z)$.

$$R_n^2(Z) = \frac{N^2(Z)}{D^2(Z)} \tag{F-38}$$

This equation can be substituted into Eq. (F-28) to form a Z-transformation loss function squared.

$$F^2(s) = 1 + \varepsilon^2 \frac{N^2(Z)}{D^2(Z)} \tag{F-39}$$

which can be written as

$$F_n^2(Z) = \frac{D^2(Z) + \varepsilon^2 N^2(Z)}{D^2(Z)} \tag{F-40}$$

Here we see the numerator and denominator polynomials of the rational function must be multiplied out and combined into a single numerator polynomial. Since we already have the factored roots of the denominator in the s-domain, we only have to factor the numerator and transform it from the Z-transformation to the s-domain.

The numerator of Eq. (F-40) will often be larger than the fourth order. Numerical methods of factoring will therefore be required. One such method is Bairstow's method. A complete discussion of this method can be found in *Applied Numerical Methods*[4]; we will only apply the method to our case.

[4]M. L. James, G. M. Smith, and J. C. Wolford, *Applied Numerical Methods for Digital Computation with Fortran and CSMP*, 2nd ed., Harper & Row, Publishers, Inc., New York, 1977, pp. 141–146.

This method is an iterative method that starts with an arbitrary quadratic, divides the polynomial by this quadratic, corrects the value of the quadratic, and then divides the polynomial by the corrected quadratic. This is repeated until the quadratic is corrected to the accuracy required. Since we are using a quadratic, we will not have complex numbers to deal with.

In our case there will be only even powers of the variable Z. Therefore,

$$D^2(Z) + \varepsilon^2 N^2(Z) = C(Z^{2I} + A_1 Z^{2(I-1)} + A_2 Z^{2(I-2)} + \cdots + A_I) \qquad \text{(F-41)}$$

where I is the degree of the polynomial
notice that A_0 is equal to 1

We need to find an a_z and b_z such that

$$C(Z^4 + a_z Z^2 + b_z)(B_0 Z^{2(I-2)} + B_1 Z^{2(I-3)} + \cdots + B_{I-2}) = D^2(Z) + \varepsilon^2 N^2(Z) \qquad \text{(F-42)}$$

We will then take the remaining B polynomial and factor out another quadratic until we have all the factors of the polynomial as

$$D^2(Z) + \varepsilon^2 N^2(Z) = C(Z^2 - a_{z0})^J \prod_{i=1}^{\text{int }(n/2)} (Z^4 + a_{zi} Z^2 + b_{zi}) \qquad \text{(F-43)}$$

where C is defined in Eq. (F-41)
$J = 0$ for even n and $J = 1$ for odd n

Using variables defined in Eqs. (F-41), (F-42), and (F-43), we can define Bairstow's method as follows:

1. Make sure that the polynomial is greater than second order, $n > 2$, since Bairstow's method will not be required for $n \le 2$.
2. Set M equal to the degree of the polynomial, n. Set i to an initial value of 1 to point to an a_{zi} and b_{zi}.
3. Factor the constant, C, out of polynomial A as shown in Eq. (F-41) so that the first coefficient is 1.
4. Choose a starting value for a_{zi} and b_{zi}. Values from -1 to 5 will usually give good results.
5. Set initial values for the starting a B polynomial and a C polynomial.

$$B_0 = 1 \qquad B_1 = A_1 - a_i$$

$$C_0 = 1 \qquad C_1 = B_1 - a_i$$

6. Initialize a counter variable. $K = 2$.
7. Find the next B coefficient.

$$B_K = A_K - B_{K-1} a_{zi} - B_{K-2} b_{zi}$$

8. If $K = M$, go to step 11; if not, continue.

9. Find the next C coefficient.

$$C_K = B_K - C_{K-1}a_{zi} - C_{K-2}b_{zi}$$

10. Increment the counter K ($K = K + 1$) and go back to step 7.
11. Calculate the correction factor for a_{zi} and b_{zi}.

$$D_a = \frac{B_n C_{n-3} - B_{n-1}C_{n-2}}{C_{n-1}C_{n-3} - C_{n-2}^2}$$

$$D_b = \frac{B_{n-1}C_{n-1} - B_n C_{n-2}}{C_{n-1}C_{n-3} - C_{n-2}^2}$$

Here we need to check if D_a's and D_b's magnitude is continuing to decrease (converge) or increase (diverge). If either one is diverging, we need to go back to step 4 and choose different starting a_{zi} and b_{zi} values.

12. Calculate the new a_{zi} and b_{zi}.

$$a_{zi} = a_{zi} + D_a$$

$$b_{zi} = b_{zi} + D_b$$

13. Determine if more accuracy is required. If $|D_a| + |D_b| >$ a desired small number, go back to step 5. Ideally, D_a and D_b should become zero.
14. The a_{zi} and b_{zi} at this point are coefficients in one of the factors.

$$Z^4 + a_{zi}Z^2 + b_{zi}$$

Or if $b_{zi} = 0$, we have found

$$Z^2 + a_{z0}$$

15. Set up to factor the remaining polynomial by moving the B coefficients to the A array, and reduce M by one factor.

$$A_j = B_j \text{ for } j = 1 \text{ to } n - 2$$

$$\text{if } b_{zi} <> 0, \text{ then } M = M - 2$$

$$\text{if } b_{zi} = 0, \text{ then } M = M - 1$$

16. Set i to point to the next a_{zi} and b_{zi} value ($i = i + 1$).
17. If $M > 2$, go to step 4 to start looking for the next a_{zi} and b_{zi} values.
18. If $M = 1$, then a_{zi} is a_{z0}.

When we finish Bairstow's method, we will have the factored Z-transformation shown in Eq. (F-43). This is the numerator of the loss function squared. We can find the inverse Z-transformation and the stable roots by the following procedure.

$$\sqrt{D^2(s) + \varepsilon^2 N^2(s)} = C(s - a_0)^J \prod_{i=1}^{\text{int }(n/2)} (s^2 + a_i s + b_i) \qquad \text{(F-44)}$$

where

$$a_i = \left(\frac{-2 - a_{zi} + 2\sqrt{1 + a_{zi} + b_{zi}}}{1 + a_{zi} + b_{zi}}\right)^{1/2}$$

$$b_i = \left(\frac{1}{1 + a_{zi} + b_{zi}}\right)^{1/2}$$

$$a_0 = \left(\frac{1}{a_{z0} - 1}\right)^{1/2}$$

$J = 0$ for even n, and $J = 1$ for odd n

The constant, C, we will not evaluate since we will evaluate the loss functions constant using Eq. (F-9).

We now have the roots for the numerator and denominator of the loss function. With the elliptic function equations we need to find L_{min}, since this function does not continue to increase in attenuation in the stop band. This value will be

$$L_{min} = 10 \log \sqrt{1 + \varepsilon^2 X^2} \tag{F-45}$$

where

$$X = \frac{1}{(\Omega_s^{-1})^n \prod_{i=0}^{\text{int }(n/2)-1} \text{sn}^4(y_i; \Omega_s^{-1})}$$

The elliptic sine function to the fourth power is calculated by using Eqs. (F-35) and (F-36) with

$$y_i = \frac{(1 + 2i)K(\Omega_s^{-1})}{n}$$

In the summary of this procedure, the equation that specifically apply to the elliptic function equations will only have the equation numbers listed since they are so complex. This will facilitate the understanding of the order of using the equations.

Following is a summary of how to find the roots for the elliptic filter equations:

1. Select an L_{max} and filter order, n.
2. Calculate ε from Eq. (F-2).

$$\varepsilon = \sqrt{10^{0.1L_{max}} - 1}$$

3. Find the Chebyshev rational function roots for Eq. (F-27).
   a. Find $K(\Omega_s^{-1})$ using Eq. (F-31).
   b. Find $K'(\Omega_s^{-1})$ using Eqs. (F-32) and (F-33).
   c. Find y_i using Eq. (F-34).

d. Find u_i using Eq. (F-35).
e. Find the r_{zi} roots of Eq. (F-29a) using Eq. (F-36).
f. Find the r_{pi} roots of Eq. (F-29b) that are also the roots of the denominator of the loss function squared.

4. Convert the Chebyshev rational function to the squared Z-transformation Chebyshev rational function using Eq. (F-37).

5. Form the Z-transformation loss function numerator of Eq. (F-40).

6. Factor the numerator squared using Bairstow's method.

7. Find the stable roots of the loss function numerator using Eq. (F-44).

8. Write the elliptic function equation of Eq. (F-8) using the roots found.

$$L(s) = \frac{(s + a_0)^J \prod_{i=1}^{\text{int }(n/2)} (s^2 + a_i s + b_i)}{C_D \prod_{i=1}^{\text{int }(n/2)} (s^2 + c_i)}$$

9. Determine the denominator constant using Eq. (F-9).

$$C_D = \left| \frac{(j + a_0)^J \prod_{i=1}^{\text{int }(n/2)} (-1 + a_i j + b_i)}{\prod_{i=1}^{\text{int }(n/2)} (-1 + c_i)} \right|_{\text{mag}} \left(\frac{1}{10^{L_{\max}/2}} \right)$$

where n is the order of the filter
int(n/2) is the integer portion of $n/2$
$J = 0$ for even n, and $J = 1$ for odd n

10. Form the completed loss function by substituting the value of C_D into Eq. (F-8).

11. Find the pass-band loss, $L_{\min}$, using Eq. (F-45).

APPENDIX G

Disk Programs

This appendix briefly explains the functions of the programs on a disk available through the address listed in the Preface of this book. These programs should be used to verify solutions and examples, and are not intended to replace the process of hand working the problems. When the disk is inserted and START is entered, the user may select form six programs.

1. TIME-GRAPH. This program allows the user to input equations of complex waveforms and display them graphically.
2. POLYNOMIAL. This program allows the user to input up to 100 polynomials, 100 sets of roots, and 100 complex numbers. These can be added subtracted, divided, manipulated, differentiated, and factored.
3. INVERSE-LAPLACE. This allows the user to input roots of the denominator and numerator of a function and find the inverse Laplace transform.
4. BODE. This allows the user to input a numerator and denominator of a transfer function and display the exact or Bode graph of the magnitude or phase.
5. FILTER-SYNTHESIS. This allows the user to input the coefficients of a first- or a second-order transfer function; then the program calculates the impedances for the filter, and allows the user to frequency and/or impedance shift the impedances.
6. FILTER-TABLE. This will calculate the roots or factors of a Butterworth, Chebyshev, or elliptic equation.

The following sections gives a more detailed explanation of these programs. Since the programs are menu driven, the information contained here is not intended to be complete, but is intended to get you started.

G-1 SYSTEM REQUIREMENTS

These programs are written in Microsoft QuickBASIC for IBM compatible computers. The system requires CGA monitor, 640k RAM (some programs will work on less memory), 360k disk drive (dual drives are required for data storage), and MS-DOS or PC-DOS version 2.1 or later. To start the system:

1. Boot from a disk in drive A, which can be used for data.
2. Insert the program disk in drive B and type B: and the Enter key.
3. To start the program, type START and the Enter key.
4. Select the program or function from the menu displayed.

G-2 TIME-GRAPH

When this program starts there are four choices: INPUT/DISPLAY, PLOT, ACCESS DISK, and EXIT. The choice is made by using the Left-arrow and Right-arrow keys. The Enter key is used to start the command.

G-2-1 INPUT/DISPLAY

This command will start another command line with the choices, NEXT, PREVIOUS, CHANGE, and NUM OF FUNCTIONS. NEXT and PREVIOUS allow the user to step through a display of the first term of each function. NUM OF FUNCTIONS allows the user to specify up of 10 different functions to be graphed at one time. CHANGE will activate another command line.

The command line choices when CHANGE is activated are NEXT, PREVIOUS, CHANGE, and NUM OF TERMS. NEXT and PREVIOUS allow the user to display the coefficients of each term of the current function. NUM OF TERMS allows the user to select the number of terms for the current function. CHANGE will allow the user to modify the current term of the current function.

To get back to the beginning command choices, select the up arrow and use the Enter key.

G-2-2 PLOT

After a function has been specified, the command will plot each function on the same axis. The command will activate the following set of questions for the user to answer.

MAKE GRAPH PAPER ONLY. A Y will cause the axis to be displayed without the functions.

DASHED GRID. An N will cause the graph's grid to be solid lines.

STARTING TIME. Enter the value for the time in seconds for the graphing to begin.

ENDING TIME. Enter the last value of time in seconds for the graphing to stop.

MARK INTERVAL. Enter the increments in seconds to be marked on the horizontal axis.

LOW MAG. Enter the lowest magnitude for the graph to display.

HIGH MAG. Enter the highest magnitude for the graph to display.

MARK INTERVAL. Enter the increments to be marked on the vertical axis.

G-2-3 ACCESS DISK

The functions entered can be saved on the data disk using this command. The filename extension is automatically specified as .TIM. The user may give only a name without an extension. This command will activate another command line with the choices SAVE ON DISK, LOAD FROM DISK, and DIRECTORY. SAVE ON DISK will allow the user to enter a filename. If a filename is not specified, the last filename used will be used. LOAD FROM DISK will ask the user for a filename and load that file from the data disk if it exists. DIRECTORY works like the DOS system currently installed on the program disk. If no filename is specified, the directory will be of *.TIM.

Filenames must conform to the DOS system being used. Failure to do so may result in a program crash and loss of data.

G-2-4 EXIT

This exits this program.

G-3 POLYNOMIAL

This program displays three different-colored blocks at the top of the screen. Each block represents 100 cell locations. Empty cells are represented in the block by blanks, and cells with values in them are represented by { }. The cells may be marked using the following system:

R or 0: usually indicates a result or destination.
S or .: A secondary result or destination.
A or 1, B or 2, C or 3, E or 4, F or 5, G or 6, I or 7, J or 8, K or 9: indicates cells to use in the process

Going from right to left, the right-most block's cells will hold one complex number in each cell, the next block's cells will hold roots of a polynomial in each cell, and the leftmost block's cells will hold one polynomial in each cell. The contents of one specified cell of each block is displayed under the blocks in the same background color as the block's color.

At the top left of the display is a location for the command blocks. There are four different command blocks: POLYNOMIAL, ROOT, COMPLEX, and DISK ACCESS. The particular command block can be changed by using the Tab key. To select a command within a command block, use the Up-, Down-, Left-, and Right-arrow keys. To start the command use the Enter key. This will move the cursor to the appropriate block of cells. To stop execution of the command, use the TAB key.

When a command is selected and the cursor is in a block of cells, the up-, down-, left-, and right-arrow keys are used to select a cell. The cell is marked by typing the appropriate letter or number. When the markers are set, the Enter key is used to continue the command or the Tab key to stop the command.

G-3-1 POLYNOMIAL COMMAND BLOCK

This controls functions that operate mainly on the polynomial block of cells. The commands and a brief description follow.

INPUT R. This allows the user to input or modify a polynomial in the cell marked R with a polynomial of up to the 30th power.

A + B. Add polynomial in cell A to cell B and store the result in cell R.

A × B. Multiply cell A by cell B and store the result in cell R.

EVAL. Evaluate the fraction A/B with the polynomial variable set to the value in the complex block of cells marked by A and store the value in cell R of the complex block of cells. If B in the polynomial block of cells is a blank cell, then only the polynomial marked A is used.

DET 2×2. Store in cell R the determinant

$$\begin{vmatrix} A & B \\ C & D \end{vmatrix}$$

COPY A. Copy cell A into cell R.

A − B. Subtract cell B from A and store the result in R.

A/B. Divide A by B and store result in cell R with the remainder in cell S.

d/dt. Find the derivative of the fraction A/B and store the result in R/S. If B is an empty cell, find the derivative of A and store the result in R.

RDC A/B. Reduce the fraction A/B and store the result in A/B. The power of the polynomial in A must be less than the power of the polynomial in B.

DET 3×3. Store in cell R the determinant

$$\begin{vmatrix} A & B & C \\ E & F & G \\ I & J & K \end{vmatrix}$$

G-3-2 ROOT COMMAND BLOCK

This controls functions that operate mainly on the root block of cells. The commands and a brief description follow.

INPUT R. This allows the user to input a set of roots into the cell marked with R.

A + B. Combine the set of roots in cell A to the set of roots in cell B and store the result in cell R.

MOVE A. Move one of the roots from cell A to cell R.

CMPLX. Move one of the roots from cell A to the complex block of cells.

COPY A. Copy a set of roots from A to cell R.

REMOVE. Copy all but one root from the set of roots in cell A to cell R.

POLY. Store the set of roots in cell A as a polynomial, in the polynomial block of cells in cell R.

G-3-3 COMPLEX COMMAND BLOCK

This controls functions that operate mainly on the complex block of cells. The commands and a brief description follow.

INPUT R. This allows the user to input complex numbers, in either polar or rectangular form, into the cell marked R.

A + B. Add the complex number in cell A to the complex number in cell B and store the result in cell R.

A × B. Multiply the complex number in cell A by the complex number in cell B and store the result in cell R.

ROOT. Store the complex number in cell A in the root block of cells in cell R.

COPY A. Copy cell A into cell R.

A − B. Subtract the complex number in cell B from the complex number in cell A and store the result in R.

A/B. Divide A by B and store the result in cell R.

G-3-4 DISK ACCESS

SAVE. Save all the values in all the blocks on the data disk with the filename extension .POL.

S. LAPL. Save specified roots from the root array on the data disk for the LAPLACE program to use.

S. BODE. Save specified polynomials on the data disk for the BODE program to use.

LOAD. Load a previously saved value from the data disk into all blocks.

DIR. Disk directory of all files.

EXIT. Exit the POLYNOMIAL program.

Filenames must conform to the DOS system being used. Failure to do so may result in a program crash and loss of data.

G-4 INVERSE-LAPLACE

When this program starts there are four choices: INPUT/DISPLAY, PLOT, ACCESS DISK, and EXIT. The choice is made by using the Left-arrow and Right-arrow keys. The Enter key is used to start the command.

G-4-1 INPUT/DISPLAY

When this command is activated, the user is asked to select one of the 10 available functions to work with. The user then enters a multiplying factor for the numerator of the selected function followed by roots of the numerator. The denominator multiplying factor followed by the roots of the denominator are then entered. The number of roots in the numerator must be equal to or less than the number of roots in the denominator. With each root a power is requested. This is the power of the factor that is currently being entered.

G-4-2 INVERSE

When this command is activated, the user is asked to select one of the 10 available functions. The function selected must have been defined previously by the INPUT/OUTPUT command or loaded from the DISK ACCESS command. The inverse of the function is then calculated and displayed.

G-4-3 ACCESS DISK

The functions entered can be saved on the data disk using this command. The filename extension is automatically specified as .LAP. The user may give only a name without an extension. This command will activate another command line with the choices SAVE ON DISK, LOAD FROM DISK, and DIRECTORY. SAVE ON DISK will allow the user to enter a filename. If a filename is not specified, the last filename used will be used. LOAD FROM DISK will ask the user for a filename and load that file from the data disk if it exists. DIRECTORY works like the DOS system currently installed on the program disk. If no filename is specified, the directory will be of *.LAP.

Filenames must conform to the DOS system being used. Failure to do so may result in a program crash and loss of data.

G-4-4 EXIT

This exits the program.

G-5 BODE

When this program starts there are five choices: INPUT/DISPLAY, PLOT MAG, PLOT PHASE, ACCESS DISK, and EXIT. The choice is made by using the Left-arrow and Right-arrow keys. The Enter key is used to start the command.

G-5-1 INPUT/DISPLAY

This command will start another command line with the choices NEXT, PREVIOUS, CHANGE, and NUM OF FUNCTIONS. The NEXT and PREVIOUS allow the user to step through a display of the first term of each transfer function. NUM OF FUNCTIONS allows the user to specify up to 10 different transfer functions. CHANGE will allow the user to modify the current transfer function being displayed.

When CHANGE is activated, the user must choose whether the function will be plotted as exact or Bode. If Bode is selected, the numerator and denominator functions must be a first- or second-order function that have complex conjugate roots. Next the number of terms for the numerator are selected, and then the numerator multiplying factor is entered. For each term the highest power of the variable is selected, followed by the coefficient of each term of the factor. After each term of the numerator is entered, the denominator is entered in a similar manner.

G-5-2 PLOT MAG and PLOT PHASE

After a function has been specified, these commands will plot each function on the same axis. These two commands are similar except that one plots the magnitude in dB and the other plots the phase in degrees. When either of these commands are activated, the following values are requested.

MAKE GRAPH PAPER ONLY. A Y will cause the axis without the functions.

DASHED GRID. An N will cause the graphs grid to be solid lines.

1-HZ 2-RD. Entering a 1 will cause the horizontal axis to be marked in hertz, and entering a 2 will cause the horizontal axis to be marked in radians.

STARTING DECADE. Enter the value of the starting frequency for the graphing to begin.

NUMBER OF DECADES. Enter the number of decades or cycles for the horizontal log axis.

LOW MAG / LOW DEG. Enter the lowest value for the vertical axis.

HIGH MAG / HIGH DEG. Enter the highest value for the vertical axis.

MAG INC. OF / DEG INC. OF. Enter the increments to be marked.

G-5-3 ACCESS DISK

The functions entered can be saved on the data disk using this command. The filename extension is automatically specified as .BOD. The user may give only a name without an extension. This command will activate another command line with the choices SAVE ON DISK, LOAD FROM DISK, and DIRECTORY. SAVE ON DISK will allow the user to enter a filename. If a filename is not specified, the last filename used will be used. LOAD FROM DISK will ask the user for a filename and load that file from the data disk if it exists. DIRECTORY works like the DOS system currently installed on the program disk. If no filename is specified, the directory will be of *.BOD. Filenames must conform to the DOS system being used. Failure to do so may result in a program crash and loss of data.

G-5-4 EXIT

This exits the program.

G-6 FILTER-SYNTHESIS

When this program starts there are five choices: DESIGN FILTER, CALC. ORDER, LP TO HP,BP,NC, SHIFT COMPT, and EXIT. The choice is made by using the Left-arrow and Right-arrow keys. The Enter key is used to start the command.

G-6-1 DESIGN FILTER

When this command is activated, another command block is displayed which allows the user to select either an INVERTING, NONINVERTING, or TWIN-T filter. When this is selected, another command block is displayed which allows the user to select either SINGLE FEEDBACK or DUAL FEEDBACK. When this is selected, another command block is displayed which allows the user to select either LOW PASS, HIGH PASS, or BAND PASS.

Once the particular filter is selected, the appropriate transfer function form is displayed, and the user will enter the coefficients of this transfer function form. The user will then select a component value, and the resulting components will be calculated and displayed. These values are referenced to the circuits in this book. After the initial calculation and display, the user is allowed to frequency shift and/or impedance shift the components.

G-6-2 CALC. ORDER

This command will ask the user for the L_{max}, L_{min}, and DESIRED STOP BAND. The filter order required to meet these specifications will then be calculated and displayed for the Butterworth and Chebyshev functions.

G-6-3 LP, TO HP, BP, NC

This command will activate another command block which allows the user to select 1st ORDER, 2nd ORDER, 2nd/(s^2 + e), or R_α R_β. The 2nd ORDER and 2nd/(s^2 + e) are the same except that the 2nd ORDER has just the s^2 term in the denominator.

When 1st ORDER, 2nd ORDER, or 2nd/(s^2 + e) are selected, another command block is activated which allows the user to select a conversion from Low Pass to either High Pass, Band Pass, or Notch. The user will then enter the center frequency, the high break frequency, and the low break frequency;

or the user will enter the center frequency and the Q. The coefficients of the low-pass equation are then entered and the appropriate coefficients for the selected equation are calculated and displayed. When R_α R_β is selected, the user has one of two choices. When the user inputs R_α and L_{shift}, the R_β will be calculated and displayed. When L_{shift} and h are input, M and h_{new} will be calculated and displayed. (See Chapter 8.)

G-6-4 SHIFT COMPT

This command allows the user to select a number of resistors and capacitors and their values. The user then may frequency shift and/or impedance shift the components.

G-6-5 EXIT

This exits this program.

G-7 FILTER-TABLE

When this program starts there are four choices: BUTTERWORTH, CHEBYSHEV, ELLIPTIC, and EXIT. The choice is made by using the Left-arrow and Right-arrow keys. The Enter key is used to start the command. When BUTTERWORTH or CHEBYSHEV is selected, the user will be asked for the maximum pass-band loss and then what orders to calculate. When ELLIPTIC is selected, the same information is required plus the normalized stop-band frequency. The equations are calculated and displayed either as roots or as polynomial factors.

Bibliography

CHRISTIAN, ERICH, and EGON EISENMANN, *Filter Design Tables and Graphs*. New York: John Wiley & Sons, Inc., 1966.

DANIELS, RICHARD W., *Approximation Methods for Electronic Filter Design*. New York: McGraw-Hill Book Company, 1974.

DARYANANI, GOBIND, *Principles of Active Network Synthesis and Design*. New York: John Wiley & Sons, Inc., 1976, by Bell Laboratories, Inc.

HUELSMAN, LAWRENCE P., and PHILLIP E. ALLEN, *Introduction to the Theory and Design of Active Filters*. New York: McGraw-Hill Book Company, 1980.

JAMES, M. L., G. M. SMITH, and J. C. WOLFORD, *Applied Numerical Methods for Digital Computation*, 2nd ed. New York: Harper & Row, Publishers, Inc., 1977.

JOHNSON, DAVID E., *Introduction to Filter Theory*. Englewood Cliffs, N.J.: Prentice-Hall, Inc., 1976.

LACROIX, ARILD, and KARL-HEINZ WITTE, *Design Tables for Discrete Time Normalized Low-Pass Filters*. Dedham, Mass.: Artech House, Inc., 1986.

LAGO, GLADWYN, and LLOYD M. BENNINGFIELD, *Circuit and System Theory*. New York: John Wiley & Sons, Inc., 1979.

LAM, HARRY Y-F., *Analog and Digital Filters: Design and Realization*. Englewood Cliffs, N.J.: Prentice-Hall, Inc., 1979.

MILNE-THOMSON, L. M., "Elliptic Integrals," in *Handbook of Mathematical Functions*. Washington, D.C.: National Bureau of Standards, March 1965.

MOSCHYTZ, GEORGE SAMSON, and P. HORN, *Active Filter Design Handbook*. New York: John Wiley & Sons, Inc., 1981.

NATARAJAN, SUNDARAM, *Theory and Design of Linear Active Networks*. New York: Macmillan Publishing Company, Inc., 1987.

RICHMOND, A. E., *Calculus for Electronics*, 2nd ed. New York: McGraw-Hill Book Company, 1972.

Stanley, William D., *Network Analysis with Applications*. Reston, Va.: Reston Publishing Co., Inc., 1985.

Stanley, William D., *Transform Circuit Analysis for Engineering and Technology*, 2nd ed. Englewood Cliffs, N.J.: Prentice-Hall, Inc., 1989.

Williams, Arthur Bernard, *Electronic Filter Design Handbook*. New York: McGraw-Hill Book Company, 1981.

Answers to Odd-Numbered Problems

CHAPTER 2

1. Fig. P2-1

3. $i(t) = 6 \times 10^{-3}u(t)$

5. Fig. P2-2

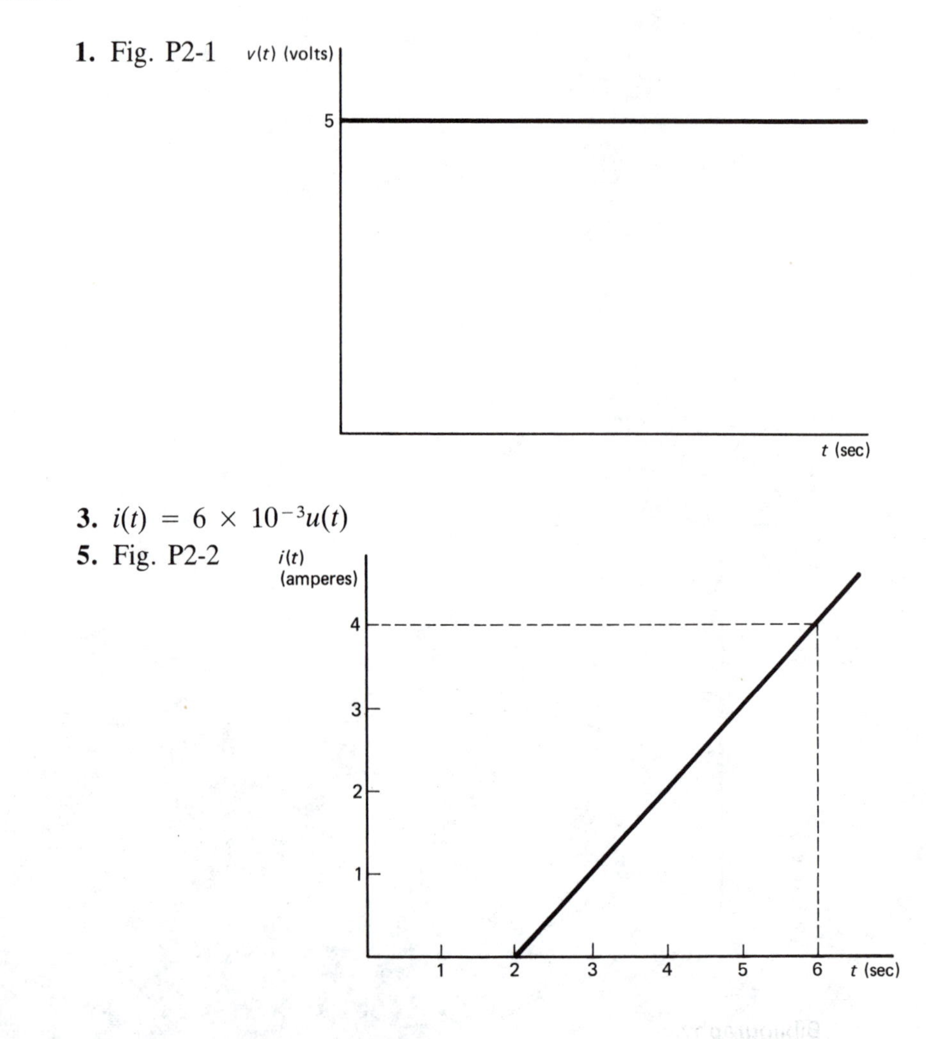

9. Fig. P2-3

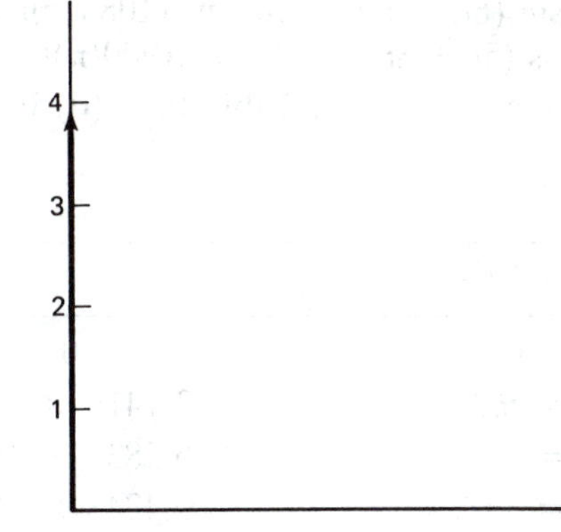

11. (a) $\tau = 5 \times 10^{-4}$ (b) $\alpha = 2 \times 10^{3}$ (c) Table P2-1 (d) Fig. P2-4

TABLE P2-1

t (time constants)	t (sec)	$v(t)$ $5e^{-2000t}u(t)$
0	0	5
0.5τ	2.5×10^{-4}	3.0327
τ	5×10^{-4}	1.8394
1.5τ	7.5×10^{-4}	1.1157
2τ	1×10^{-3}	6.6767×10^{-1}
2.5τ	1.25×10^{-3}	4.1043×10^{-1}
3τ	1.5×10^{-3}	2.4894×10^{-1}
3.5τ	1.75×10^{-3}	1.5099×10^{-1}
4τ	2×10^{-3}	9.1578×10^{-2}
4.5τ	2.25×10^{-3}	5.5545×10^{-2}
5τ	2.5×10^{-3}	0

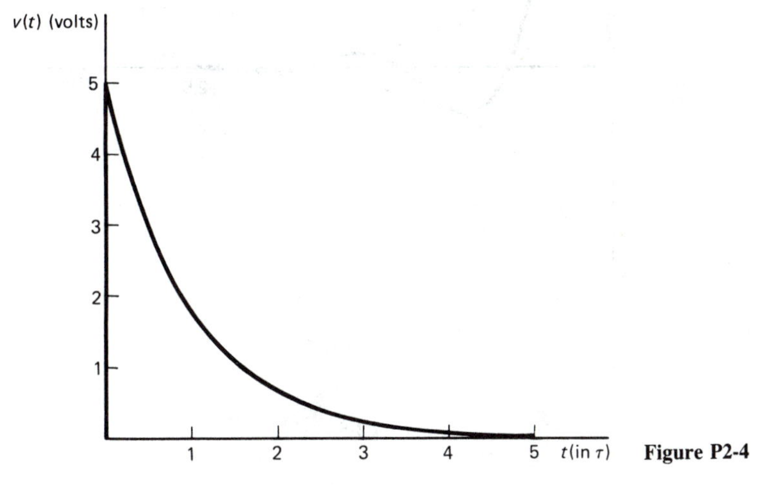

Figure P2-4

13. (a) $14 \sin (6t + \pi/6)$ $14 \sin [(1080/\pi)t + 30°]$
(b) $9 \cos (5t + \pi/12)$ $9 \cos [(900/\pi)t + 15°]$

15. (a) 2.5 sec (b) 0 A (c) Table P2-2 (d) Fig. P2-5

TABLE P2-2

Point	$\omega t + \theta$	t	$4e^{-2t} \sin (5t)\, u(t)$
A	$5t = 0$	0	0
B	$5t = \pi/2$	3.1416×10^{-1}	2.1340
C	$5t = \pi$	6.2832×10^{-1}	0
D	$5t = 3\pi/2$	9.4248×10^{-1}	-6.0734×10^{-1}
A or E	$5t = 2\pi$	1.2566	0
B	$5t = 2\pi + (\pi/2)$	1.5708	1.7286×10^{-1}
C	$5t = 2\pi + \pi$	1.8850	0
D	$5t = 2\pi + (3\pi/2)$	2.1991	-4.9196×10^{-2}
E	$5t = 2\pi + 2\pi$	2.5133	0

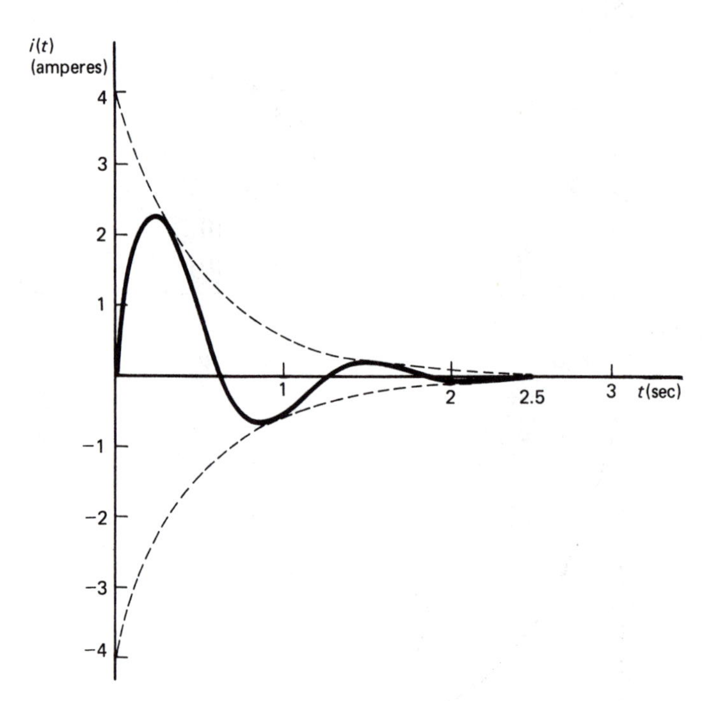

Figure P2-5

17. $5.4464 \sin(4\pi t) + 8.3867 \cos(4\pi t)$

19. $-2.5712 \sin(3\pi t) + 3.0642 \cos(3\pi t)$

21. $7.8102 \sin(4\pi t + 39.806°)$

23. $4(t - 2)u(t - 2)$

25. $7e^{-2(t - 3 \times 10^{-6})} \sin[3(t - 3 \times 10^{-6}) + 40°]u(t - 3 \times 10^{-6})$

27. $4e^{-2t} \sin(4t + 10°)u(t)$

29. Fig. P2-6

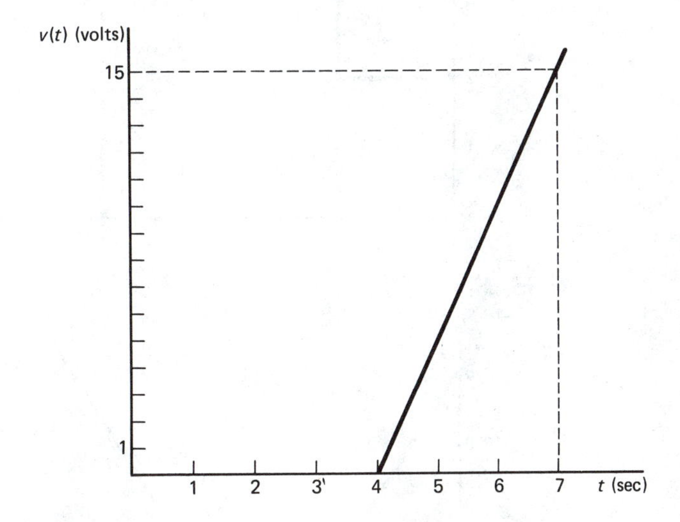

31. Fig. P2-7

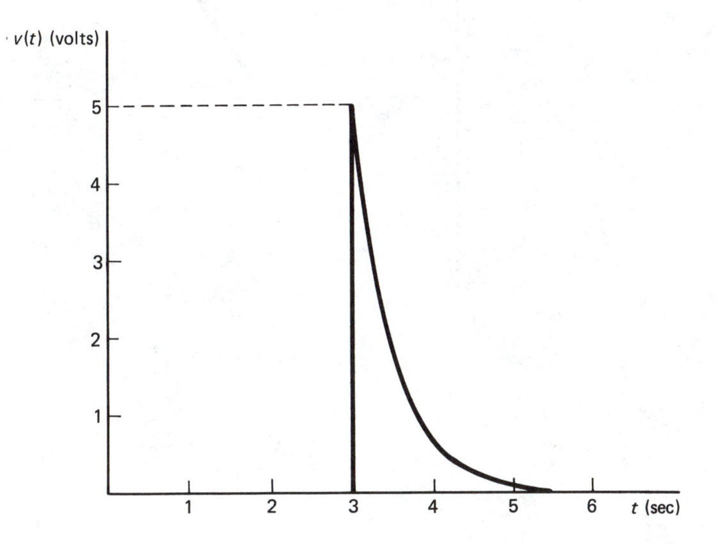

33. Fig. P2-8

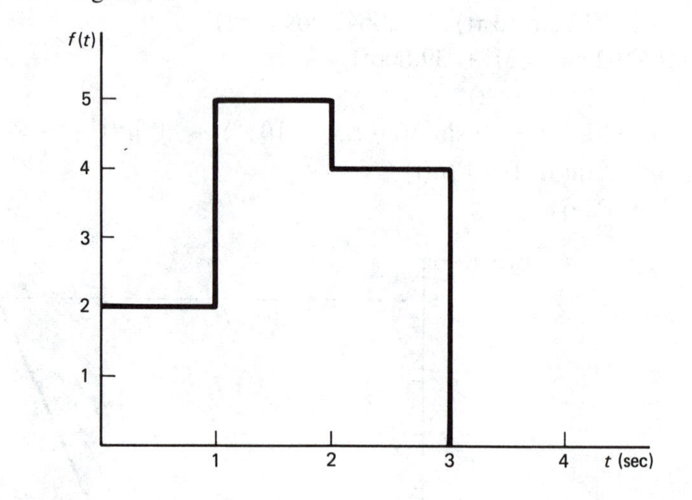

35. Fig. P2-9

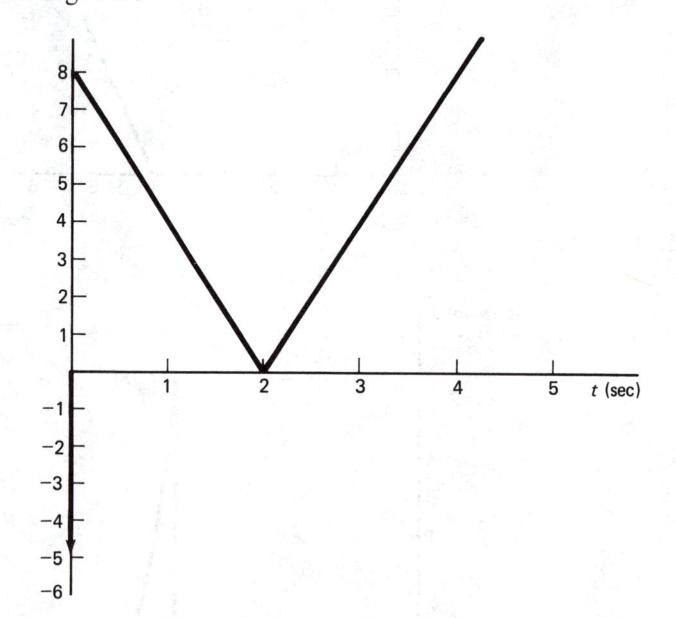

37. Fig. P2-10

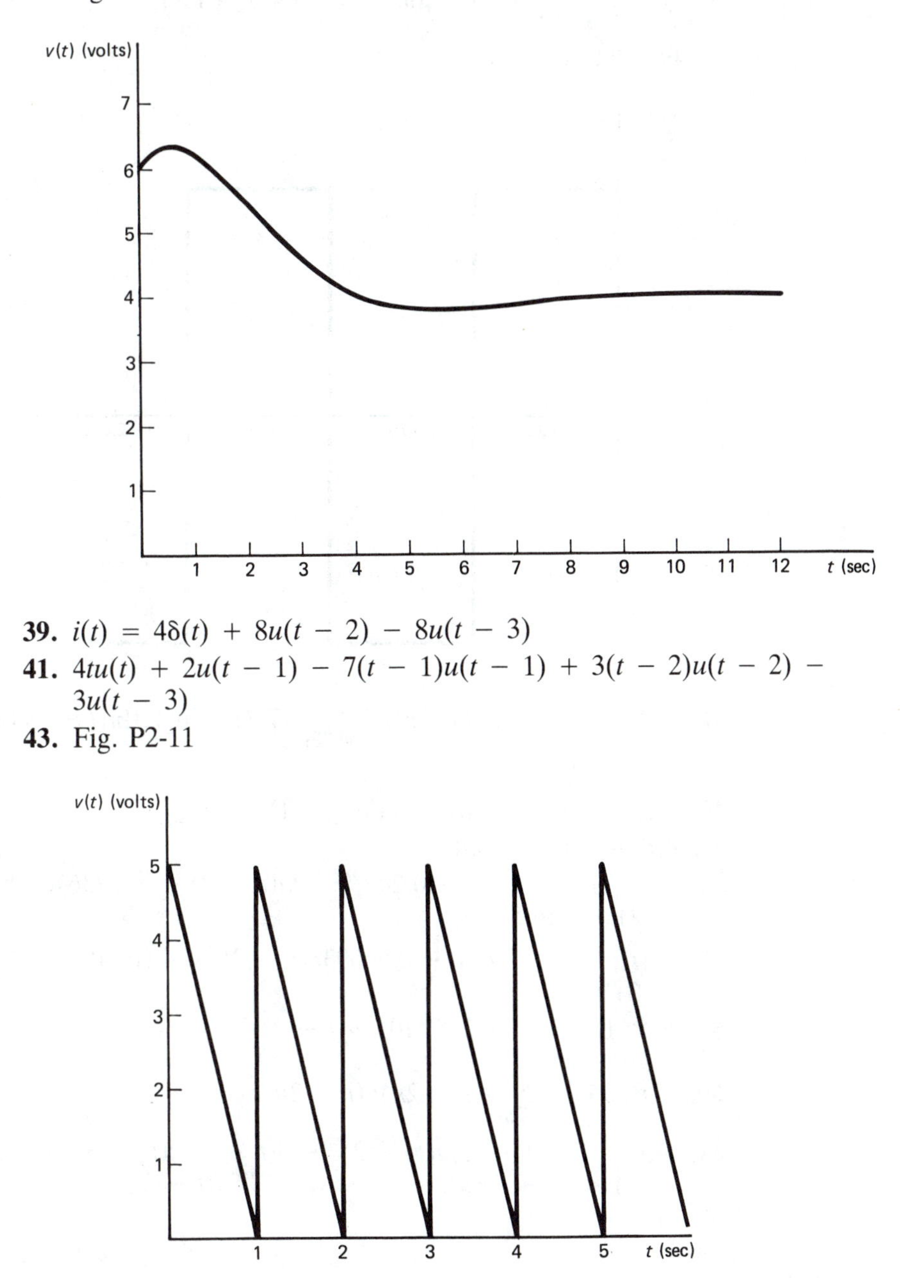

39. $i(t) = 4\delta(t) + 8u(t-2) - 8u(t-3)$

41. $4tu(t) + 2u(t-1) - 7(t-1)u(t-1) + 3(t-2)u(t-2) - 3u(t-3)$

43. Fig. P2-11

45. $20u(t) + \sum_{n=1}^{\infty} (-1)^n 40u(t - n(0.5 \times 10^{-3}))$

Fig. P2-12

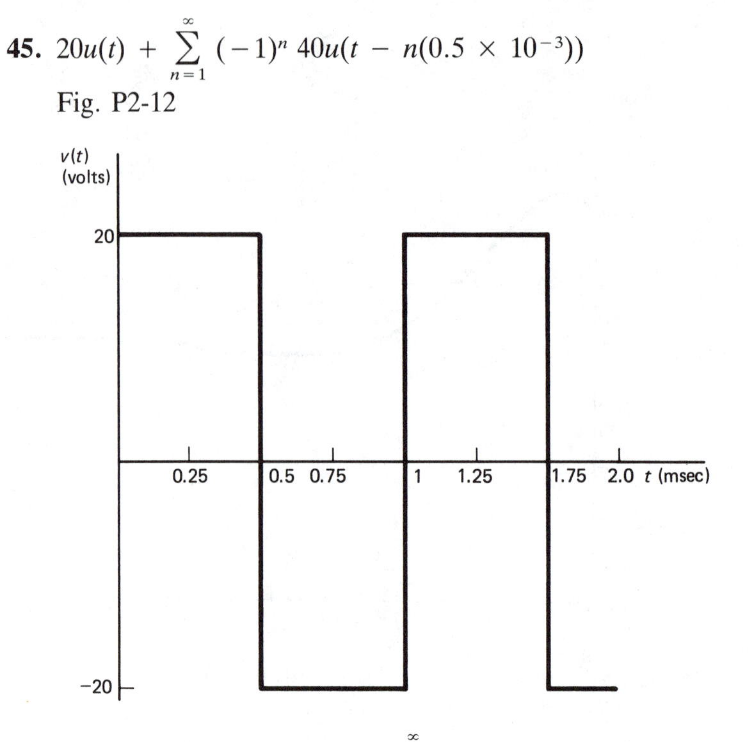

47. $v(t) = 8.9095 \sin(120\pi t) + \sum_{n=1}^{\infty} 17.819 \sin[120\pi(t - n/120)]u(t - n/120)$

49. $u(t) + 7tu(t) - 4(t - 1)^2u(t - 1)$

51. $\delta(t) + 10(t - 2)u(t - 2)$

53. $5(t - 1)u(t - 1) + 0.2e^{-2(t-2)}u(t - 2) + 0.22361e^{-2(t-2)} \sin[4(t - 2) - 116.57°]u(t - 2)$

55. $6\delta(t) - 18e^{-3t}u(t) + 0.96593\delta(t - 3) + 4 \sin[4(t - 3) + 165°)u(t - 3)$

57. $5 \times 10^{-3}\delta(t) + 2 \times 10^{-2}u(t - 3)$

59. $-t^2u(t) + \sum_{n=0}^{\infty} 4(t - 2n)u(t - 2n)$

61. $8tu(t) - 8(t - 2)u(t - 2) + 8(t - 3)u(t - 3) - 12(t - 5)u(t - 5) + (t - 5)^2u(t - 5) - (t - 7)^2u(t - 7)$

CHAPTER 3

3. $\dfrac{0.86603(s - 1.7321)}{s^2 + 9}$

5. $5.6569 \sin(2t + 45°)$

7. (a) $\dfrac{(s^2 + 1)\, e^{-2s}}{s^2}$ (b) $\dfrac{0.5794e^{-s}(s - 2.8134)}{s^2 + 4}$

9. (a) $\dfrac{e^{-s}}{(s + 2)^2} + \dfrac{e^{-2s}}{s + 3}$ (b) $\dfrac{4e^{-3s}}{(s + 1)^2 + 16}$

11. (a) $\dfrac{1}{s}$ (b) $\dfrac{2.8284(s - 4)}{s^2 + 16}$ (c) $\dfrac{-6}{s + 6}$ (d) $\dfrac{-4}{s^2 + 4}$

13.

$$R(s)_{-3} = \frac{s^2(s + 1)}{(s + 2)\,[(s + 4)^2 + 9]}$$

$$R(s)_{-2} = \frac{s^2(s + 1)}{(s + 3)^2\,[(s + 4)^2 + 9]}$$

$$R(s)_{-4 + 3j} = \frac{s^2(s + 1)}{(s + 2)(s + 3)^2}$$

15. (a) $-0.16667e^{-4t}u(t) + 0.27273e^{-5t}u(t) - 0.10606e^{-16t}u(t)$
(b) $\delta(t) + 15e^{-3t}u(t) - 4e^{-2t}u(t) - 16e^{-4t}u(t)$

17. (a) $\delta(t) + 12tu(t) - u(t) + 6e^{-t}u(t)$
(b) $6e^{-t}u(t) - t^2e^{-2t}u(t) - 5te^{-2t}u(t) - 6e^{-2t}u(t)$

19. (a) $0.625t^2e^{-2t}\sin(t + 126.87°)\, u(t) + 1.7002te^{-2t}\sin(t - 107.10°)$ $u(t) + 1.625e^{-2t}\sin(t)\, u(t)$
(b) $2.2535te^{-t}\sin(2t + 176.82°)\, u(t) + 1.0915e^{-t}\sin(2t + 66.371°)$ $u(t)$

CHAPTER 4

1. Fig. P4-1

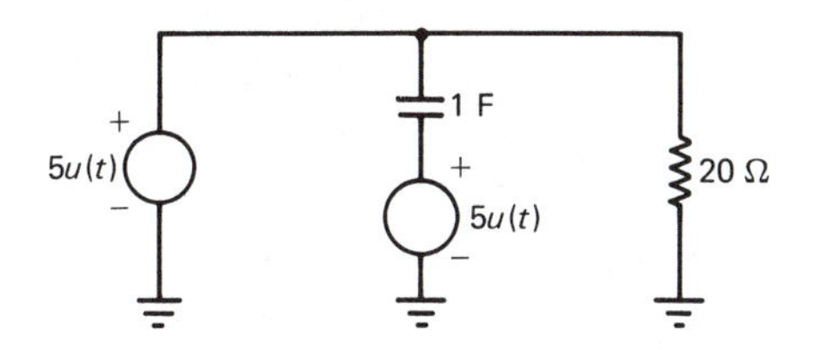

3. Fig. P4-2

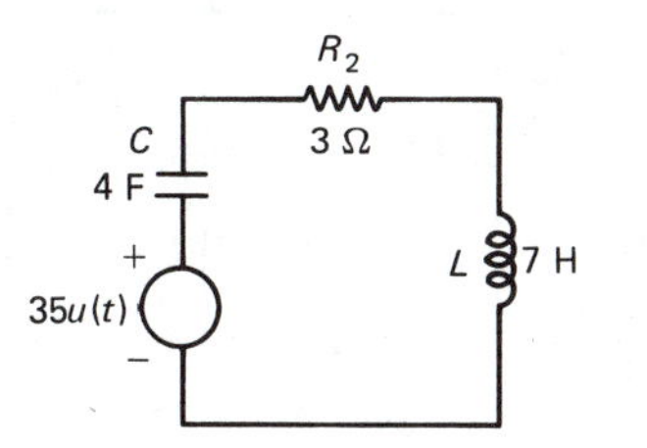

5. Fig. P4-3

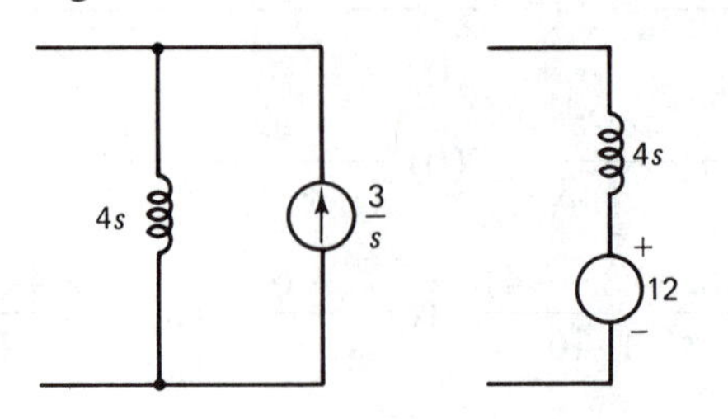

7. Fig. P4-4

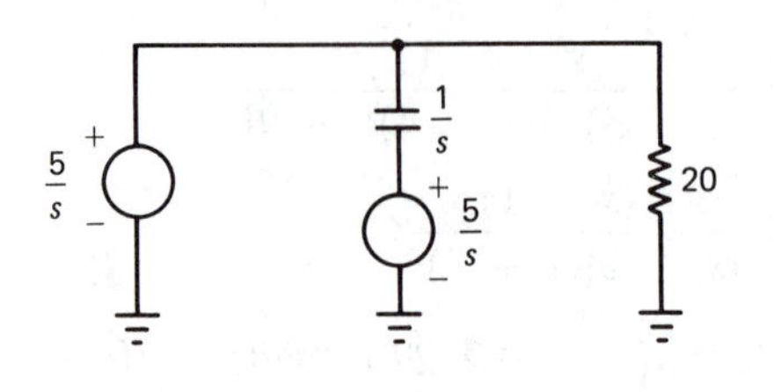

9. Fig. P4-5

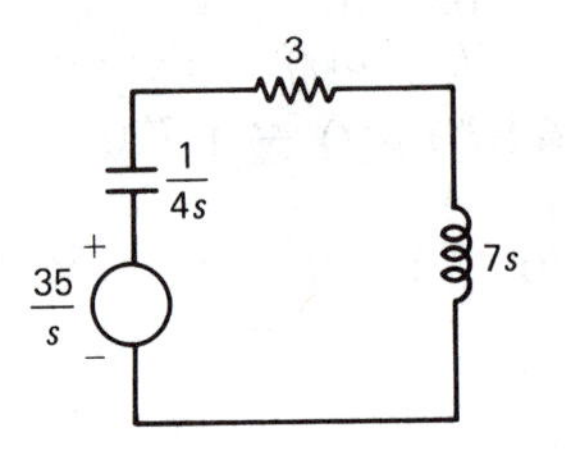

11. Fig. P4-6

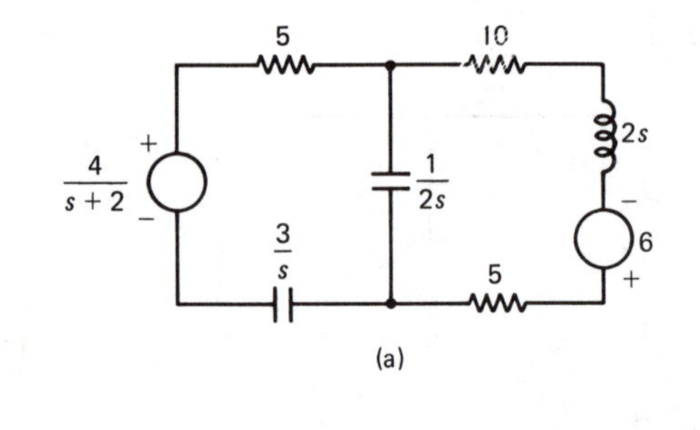

(a)

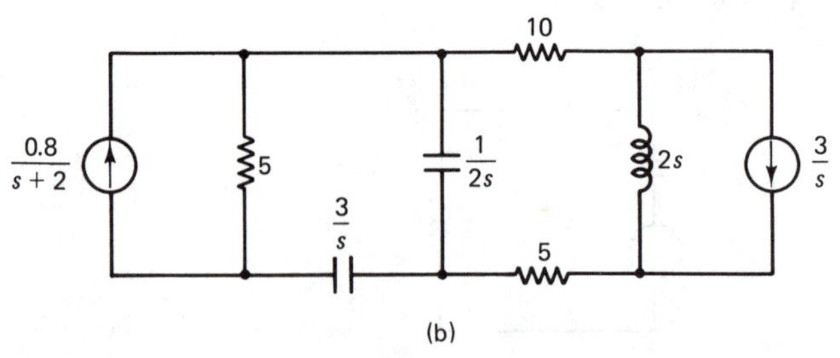

(b)

13. (a) $\dfrac{4}{(s^2 + 100)(s + 0.1)}$

(b) $3.9996 \times 10^{-2} e^{-0.1t} u(t) + 3.9998 \times 10^{-2} \sin(10t - 89.427°)\, u(t)$

(c) Transient: $3.9996 \times 10^{-2} e^{-0.1t} u(t)$; steady state $3.9998 \times 10^{-2} \sin(10t - 89.427°)\, u(t)$

15. (a) $\dfrac{-10s}{[(s + 1)^2 + 99]}$

(b) $10.050 e^{-t} \sin(9.9499t - 84.261°)\, u(t)$

(c) The whole function is transient.

17. $\dfrac{8[(s + 0.225)^2 + 1.0972]}{s(s + 0.9)[(s + 0.3)^2 + 0.91]}$

19. (a) $\dfrac{-2(s^2 + 16.5)}{(s + 1.25)(s^2 + 16)}$, $\dfrac{-4[(s + 0.3125)^2 + (2.8552)^2]}{(s + 1.25)(s^2 + 16)}$

(b) $-2.0569 e^{-1.25t} u(t) + 5.9655 \times 10^{-2} \sin(4t + 107.35°)\, u(t)$
$-2.0569 e^{-1.25t} u(t) + 1.9431 \sin(4t - 90.525°)\, u(t)$

21. (a) $\left(\dfrac{1}{2} + \dfrac{1}{5} + \dfrac{1}{2s}\right) V_1 - \dfrac{1}{2s} V_2 = \dfrac{2}{s^2 + 16}$

$$\frac{-1}{2s} V_1 + \left(\frac{1}{2s} + \frac{1}{3/s}\right) V_2 = \frac{-4}{s}$$

(b) $0.62492 \sin(4t + 28.968°)\, u(t) + 11.697 e^{-2.2143t} u(t)$

23. (a) Same as Problem 21a

(b) $2.7097u(t) - 2.6413 e^{-2.2143t} u(t) + 0.078115 \sin(4t - 61.032°)\, u(t)$

25. (a) $8I_1(s)$ (b) $\dfrac{0.4}{s} V_1(s)$ (c) $0.8I_1(s)$

27. $\dfrac{20(s - 0.16667)}{s[(s + 0.33333)^2 + (1.7951)^2]}$

29. $Z_{TH} = \dfrac{10(s + 0.83333)}{(s + 1)(s + 5.8333)}$, $V_{TH} = \dfrac{10(s + 1.83333)}{(s + 1)(s + 5.8333)}$

31. (a) $Z_{TH} = \dfrac{2(s + 2.2222)}{(s + 0.62679)(s + 1.5954)}$

$$V_{TH} = \frac{0.22222}{(s^2 + 4)(s + 0.62679)(s + 1.5954)}$$

(b)
$$\frac{0.22222}{(s^2 + 4)[(s + 1.6111)^2 + (0.79154)^2]}$$

$$\frac{0.22222}{(s^2 + 4)(s + 1.0616)(s + 1.3606)}$$

33. 5th

35. $-0.5\delta(t) + e^{-2t}u(t)$

37. $\dfrac{0.5}{s^2 + 0.2s + 0.0625}$ underdamped

39. $\dfrac{0.5s}{s^2 + 0.2s + 1}$ underdamped

CHAPTER 5

1. (a) $\dfrac{s}{s + 0.5}$ (b) $\delta(t) - 0.5e^{-0.5t}u(t)$

3. (a) $\dfrac{6.6667s^2}{(s + 0.16667)(s + 1)}$

(b) $6.6667\delta(t) + 0.22222e^{-0.16667t}u(t) - 8e^{-t}u(t)$

5. (a) Fig. P5-1

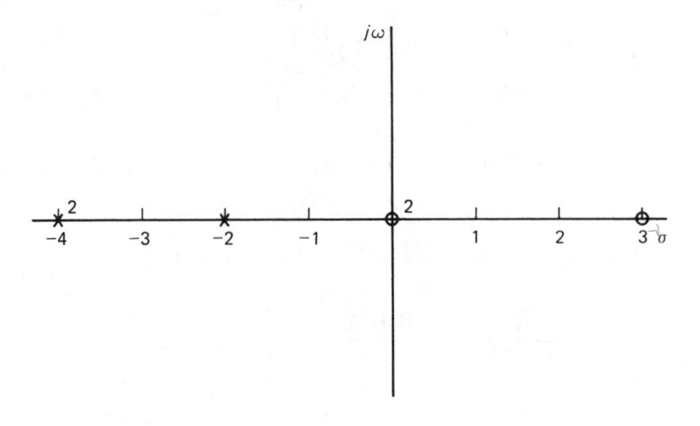

(b) Transient: $s = -2, -4, -4$; steady state: none
(c) Stable (d) $s = -4, -4, -2$

7. (a) Fig. P5-2

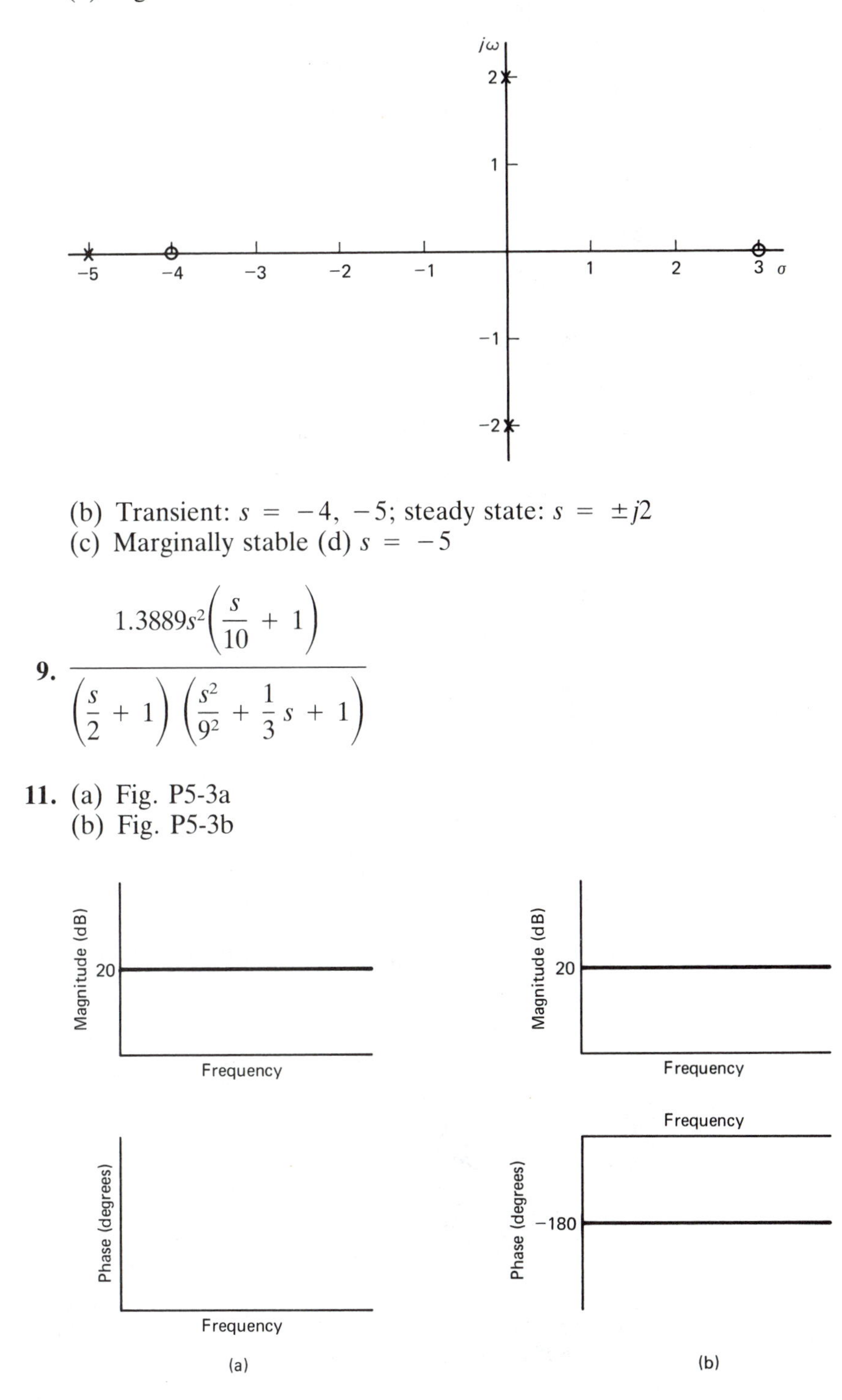

(b) Transient: $s = -4, -5$; steady state: $s = \pm j2$
(c) Marginally stable (d) $s = -5$

9. $$\frac{1.3889s^2\left(\dfrac{s}{10} + 1\right)}{\left(\dfrac{s}{2} + 1\right)\left(\dfrac{s^2}{9^2} + \dfrac{1}{3}s + 1\right)}$$

11. (a) Fig. P5-3a
(b) Fig. P5-3b

13. (a) Fig. P5-4
(b) -1.4451 dB, $-32.142°$

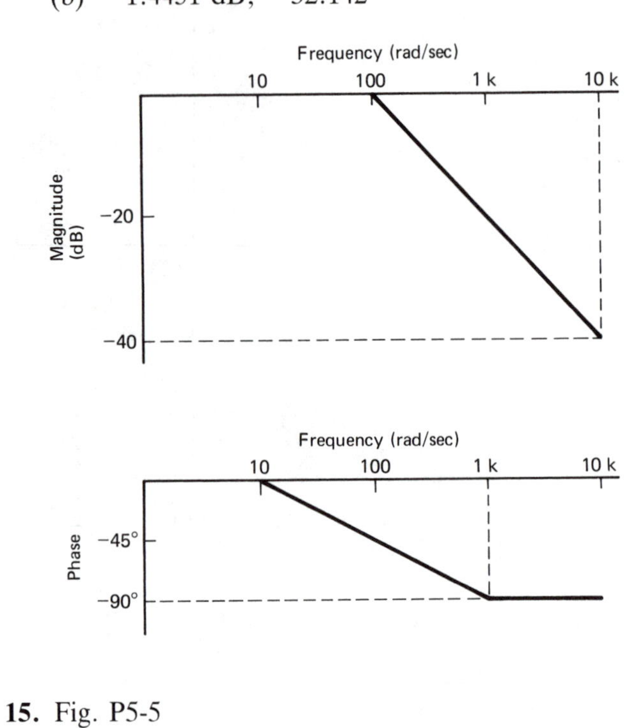

15. Fig. P5-5

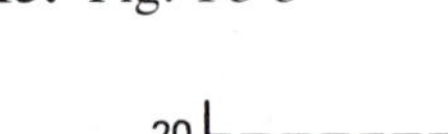

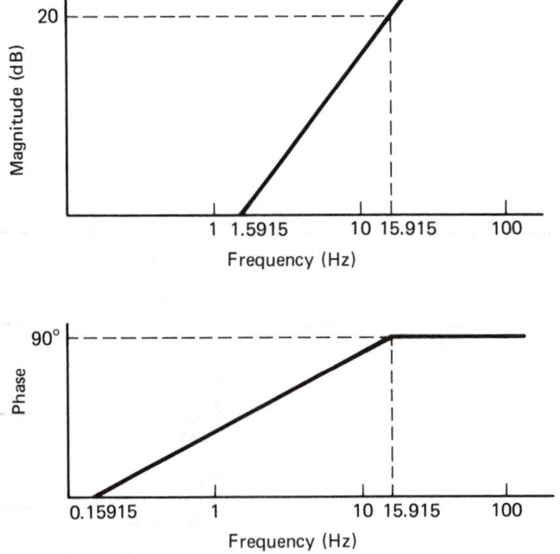

17. (a) Fig. P5-6
(b) 2.6954 dB, 16.4924 rad/sec

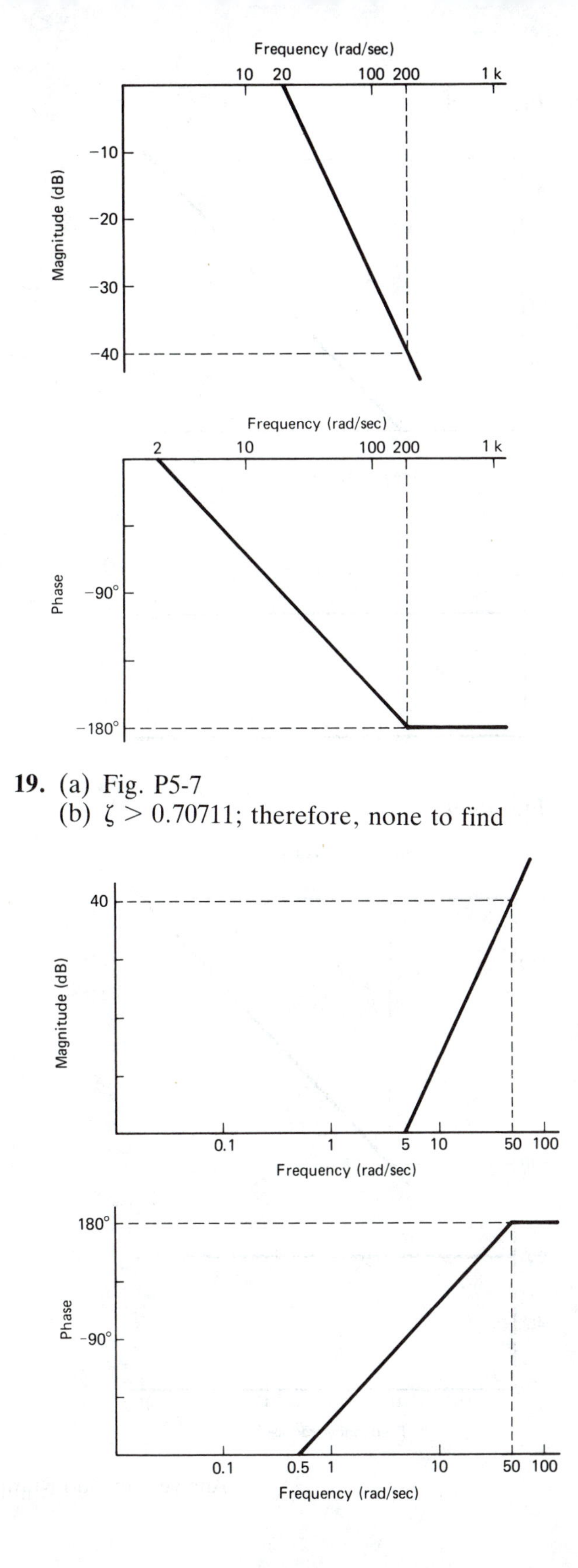

19. (a) Fig. P5-7
(b) $\zeta > 0.70711$; therefore, none to find

21. (a) Fig. P5-8

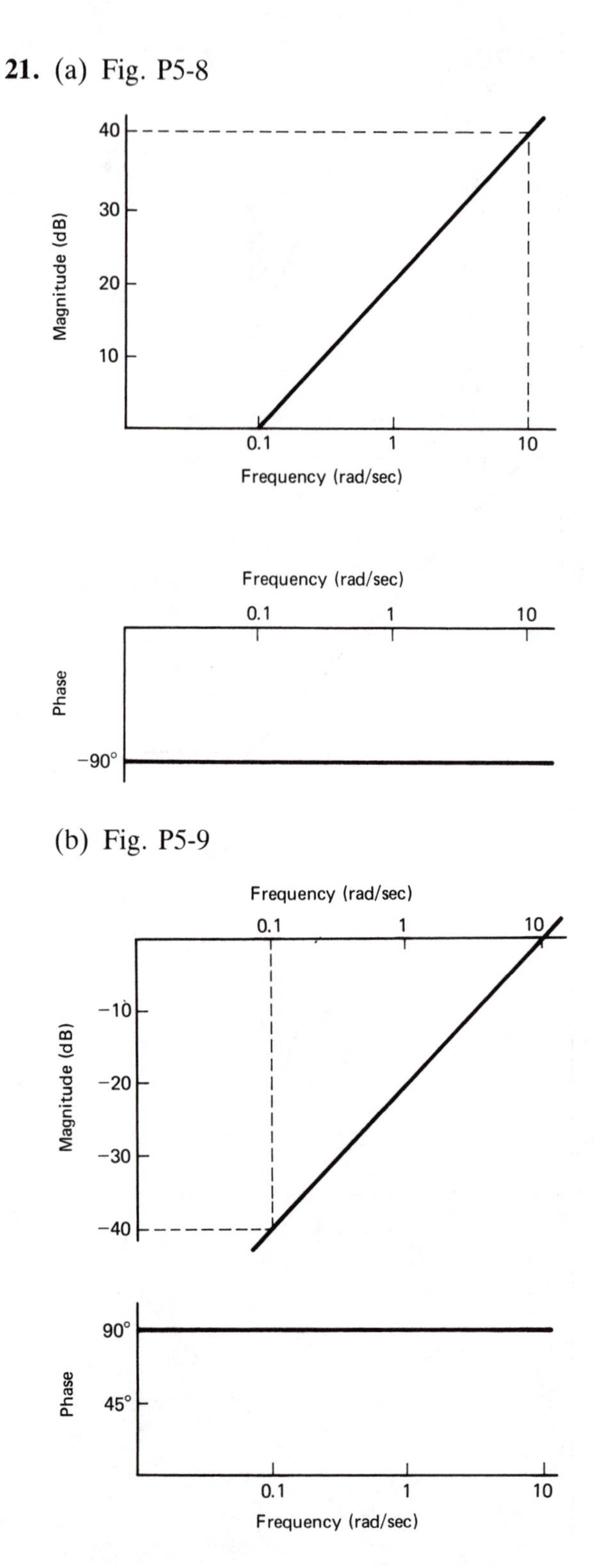

40
30
20
10
Magnitude (dB)
0.1
1
10
Frequency (rad/sec)
Frequency (rad/sec)
0.1
1
10
Phase
−90°
(b) Fig. P5-9
Frequency (rad/sec)
0.1
1
10
−10
−20
−30
−40
Magnitude (dB)
90°
45°
Phase
0.1
1
10
Frequency (rad/sec)

23. (a) 31.956 dB, 31.4960 rad/sec
Fig. P5-10

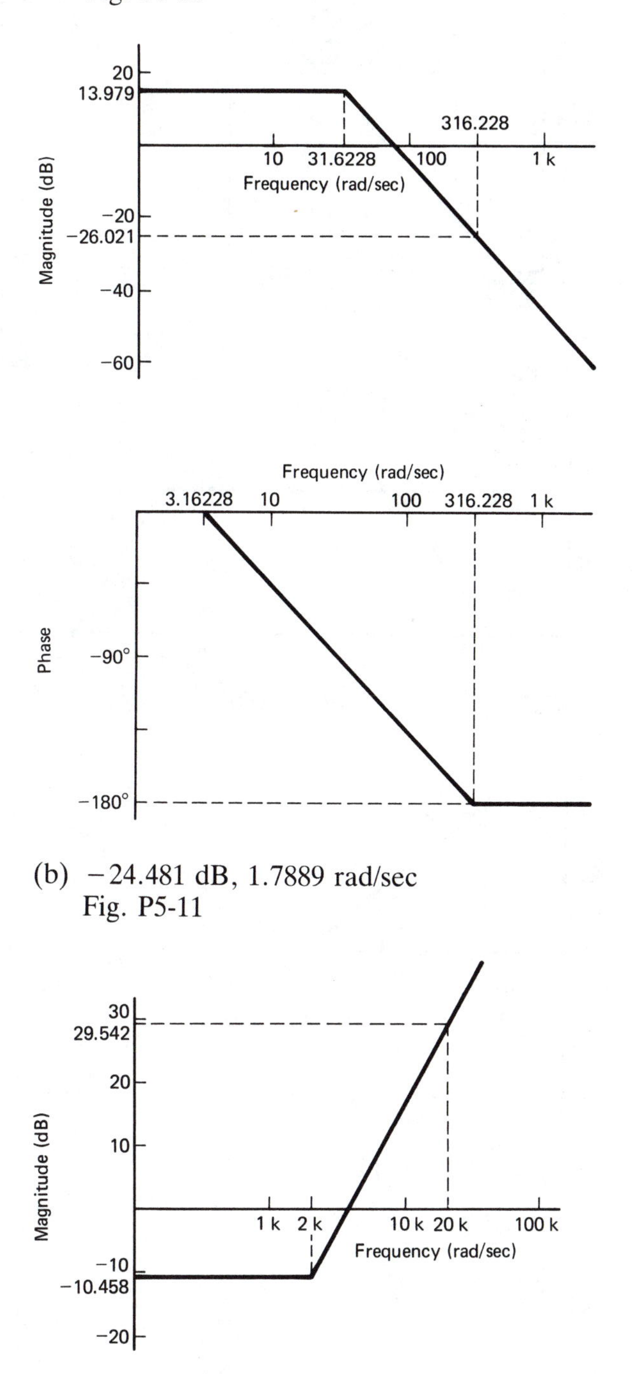

(b) −24.481 dB, 1.7889 rad/sec
Fig. P5-11

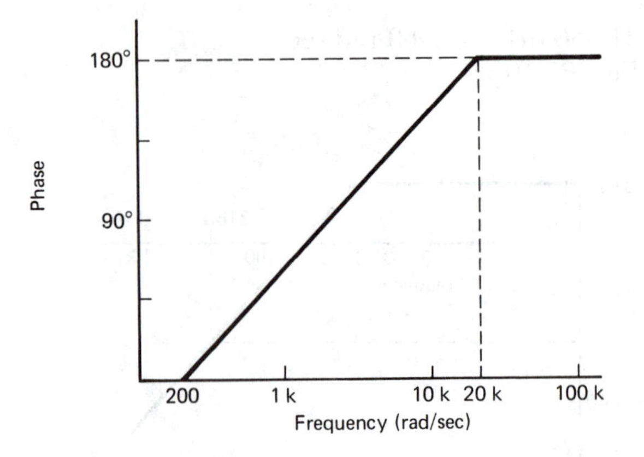

25. (a) Fig. P5-12

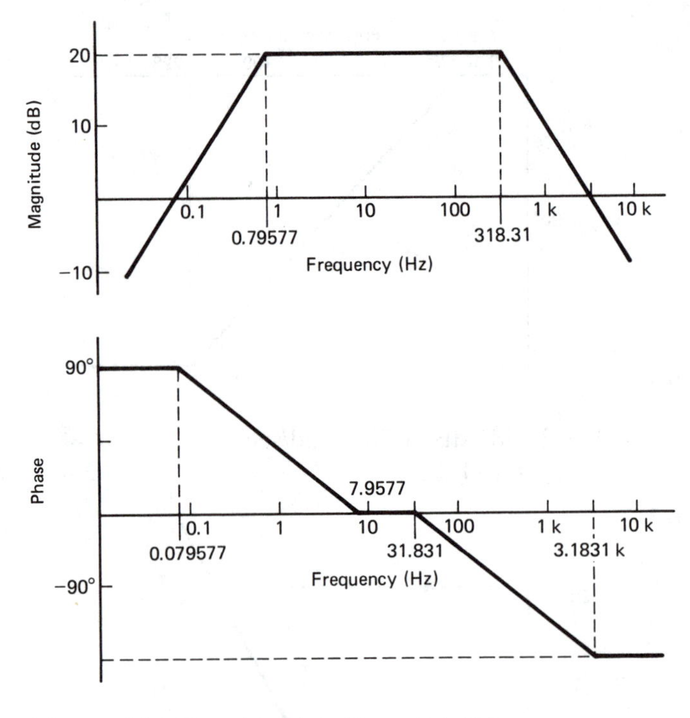

(b) 3.1831 kHz (c) 14.6 dB, −57.427°

27. Fig. P5-13

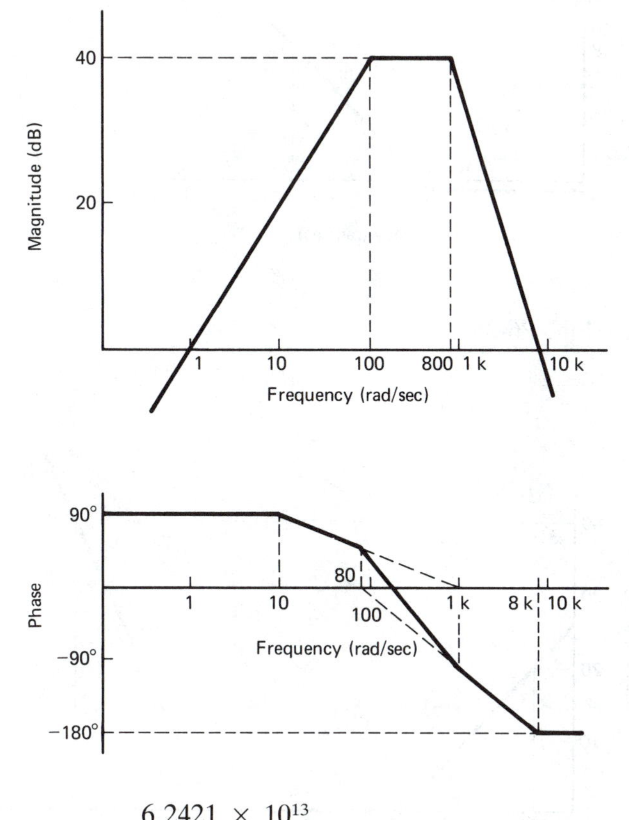

29. $\dfrac{6.2421 \times 10^{13}}{(s + 31.416)\,(s + 6.2832 \times 10^{6})}$

CHAPTER 6

1. (a) Amplified (b) +5 dB

3. (a) Fig. P6-1a

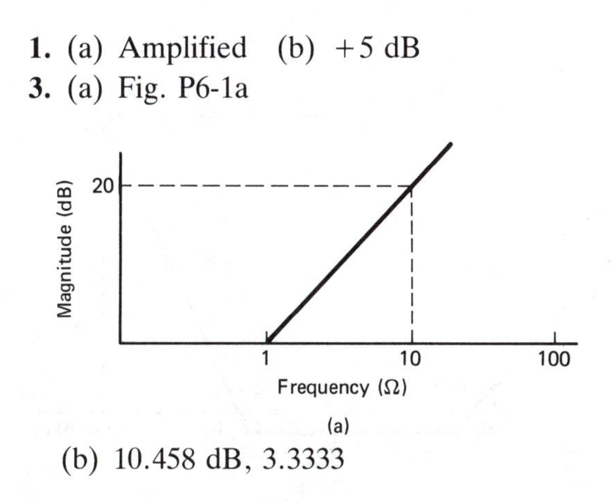

(a)

(b) 10.458 dB, 3.3333

(c) Fig. P6-1b

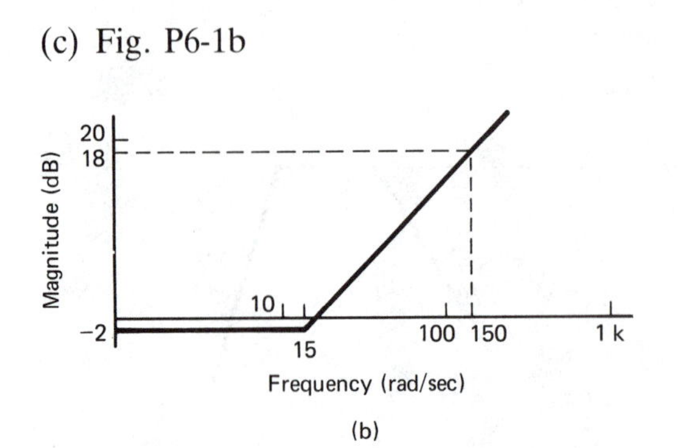

(b)

5. (a) Fig. P6-2a

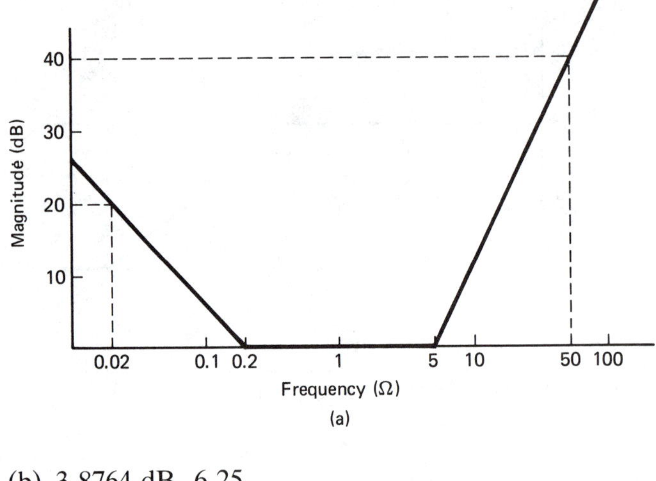

(a)

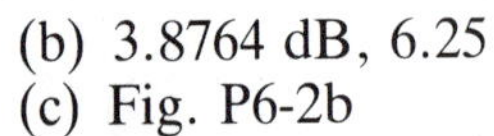
(b) 3.8764 dB, 6.25
(c) Fig. P6-2b

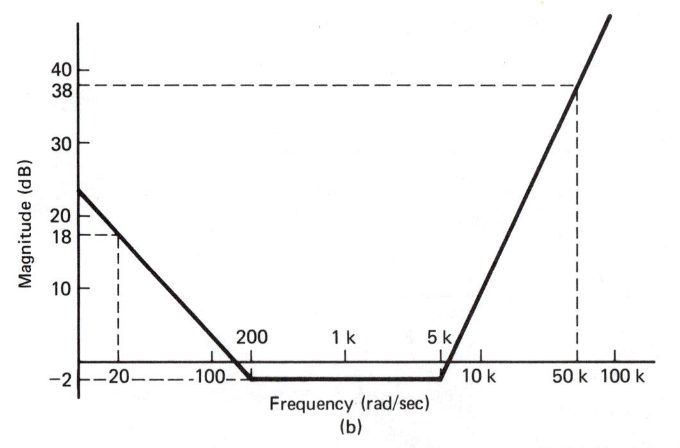

(b)

7. Low pass; Fig. P6-3

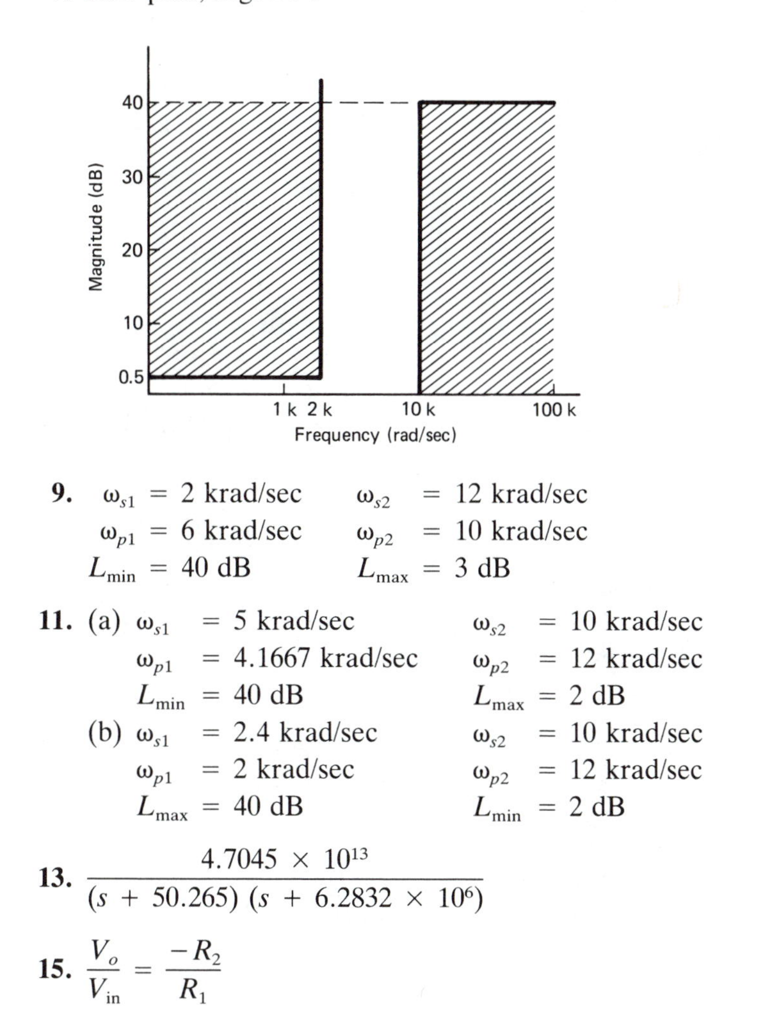

9. $\omega_{s1} = 2$ krad/sec $\quad \omega_{s2} = 12$ krad/sec
$\omega_{p1} = 6$ krad/sec $\quad \omega_{p2} = 10$ krad/sec
$L_{\min} = 40$ dB $\quad L_{\max} = 3$ dB

11. (a) $\omega_{s1} = 5$ krad/sec $\quad \omega_{s2} = 10$ krad/sec
$\omega_{p1} = 4.1667$ krad/sec $\quad \omega_{p2} = 12$ krad/sec
$L_{\min} = 40$ dB $\quad L_{\max} = 2$ dB
(b) $\omega_{s1} = 2.4$ krad/sec $\quad \omega_{s2} = 10$ krad/sec
$\omega_{p1} = 2$ krad/sec $\quad \omega_{p2} = 12$ krad/sec
$L_{\max} = 40$ dB $\quad L_{\min} = 2$ dB

13. $$\frac{4.7045 \times 10^{13}}{(s + 50.265)(s + 6.2832 \times 10^6)}$$

15. $$\frac{V_o}{V_{in}} = \frac{-R_2}{R_1}$$

CHAPTER 7

5. (a) $Y_A = sC_A$, $Y_B = 1/R_B$, $Y_C = sC_C$, $Y_D = 1/R_D$, $Y_E = 0$
(b) $Y_A = sC_A$, $Y_B = sC_B$, $Y_C = sC_C$, $Y_D = 1/R_D$, $Y_E = 1/R_E$
7. $R_A = R_C = 1\ \Omega$, $C_B = C_D = 1$ F, $R_1 = 0.5858$, $h = 1.5858$
9. (a) Second-order low pass (b) Second-order band pass
(c) First-order high pass
11. 16.454 dB
13. (a) $C_{D1} = 0.1$ F, $C_{E1} = 1$ F
$R_{A1} = 0.73664\ \Omega$, $R_{B1} = R_{C1} = 2.8309\ \Omega$
$C_{D2} = 0.1$ F, $C_{E2} = 1$ F
$R_{A2} = 6.7360\ \Omega$, $R_{B2} = R_{C2} = 2.8309\ \Omega$

(b) $C_{B1} = C_{D1} = 1$ F
$R_{A1} = R_{C1} = 0.89522\ \Omega$, $R_1 = 0.1523\ \Omega$
$C_{B2} = C_{D2} = 1$ F
$R_{A2} = R_{C2} = 0.89522\ \Omega$, $R_1 = 1.2346\ \Omega$
$R_{F3} = 0.38833\ \Omega$, $R_{13} = 1\Omega$ (inverting amplifier)

15. (a) $C_D = 0.1$ F, $C_E = 1$ F, $h = 0.70711$
$R_A = 8.9478\ \Omega$, $R_B = R_C = 3.7606\ \Omega$
(b) $C_D = C_B = 1$ F, $h = 1.5801$
$R_A = R_C = 1.1892\ \Omega$, $R_1 = 1.2346\ \Omega$

17. (a) $1.0264 \underline{/-152.29^\circ}$ or 0.22618 dB
(b) $6.1506 \underline{/-42.395^\circ}$ or 15.778 dB

19. $C_{B1} = 1$ F, $R_{A1} = 66.839\ \Omega$, $R_{B1} = 3.3183\ \Omega$
$C_{A2} = C_{E2} = 1$ F, $C_{C2} = 2$ F, $C_{G2} = 20$ F, $R_{12} = 19.975\ \Omega$
$R_{B2} = R_{F2} = 0.17364\ \Omega$, $R_{D2} = 0.086822\ \Omega$, $R_{G2} = 8.4064\ \Omega$

CHAPTER 8

1. $C_1 = 2$ F, $C_2 = 0.2$ F
$R_1 = 7.9577 \times 10^{-5}\ \Omega$, $R_2 = 1.5915 \times 10^{-4}\ \Omega$

3. $C_1 = 0.002$ F, $C_2 = 0.02$ F, $R_1 = 3\ \Omega$, $R_2 = 2.2\ \Omega$

5. (a) $C_1 = 10\ \mu$F, $C_2 = 1\ \mu$F, $R_1 = 4$ MΩ, $R_2 = 200$ kΩ
(b) $C_1 = 100\ \mu$F, $C_2 = 10\ \mu$F, $R_1 = 400$ kΩ, $R_2 = 20$ kΩ

7. $C_A = C_E = 0.001\ \mu$F, $C_C = 0.002\ \mu$F, $C_G = 0.1\ \mu$F, $h = 0.46679$
$R_B = R_F = 5.3863$ kΩ, $R_D = 2.6931$ kΩ, $R_G = 5.6676$ kΩ
$R_1 = 249.99$ kΩ

9. (a) $\dfrac{s + 1.5569}{2.7686}$ (b) $\dfrac{s + 1.5569}{0.87551}$

11. (a) $R_\beta = 9.1127$ kΩ; output between R_α and R_β
(b) $R_\beta = 7.3375$ kΩ; feedback between R_α and R_β

13. (a) $R_\beta = 19.1432$ kΩ; output between R_α and R_β
(b) $R_\beta = 15.646$ kΩ; feedback between R_α and R_β

15. $\dfrac{(s + 2.3803)\,(s^2 + 0.45344s + 1.0793)}{s^3}$

17. $C_A = C_C = 0.22\ \mu$F, $R_B = R_D = 20.939$ kΩ, $R_1 = 12.265$ kΩ, $R_\beta = 5.4055$ kΩ; feedback between R_α and R_β

19. 2.2222, 1.35 krad/sec, 214.86 Hz, 477.46 Hz, 3 krad/sec

21. $C_{A1} = C_{B1} = C_{C1} = 0.022\ \mu$F, $R_{D1} = 30.242$ kΩ, $R_{E1} = 3.0887$ kΩ
$R_{\beta 1} = 1.8851$ kΩ, feedback between R_α and R_β
$C_{D2} = 0.0022\ \mu$F, $C_{E2} = 0.022\ \mu$F
$R_{A2} = 5.1225$ kΩ, $R_{B2} = R_{C2} = 5.2852$ kΩ
$R_{\beta 2} = 1.8851$ kΩ, feedback between R_α and R_β

23. $$\frac{(s^2 + 0.1159s + 1)(s^2 + 5.5044 \times 10^{-2}s + 0.90450)(s^2 + 6.0856 \times 10^{-2}s + 1.1056)}{1.5569 \times 10^{-3}s^3}$$

25. $$\frac{(s^2 + 0.18496s + 0.52641)(s^2 + 0.35136s + 1.89965)}{0.34722s^2}$$

27. $C_B = C_C = 1\ \mu F$, $R_A = R_E = 20.541\ k\Omega$, $R_D = 2.4342\ k\Omega$
$R_F = 100\ k\Omega$, $R_1 = 3.3321\ k\Omega$, $R_2 = 56.234\ k\Omega$

29. $$\frac{(s^2 + 0.67105s + 1)(s^2 + 3.9337 \times 10^{-2}s + 1.24)(s^2 + 3.1724 \times 10^{-2}s + 0.80645)}{0.99998(s^2 + 1)^3}$$

31. $C_{A1} = C_{E1} = 0.1\ \mu F$, $C_{C1} = 0.2\ \mu F$, $C_{G1} =$ none
$R_{B1} = R_{F1} = 2.3216\ k\Omega$, $R_{D1} = 1.1608\ k\Omega$, $R_{G1} = 44.969\ k\Omega$
$R_{11} = 2.3740\ k\Omega$, $R_{21} = 2.3216\ k\Omega$
$C_{A2} = C_{E2} = 0.1\ \mu F$, $C_{C2} = 0.2\ \mu F$, $C_{G2} = 0.1\ \mu F$
$R_{B2} = R_{F2} = 2.4419\ k\Omega$, $R_{D2} = 1.2209\ k\Omega$, $R_{G2} = 2.8404\ k\Omega$
$R_{12} = 6.7901\ k\Omega$, $R_{22} = 2.4419\ k\Omega$

Index

D

E

F

G

M

N

O

P

U

V

W

Z